Neuropeptide Protocols

METHODS IN MOLECULAR BIOLOGY™

John M. Walker, SERIES EDITOR

73. **Neuropeptide Protocols,** edited by *G. Brent Irvine and Carvell H. Williams, 1997*
72. **Neurotransmitter Methods,** edited by *Richard C. Rayne, 1997*
71. **PRINS and *In Situ* PCR Protocols,** edited by *John R. Gosden, 1997*
70. **Sequence Data Analysis Guidebook,** edited by *Simon R. Swindell, 1997*
69. **cDNA Library Protocols,** edited by *Ian G. Cowell and Caroline A. Austin, 1997*
68. **Gene Isolation and Mapping Protocols,** edited by *Jacqueline Boultwood, 1997*
67. **PCR Cloning Protocols:** *From Molecular Cloning to Genetic Engineering,* edited by *Bruce A. White, 1996*
66. **Epitope Mapping Protocols,** edited by *Glenn E. Morris, 1996*
65. **PCR Sequencing Protocols,** edited by *Ralph Rapley, 1996*
64. **Protein Sequencing Protocols,** edited by *Bryan J. Smith, 1996*
63. **Recombinant Proteins:** *Detection and Isolation Protocols,* edited by *Rocky S. Tuan, 1996*
62. **Recombinant Gene Expression Protocols,** edited by *Rocky S. Tuan, 1996*
61. **Protein and Peptide Analysis by Mass Spectrometry,** edited by *John R. Chapman, 1996*
60. **Protein NMR Protocols,** edited by *David G. Reid, 1996*
59. **Protein Purification Protocols,** edited by *Shawn Doonan, 1996*
58. **Basic DNA and RNA Protocols,** edited by *Adrian J. Harwood, 1996*
57. **In Vitro Mutagenesis Protocols,** edited by *Michael K. Trower, 1996*
56. **Crystallographic Methods and Protocols,** edited by *Christopher Jones, Barbara Mulloy, and Mark Sanderson, 1996*
55. **Plant Cell Electroporation and Electrofusion Protocols,** edited by *Jac A. Nickoloff, 1995*
54. **YAC Protocols,** edited by *David Markie, 1995*
53. **Yeast Protocols:** *Methods in Cell and Molecular Biology,* edited by *Ivor H. Evans, 1996*
52. **Capillary Electrophoresis:** *Principles, Instrumentation, and Applications,* edited by *Kevin D. Altria, 1996*
51. **Antibody Engineering Protocols,** edited by *Sudhir Paul, 1995*
50. **Species Diagnostics Protocols:** *PCR and Other Nucleic Acid Methods,* edited by *Justin P. Clapp, 1996*
49. **Plant Gene Transfer and Expression Protocols,** edited by *Heddwyn Jones, 1995*
48. **Animal Cell Electroporation and Electrofusion Protocols,** edited by *Jac A. Nickoloff, 1995*
47. **Electroporation Protocols for Microorganisms,** edited by *Jac A. Nickoloff, 1995*
46. **Diagnostic Bacteriology Protocols,** edited by *Jenny Howard and David M. Whitcombe, 1995*
45. **Monoclonal Antibody Protocols,** edited by *William C. Davis, 1995*
44. ***Agrobacterium* Protocols,** edited by *Kevan M. A. Gartland and Michael R. Davey, 1995*
43. **In Vitro Toxicity Testing Protocols,** edited by *Sheila O'Hare and Chris K. Atterwill, 1995*
42. **ELISA:** *Theory and Practice,* by *John R. Crowther, 1995*
41. **Signal Transduction Protocols,** edited by *David A. Kendall and Stephen J. Hill, 1995*
40. **Protein Stability and Folding:** *Theory and Practice,* edited by *Bret A. Shirley, 1995*
39. **Baculovirus Expression Protocols,** edited by *Christopher D. Richardson, 1995*
38. **Cryopreservation and Freeze-Drying Protocols,** edited by *John G. Day and Mark R. McLellan, 1995*
37. **In Vitro Transcription and Translation Protocols,** edited by *Martin J. Tymms, 1995*
36. **Peptide Analysis Protocols,** edited by *Ben M. Dunn and Michael W. Pennington, 1994*
35. **Peptide Synthesis Protocols,** edited by *Michael W. Pennington and Ben M. Dunn, 1994*
34. **Immunocytochemical Methods and Protocols,** edited by *Lorette C. Javois, 1994*
33. ***In Situ* Hybridization Protocols,** edited by *K. H. Andy Choo, 1994*
32. **Basic Protein and Peptide Protocols**, edited by *John M. Walker, 1994*
31. **Protocols for Gene Analysis,** edited by *Adrian J. Harwood, 1994*
30. **DNA–Protein Interactions,** edited by *G. Geoff Kneale, 1994*
29. **Chromosome Analysis Protocols,** edited by *John R. Gosden, 1994*
28. **Protocols for Nucleic Acid Analysis by Nonradioactive Probes,** edited by *Peter G. Isaac, 1994*
27. **Biomembrane Protocols:** *II. Architecture and Function,* edited by *John M. Graham and Joan A. Higgins, 1994*
26. **Protocols for Oligonucleotide Conjugates:** *Synthesis and Analytical Techniques,* edited by *Sudhir Agrawal, 1994*
25. **Computer Analysis of Sequence Data:** *Part II,* edited by *Annette M. Griffin and Hugh G. Griffin, 1994*
24. **Computer Analysis of Sequence Data:** *Part I,* edited by *Annette M. Griffin and Hugh G. Griffin, 1994*
23. **DNA Sequencing Protocols**, edited by *Hugh G. Griffin and Annette M. Griffin, 1993*
22. **Microscopy, Optical Spectroscopy, and Macroscopic Techniques**, edited by *Christopher Jones, Barbara Mulloy, and Adrian H. Thomas, 1993*
21. **Protocols in Molecular Parasitology**, edited by *John E. Hyde, 1993*
20. **Protocols for Oligonucleotides and Analogs:** *Synthesis and Properties,* edited by *Sudhir Agrawal, 1993*
19. **Biomembrane Protocols:** *I. Isolation and Analysis,* edited by *John M. Graham and Joan A. Higgins, 1993*

METHODS IN MOLECULAR BIOLOGY™

Neuropeptide Protocols

Edited by

G. Brent Irvine

Carvell H. Williams

The Queen's University of Belfast, Northern Ireland, UK

Humana Press **Totowa, New Jersey**

999 Riverview Drive, Suite 208
Totowa, New Jersey 07512

This publication is printed on acid-free paper. ∞
ANSI Z39.48-1984 (American Standards Institute) Permanence of Paper for Printed Library Materials.

Cover illustration: Fig. 3 from Chapter 25, "Immunocytochemical Methods for Regulatory Peptides," by Kathy M. Pogue and Colin F. Johnston.

Cover design by Patricia F. Cleary.

For additional copies, pricing for bulk purchases, and/or information about other Humana titles, contact Humana at the above address or at any of the following numbers: Tel: 201-256-1699; Fax: 201-256-8341; E-mail: humana@interramp.com

Printed in the United States of America. 10 9 8 7 6 5 4 3 2 1

Library of Congress Cataloging in Publication Data

Main entry under title:

Methods in molecular biology™.

Neuropeptide protocols/edited by G. Brent Irvine and Carvell H. Williams.
p. cm.—(Methods in molecular biology™; vol. 73)
Includes index.
ISBN 0-89603-399-6 (alk. paper)
1. Neuropeptides—Analysis—Laboratory manuals. I. Irvine, G. Brent. II. Williams, Carvell H. III. Series: Methods in molecular biology (Totowa, NJ); 73
[DNLM: 1. Neuropeptides—analysis—laboratory manuals. 2. Neuropeptides—chemical synthesis—laboratory manuals. W1 ME9616J v. 73 1997/ WL 25 N4947 1997]
QP552.N39N4742 1997
612.8'042—dc20
DNLM/DLC
for Library of Congress

96-36390
CIP

Preface

After 20 years of intensive effort, novel neuropeptides continue to be discovered, and the field of neuropeptide research is still expanding. As new analytical techniques become available, their applicability to the study of neuropeptides brings fresh insights into the properties and functions of these ubiquitous chemical messengers. Presented in this single volume, *Neuropeptide Protocols,* are 33 chapters covering these new techniques, together with more established methods.

Each contributor is actively engaged in neuropeptide research and so brings to his or her description an awareness of the practical problems inherent in the method, and provides sound advice on how to overcome them. The format conforms to the style of previous books in the *Methods in Molecular Biology* series. Each chapter provides an instruction to the technique, and itemized list of equipment and reagents, and a step-by-step set of instructions to enable practitioners to reproduce the method. The Notes section gives insights into pitfalls or critical stages, tips to overcome these obstacles, and suggestions for extensions or modifications of the basic protocol.

Neuropeptide Protocols is intended as a benchtop manual providing the entire gamut of techniques that form the essential tool kit of the practicing neuropeptide researcher. It will be useful for those new to the field, as well as for established workers who wish to try a new technique for the first time.

G. Brent Irvine
Carvell H. Williams

Contents

Contributors

PETER W. ABEL • *Department of Pharmacology, School of Medicine, Creighton University, Omaha, NE*
THOMAS E. ADRIAN • *Department of Biomedical Sciences, School of Medicine, Creighton University, Omaha, NE*
JUAN P. ADVIS • *Department of Animal Sciences, Rutgers University, New Brunswick, NJ*
JAMES M. ALLEN • *Department of Biomedical Sciences, University of Ulster, Coleraine, UK*
GRAEME J. ANDERSON • *Department of Chemistry, Manchester Metropolitan University, Manchester, UK*
WILLIAM A. BANKS • *VA Medical Center, New Orleans, LA*
ANNETTE G. BECK-SICKINGER • *Department of Pharmacy, ETH Zurich, Switzerland*
GERARD P. BRENNAN • *School of Biology and Biochemistry, The Queen's University of Belfast, UK*
J. MICHAEL CONLON • *Department of Biomedical Sciences, School of Medicine, Creighton University, Omaha, NE*
GLENN CRITCHLEY • *Micromass UK Limited, Manchester, UK*
ANDREW R. L. DAVIES • *Department of Biochemistry, University of Birmingham, UK*
MELITA B. FASOLD • *VA Medical Center, New Orleans, LA*
TIMOTHY G. GEARY • *Pharmacia & Upjohn Inc., Kalamazoo, MI*
WIJNAND P. M. GERAERTS • *Faculteit der Biologie, Vrije Universiteit, Amsterdam, The Netherlands*
DAVID J. S. GUTHRIE • *School of Biology and Biochemistry, The Queen's University of Belfast, UK*
NORBERTO A. GUZMAN • *The R. W. Johnson Pharmaceutical Research Institute (a Johnson & Johnson company), Raritan, NJ*
DAVID W. HALTON • *School of Biology and Biochemistry, The Queen's University of Belfast, UK*
ANN M. HANLY • *Department of Biomedical Sciences, School of Medicine, Creighton University, Omaha, NE*

Rickey P. Hicks • *Department of Chemistry, Mississippi State University, Oktibbeha, MS*
Nigel M. Hooper • *Department of Biochemistry and Molecular Biology, University of Leeds, UK*
John Howl • *Department of Biochemistry, University of Birmingham, UK*
Wei-Lun Hung • *Department of Animal Sciences, Rutgers University, New Brunswick, NJ*
G. Brent Irvine • *School of Biology and Biochemistry, The Queen's University of Belfast, UK*
William B. Jeffries • *Department of Pharmacology, School of Medicine, Creighton University, Omaha, NE*
Colin F. Johnston • *Department of Medicine, Royal Victoria Hospital, The Queen's University of Belfast, UK*
Abba J. Kastin • *VA Medical Center, New Orleans, LA*
Chris Kowalczyk • *Sussex Centre for Neuroscience, School of Biology, University of Sussex, Brighton, UK*
Ka Wan Li • *Faculteit der Biologie, Vrije Universiteit, Amsterdam, The Netherlands*
Sándor Lovas • *Department of Biomedical Sciences, School of Medicine, Creighton University, Omaha, NE*
Aaron G. Maule • *School of Biology and Biochemistry, The Queen's University of Belfast, UK*
Mark Murnin • *Department of Biomedical Sciences, School of Medicine, Creighton University, Omaha, NE*
Richard F. Murphy • *Department of Biomedical Sciences, School of Medicine, Creighton University, Omaha, NE*
Michael O'Shea • *Sussex Centre for Neuroscience, School of Biology, University of Sussex, Brighton, UK*
Ferenc Ötvös • *Biological Research Center, Hungarian Academy of Sciences, Szeged, Hungary*
Laszlo Otvos, Jr. • *The Wistar Institute, Philadelphia, PA*
SungAe S. Park • *The R. W. Johnson Pharmaceutical Research Institute (a Johnson & Johnson company), Raritan, NJ*
Rosemary A. Parslow • *Department of Biochemistry, University of Birmingham, UK*
Kathy M. Pogue • *Department of Medicine, The Queen's University of Belfast, Royal Victoria Hospital, Belfast, UK*
Daniel E. Schaufelberger • *The R. W. Johnson Pharmaceutical Research Institute (a Johnson & Johnson company), Raritan, NJ*

Chris Shaw • *Department of Medicine, Royal Victoria Hospital, The Queen's University of Belfast, UK*
Jerzy Silberring • *Department of Drug Dependence Research, Karolinska Institute, Stockholm, Sweden*
D. David Smith • *Department of Biomedical Sciences, School of Medicine, Creighton University, Omaha, NE*
Gyorgyi I. Szendrei • *The Wistar Institute, Philadelphia, PA*
Géza Tóth • *Biological Research Center of the Hungarian Academy of Sciences, Szeged, Hungary*
David Waugh • *Department of Pharmacology, School of Medicine, Creighton University, Omaha, NE*
Mark Wheatley • *Department of Biochemistry, University of Birmingham, UK*
Carvell H. Williams • *School of Biology and Biochemistry, The Queen's University of Belfast, UK*
G. Brian Wisdom • *School of Biology and Biochemistry, The Queen's University of Belfast, UK*
Belinda Worster • *Sussex Centre for Neuroscience, School of Biology, University of Sussex, Brighton, UK*
Nicola J. Yarwood • *Department of Biochemistry, University of Birmingham, UK*

1

Preparation of Neuropeptide-Containing Fractions from Biological Materials

J. Michael Conlon

1. Introduction

Neuropeptides vary appreciably in terms of their molecular mass, charge, and hydrophobicity so that there is no single optimum method for their extraction from biological materials such as tissues, cultured neurons, plasma, or cerebrospinal fluid (CSF). As all neuropeptides are rapidly degraded by a range of relatively nonspecific peptidases *(1)*, the extraction procedure must release the peptides from storage vesicles into an environment in which the enzymes are inactive. Several general methods have been used to inactivate the neuropeptide-degrading enzymes while efficiently releasing the neuropeptides into the extraction medium. These include the use of boiling aqueous solvents at low or neutral pH; organic solvents, or a mixture of aqueous and organic solvents, at low temperature; and aqueous solutions of chaotropic agents such as 6*M* guanidine hydrochloride containing a cocktail of protease inhibitors. This article will describe protocols using these different approaches.

Small ($M_r < 5000$) and relatively hydrophilic peptides such as substance P, neurokinin A, neuropeptide-γ, gastrin-releasing peptide, angiotensin II, vasoactive intestinal polypeptide, somatostatin-14, galanin, neuropeptide Y, and calcitonin gene-related peptide are efficiently extracted by dilute acids at high temperatures. Acidic peptides, however, such as gastrin and the octapeptide of cholecystokinin, are more effectively extracted by boiling water than by dilute acids *(2)*. Consequently, this chapter describes a procedure of general applicability that involves sequential extraction of tissue by boiling water followed by homogenization in dilute acid. For very hydrophobic neuroendocrine peptides such as neurokinin B, neuropeptide K, corticotropin-releasing hormone, and urotensin I and for thermally unstable large peptides/proteins such as insulin

From: *Methods in Molecular Biology, Neuropeptide Protocols*
Edited by: G. B. Irvine and C. H. Williams Humana Press Inc., Totowa, NJ

and neuronal growth factors, the use of boiling aqueous solvents results in low extraction yields and loss of biological activity so that methods employing organic solvents at low temperature are preferred. Chaotropic solvents containing protease inhibitors have been used successfully to isolate precursor forms of the neuropeptides (prohormones) that contain peptide bonds that are particularly susceptible to proteolytic cleavage, e.g., prosomatostatin *(3)*.

It is important to realize, however, that all extraction procedures have the potential to generate unwanted artifacts. Extraction at high temperature and low pH can result in hydrolysis of labile peptide bonds, particularly Asp-Pro *(4),* and oxidation of sensitive residues, notably methionine, tryptophan, and cysteine *(5)*. The use of ethanolic or methanolic solvents even at low temperature can result in esterification of aspartic and glutamic acid residues *(6)*. In addition, neuropeptides have been isolated from natural sources in which rearrangements of the peptide backbone have occurred, probably during the extraction/purification procedure, e.g., a β-aspartyl shift in porcine GRP *(7)*. Extraction methods are formulated to minimize, but cannot completely eliminate, these artifacts.

Irrespective of the method of extraction, it is necessary to concentrate the neuropeptide-containing fraction, prior to further purification by gel permeation chromatography and/or reversed-phase HPLC. Selective precipitation methods (e.g., acetone or ammonium sulfate) are generally not applicable to small peptides, and the use of ultrafiltration has the disadvantage that many peptides bind tightly to the membrane filters. Lyophilization is time-consuming and frequently results in low recoveries of neuropeptide owing to sequestration within insoluble material. Immunoaffinity chromatography, using the γ-globulin fraction of a specific antiserum immobilized on CNBr-activated sepharose, is a powerful method for simultaneously concentrating and purifying a neuropeptide from tissue extracts or biological fluids. The method, however, requires the availability of large volumes of high-titer antiserum to prepare a column of high capacity and so the procedure is of restricted applicability. In addition, recoveries of tightly-bound antigens from immunoaffinity columns are often low *(8)*. In the classical isolation studies of Mutt and coworkers, a (neuro)peptide-containing fraction was prepared from an extract of pig intestine by batch adsorption onto alginic acid. Later workers have substituted cation-exchange resins, such as SP-sephadex and CM-cellulose for naturally occurring materials, but recoveries of peptides from charged matrices are usually quite low. In recent years, these methods have been superseded by the introduction of high capacity reversed-phase supports based on octadecylsilylsilica beads. These materials are available in a convenient cartridge form that permits fast flow rates and, consequently, short isolation times and good recoveries of the adsorbed peptides. The particle size of the packing material in the cartridges (25–40 μm) is larger than in conventional high-performance liquid chromatography (HPLC) columns so that the cartridges may be eluted at low pressure. This article describes the use of Sep-Pak C_{18} cartridges, supplied by Millipore Waters Chromatography (Milford, MA), to prepare neuropeptide fractions from tissue extracts, plasma, and CSF.

2. Materials

2.1. Apparatus

2.1.1. Tissue Extraction

For analytical studies involving small amounts of tissue, extractions are carried out in either 16 × 100 mm borosilicate glass tubes or 17 × 100 mm polypropylene tubes (*see* Note 1). For procedures involving high temperature, the tubes containing the aqueous solvents are immersed in a boiling water bath for at least 15 min before adding the tissues. For procedures using acidified ethanol or guanidine hydrochloride, the tubes are immersed in an ice bath. Homogenization of tissue is performed using a rotor/stator-type instrument with stainless steel probe, e.g., Tissue-Tearor homogenizer (Fisher Scientific, Pittsburgh, PA), PowerGen model 35 (Fisher), or Tissumizer model SDT-100EN (Tekmar, Cincinnati, OH).

For preparative studies, larger amounts of tissue are extracted in a borosilicate glass beaker of appropriate volume. Heating is provided by a hot plate, e.g., Corning (Corning, NY) model PC-300. Homogenization is carried out using a Waring blender with either a glass or stainless steel container. Homogenates are stirred using a variable speed motorized stirrer with stainless-steel propeller, e.g., Dyna-Mix stirrer (Fisher). Centrifugation is carried out using polypropylene centrifuge bottles (500 or 1000 mL) in a low speed instrument equipped with a high-capacity swinging bucket or angle-head rotor. Ethanol is removed from the homogenate supernatant under reduced pressure using a rotary evaporation system, e.g., Buchi Rotavapor model RE-111A (Flawil, Switzerland) equipped with water bath and water-cooled condenser (*see* Note 2).

2.1.2. Sep-Pak Isolation

For processing of small volumes of tissue extract, plasma, or CSF for analytical purposes, Sep-Pak Classic C_{18} or Sep-Pak Plus C_{18} cartridges may be used. The latter have the advantage that they may be stacked together for greater sample capacity without the need for connecting tubing. The samples are applied and the cartridges eluted with a manually operated 5- or 10-mL disposable polypropylene syringe. When large numbers of samples need to be processed at the same time, C_{18} cartridges containing 100, 200, or 500 mg of sorbent incorporated into the base of polypropylene syringe barrel can be used in conjunction with a vacuum manifold apparatus with a capacity of up to 24 cartridges (Millipore). Samples of larger volume may processed with the vacuum manifold by using C_{18} Sep-Pak Vac RC cartridges, which contain 100 mg of sorbent connected to a 20-mL polypropylene reservoir. Prior to analysis by radioimmunoassay or ELISA, Sep-Pak-concentrated samples may be conveniently dried in a Savant Speed Vac Concentrator (Savant Instruments, Hicksville, NY).

For preparative studies involving processing of a large volume of tissue extract, up to 12 Sep-Pak Plus cartridges may be stacked together in series and connected to a peristaltic pump, e.g., Pump P-1 (Pharmacia Biotech, Uppsala, Sweden). A leak-free connection between the pump and the array of Sep-Paks may be made using Teflon or Tefzel tubing (external diameter, 1.8 mm) and two Pharmacia tubing connectors. Alternatively, large capacity Sep-Pak Vac C_{18} cartridges containing 2, 5, or 10 g of sorbent may be used together with the vacuum manifold apparatus.

2.2. Chemicals

2.2.1. Solvents for Tissue Extraction

1. Water (*see* Note 3).
2. 1*M* Acetic acid (*see* Note 4).
3. Ethanol/0.8*M* hydrochloric acid (3:1, v/v). Mix 750 mL ethanol (96% non-denatured) with 250 mL water and add 18 mL concentrated hydrochloric acid (Fisher). The solvent is stored at –20°C.
4. 6*M* Guanidine hydrochloride/protease inhibitor solution: Dissolve to a concentration of 6*M*, guanidine hydrochloride (Sigma, St Louis, MO) in water containing 1 m*M* phenylmethylsulfonyl fluoride (PMSF; Sigma), 10 µg/mL pepstatin A (Isovaleryl-Val-Val-Sta-Ala-Sta; Sta = [3S,4S]-4-amino-3-hydroxy-6-methylheptanoic acid) (Sigma), 10 µg/mL E-64 (*trans*-epoxysuccinyl-L-Leucylamido-[4-guanidino]butane) (Sigma), and 100 µg/mL bacitracin (Sigma, *see* Note 5). The solvent is stored at 4°C.

2.2.2. Reagents for Sep-Pak Isolation

1. Prewetting solvent: acetonitrile (Fisher).
2. Solvent A: Add 1 mL trifluoroacetic acid (Pierce; HPLC/spectro grade) to 1000 mL water.
3. Solvent B: Mix 800 mL acetonitrile (HPLC grade) with 200 mL water and add 1 mL trifluoroacetic acid.

The solvents are stored at room temperature. Degassing and filtration are not necessary.

3. Methods

3.1. Extraction Procedures

3.1.1. Extraction of Tissue Using Boiling Water/1M Acetic Acid

The procedure described is applicable to processing a large number of samples for measurement of neuropeptide concentrations by radioimmunoassay or related techniques.

1. Rapidly weigh the tissue samples and cut into small (2–4 mm) pieces while still frozen.
2. Add the tissue to tubes containing water maintained at a temperature of at least 90°C on a boiling water bath. A volume of 10 mL/g tissue is used.
3. Leave the tubes immersed in the boiling water bath for 10 min.
4. After cooling to room temperature, centrifuge the tubes (1600*g* for 30 min) and remove the supernatant using a Pasteur pipet.
5. Add 10 mL 1*M* acetic acid to the precipitate and homogenize for 30 s using a rotor/stator-type homogenizer at maximum speed.
6. Centrifuge the tubes (1600*g* for 30 min) and remove the supernatant.
7. Combine the supernatants from the boiling water and acetic acid extractions and add trifluoroacetic acid to give a final concentration of 0.1% (*see* Note 6). The samples are ready for concentration using Sep-Pak cartridges.

3.1.2. Extraction of Tissue Using Acidified Ethanol

The procedure described is applicable to the processing of large amounts of tissue for preparative studies (*see* Note 7).

1. Weigh the tissue and, while still frozen, add to the chilled acid/ethanol solvent in a blender. A volume of 8 mL/g tissue is used.
2. Homogenize at maximum speed for 1 min.
3. Transfer the homogenate to a glass or polypropylene beaker immersed in an ice bath and stir for 1 h (*see* Note 8).
4. Transfer the homogenate to polypropylene bottles and centrifuge (1600*g* for 30 min).
5. Transfer the supernatant to an evaporation flask and remove the ethanol under reduced pressure using a rotary evaporator. A water bath temperature of 40°C is sufficient for the rapid removal of solvent provided that an efficient pump is used.
6. Centrifuge the remaining extract (1600*g* for 30 min) and remove the supernatant.
7. Dilute the supernatant with an equal volume of 0.2% (v/v) trifluoroacetic acid/water. The solution is now ready for further purification using Sep-Pak cartridges. Defatting of the extract with diethylether or hexane is not usually required in this procedure.

3.1.3. Extraction of Tissue Using Guanidine Hydrochloride/Protease Inhibitor Solution

1. Weigh the tissue and cut into small (2–4 mm) pieces while still frozen.
2. Add to a glass or polypropylene tube containing the chilled guanidine hydrochloride/protease inhibitor solution. A volume of 5 mL/g tissue is used.
3. Homogenize for 30 s using the rotor/stator-type homogenizer at maximum speed.
4. Centrifuge (4000*g* for 60 min at 4°C) and remove the supernatant.
5. Dilute the supernatant with an equal volume of 0.2% (v/v) trifluoroacetic acid/water prior to concentration using Sep-Pak cartridges.

3.2. Preparation of a Neuropeptide-Containing Fraction Using Sep-Pak Cartridges

3.2.1. Tissue Extracts

1. Using a 10-mL polypropylene syringe, precondition the Sep-Pak cartridges by passing acetonitrile (5 mL/cartridge) followed by solvent A (10 mL/cartridge). Fast flow rates (>10 mL/min) can be used.
2. Pass the tissue extract through the Sep-Pak cartridge(s) at a flow rate that does not exceed 2 mL/min. In the case of small volumes, this may be accomplished manually using a polypropylene syringe. For larger volumes, a peristaltic pump or vacuum manifold is used (*see* Note 9).
3. Irrigate the Sep-Pak cartridge(s) with solvent A (10 mL/cartridge). A fast flow rate (>10 mL/min) can be used.
4. Elute the neuropeptide fraction into a polypropylene tube by irrigating with solvent B (4 mL/cartridge) at a flow rate not greater than 1 mL/min (*see* Note 10).
5. Dry the samples in a Savant Speed-Vac concentrator (*see* Note 11).
6. The Sep-Pak cartridges may be regenerated for further use by washing with acetonitrile (4 mL/cartridge) followed by solvent A (10 mL/cartridge) (*see* Note 12).

3.2.2. Plasma and CSF

1. Mix plasma or CSF sample with an equal volume of 2% (v/v) trifluoroacetic/water.
2. Centrifuge (1600*g* for 30 min at 4°C) and remove the supernatant.
3. Repeat step 1 in Section 3.2.1.
4. Pass the acidified plasma or CSF sample through the Sep-Pak cartridge at a flow rate not exceeding 1 mL/min. The flowthrough is collected into polypropylene tubes and passed through the Sep-Pak cartridge a second time at the same flow rate.
5. Repeat steps 3–6 in Section 3.2.1.

4. Notes

1. Irreversible binding of neuropeptides in biological materials to glass or plastic surfaces is much less of a problem than with purified peptides in aqueous solutions, but polystyrene or siliconized glass tubes should not be used.

2. In laboratories with good water pressure, a water pump is adequate for removal of ethanol. If a vacuum pump is used, a rotary evaporation system with a cold trap (dry ice/acetone) instead of water cooled condenser, e.g., Buchi Rotavopor model RE-111C should be used.
3. Suitable water can be obtained using a Milli-Q purification system (Millipore) supplied with water that has been partially purified by single distillation or with a deionization resin.
4. 1*M* may not be the optimum concentration for extracting all neuropeptides. For example, extraction yields using 0.5*M* acetic acid and 2*M* acetic should be compared for the peptide and tissue under investigation.
5. The mixture of protease inhibitors is designed to provide specific protection against active site serine- (PMSF), aspartyl- (pepstatin), and cysteine- (E-64) endopeptidases. Bacitracin is a broad spectrum protease inhibitor. The addition of 10 µg/mL amastatin ([{2S,3R}-3-amino-2-hydroxy-5-methylhexanoyl]-Val-Val-Asp) and 10 µg/mL bestatin ([{2S,3R}-3-amino-2-hydroxy-4-phenylbutanoyl]-L-leucine) to the "cocktail" will provide additional protection against aminopeptidases. The PMSF should first be dissolved in the minimum volume of ethanol.
6. Addition of trifluoroacetic acid to the sample and the elution solvents is extremely important in increasing both the efficiency of binding and the recovery of adsorbed peptide.
7. The use of acidified ethanol is particularly advantageous in the extraction of neuropeptides from gastrointestinal tissues. Mucous and other gelatinous components are extracted by boiling aqueous solvents and these materials may seriously interfere in subsequent chromatographic purifications.
8. In order to avoid artifactual peptide modification *(6)*, it is very important to keep the temperature of the extraction solvent low (approx 0°C) and not to prolong the duration of the extraction.
9. The maximum flow rate consistent with efficient adsorption is dependent on the nature of the neuropeptide, and must be determined empirically. Hydrophilic peptides bind less efficiently to the Sep-Pak cartridges and should be passed through at lower flow rates. In a model study using substance P and met-enkephalin, it was found that a slow rate of elution was more important than a slow rate of application in order to obtain high (>90%) recovery of peptide *(9)*. The study emphasized the importance of using the same flow rates for application and elution of all samples in analytical studies.
10. For certain applications, it may be advantageous to prepare multiple neuropeptide-containing fractions by differential elution of the cartridge. For example, four fractions, containing a different distribution of peptides, may be obtained by sequential elution with 20% (v/v), 40% (v/v), 60% (v/v), and 80% (v/v) acetonitrile/water containing 0.1% trifluoroacetic acid (4 mL/cartridge).
11. For analytical studies, the Sep-Pak concentrated samples can be dried completely and reconstituted in an appropriate volume of assay buffer. For preparative work, however, it is strongly recommended that the drying process be stopped after

approx 75% of the solvent is removed. The residual solution, after centrifugation, can be applied directly to a gel permeation column or reversed-phase HPLC column.

12. The manufacturer recommends that Sep-Pak cartridges be used for single sample application only, but it is the author's experience that at least three plasma samples can be extracted using the same cartridge without loss of efficiency, provided that the samples are free from particulate matter.

References

1. Conlon, J. M. (1993) Proteolytic inactivation of neurohormonal peptides in the gastrointestinal tract. *Handbook Exp. Pharmacol.* **106,** 177–198.
2. Turkelson, C. M. and Solomom, T. E. (1990) Molecular forms of cholecystokinin in rat intestine. *Am. J. Physiol.* **259,** G364–G371.
3. Spiess, J. and Vale, W. (1980) Multiple forms of somatostatin-like activity in rat hypothalamus. *Biochemistry* **19,** 2861–2866.
4. Marcus, F. (1985) Preferential cleavage of aspartyl-prolyl bonds in dilute acid. *Int. J. Peptide Protein Res.* **25,** 542–546.
5. Floor, E. and Leeman, S. E. (1980) Substance P sulfoxide: separation from substance P by high pressure liquid chromatography, biological and immunological activities and chemical reduction. *Anal. Biochem.* **101,** 498–503.
6. Henry, J. S., Lance, V. A., and Conlon, J. M. (1993) Purification and characterization of insulin and the C-peptide of proinsulin from Przewalski's horse, zebra, rhino and tapir (Perissodactyla). *Gen. Comp. Endocrinol.* **89,** 299–308.
7. McDonald, T. J., Jornvall, H., Tatemoto, K., and Mutt, V. (1983) Identification and characterization of variant forms of the gastrin-releasing peptide (GRP). *FEBS Lett.* **156,** 349–356.
8. Murphy, R. F., Imam, A., McGucken, J. J., Hughes, A., Conlon, J. M., Buchanan, K. D., and Elmore, D. T. (1976) Avoidance of strongly chaotropic eluents for immunoaffinity chromatography by chemical modification of immobilized ligand. *Biochem. Biophys. Acta* **420,** 87–96.
9. Higa, T. and Desiderio, D. M. (1989) Optimizing recovery of peptides from an octadecylsilyl (ODS) cartridge. *Int. J. Peptide Protein Res.* **33,** 250–255.

2

Purification of Extracted Peptides for Structural Analysis

Chris Shaw

1. Introduction

The continued modifications in peptide/protein sequencer hardware, derivatization and coupling chemistry, reagent delivery, and component detection provide the protein/peptide chemist with the tools to determine primary structural information on subpicomole quantities of material. This dramatic quantum leap in sequencer sensitivity has made the preparation of samples even more critical with respect to purity. The task of isolating several hundred femtomoles or a few picomoles of a peptide of interest from a crude extract of cells or tissues seems daunting at first but is readily achievable if several important criteria are met and the requisite chromatographic hardware is available. A critical factor in initiating an isolation is the availability of a detection system for the peptide of interest. This may be a bioassay when the peptide of interest displays a novel biological activity, an immunoassay when the peptide displays cross-reactivity with antisera generated to a known peptide, or a chemical assay when structural attributes such as the presence of a C-terminal amide, sulfhydryl-containing, or aromatic residues are determined. Each of these detection systems has inherent advantages and disadvantages but, when possible, the system employed should be rapid and discriminating. Once a detection system has been chosen, and often the choice is dictated and limited by the individual study, a suitable quantity of starting material should be amassed and stored under conditions in which the peptide of interest is stable. The quantities required are a reflection of the relative abundance of the component of interest, but often the detection system employed plays a key role in determining this amount. Generally, bioassays require more starting material than chemical assays, which in turn require more than immunoassay. The value

From: *Methods in Molecular Biology, Neuropeptide Protocols*
Edited by: G. B. Irvine and C. H. Williams Humana Press Inc., Totowa, NJ

of performing pilot studies to determine the relevant detection system, quantity of starting material required, and optimal extraction medium to be employed should be stressed at this point. Once these parameters have been determined, the isolation procedure can be initiated. Where large quantities of tissues or cells are required, the initial extract, once tissue or cell debris has been removed by centrifugation, is often of considerable volume, may contain a high concentration of organic solvent, and is often turbid owing to the presence of microparticulates. If of an aqueous nature, the overall volume of the crude extract can be significantly reduced by lyophilization, but this procedure often incurs a large loss of peptide. The method of choice with such extracts would be, where appropriate, acidification followed by high-speed centrifugation to pellet microparticulates. Peptides present in the resultant clarified supernatant can be concentrated using disposable solid-phase extraction cartridges arranged in series, the number being related to the volume or density of the extract. An extract containing organic solvent can be treated in a similar manner after removal of solvent by prior rotary evaporation. The cartridges can be eluted step-wise with ascending concentrations of acetonitrile and each eluate can be assayed for the peptide of interest. The preparation of neuropeptide-containing fractions from tissues is dealt with in Chapter 1.

Once the eluate containing the peptide has been identified, it can be subjected to the initial high capacity chromatographic fractionation. This will most often be performed at low pressure using hydrophilic gel permeation or ion exchange resins. Used in tandem, both of these fractionations can effect a high degree of peptide purification prior to the application of reverse-phase high performance liquid chromatography (RP-HPLC). They will also provide some useful physiochemical information, such as approximate molecular mass and charge characteristics, respectively, of the peptide of interest. The RP-HPLC column chemistries employed downstream of this point are, in part, determined by this information. For peptides of molecular mass <3 kDa, small pore, high-carbon loaded columns would be chosen, but for peptides larger than this, wider pore columns with lower carbon loading would provide better resolution and elution at lower acetonitrile concentrations. It is important to note that columns of apparent identical specification, for example C-8 with 300 Å pores, may differ greatly in component discrimination, especially if manufactured from different types of silica or synthetic polymer matrices. In cases in which component resolution presents a problem, columns from different manufacturers should be evaluated. This often produces the desired effect. Whereas many workers employ sophisticated diode array detection to assess component purity, much information on sample purity can be gleaned by arranging two, fixed wavelength (214 and 280 nm) detectors in series or by simultaneous screening at these wavelengths on a two channel detector. In addition, with practice, the presence and number of tyrosyl and/or tryptophanyl residues in a peptide can be estimated. Structural analysis should ideally involve a combination of amino acid composition, mass spectroscopy, and automated Edman degradation. In most instances, the latter two techniques are all that is required for full characterization. Contemporary mass spectrometers, such as those that employ electrospray or laser desorption, are highly sensitive and can establish accurate (<0.01% error) mass values on as little as several tens of femtomoles of peptide. Likewise, automated Edman degradative chemistry can now be performed on new generation microsequencers, such as the ABI Procise series, using several hundreds of femtomoles of peptide. The limitations of the latter technique, however, remain essentially unaltered, with sample purity and amino acid sequence *per se,* determining if full structures can be obtained by direct sequencing (*see* Chapter 3). In some instances, the structure of a peptide will not permit direct sequencing. In such situations, enzymatic fragmentation will have to be resorted to. This prospect is often met with feelings of despondency, especially if small quantities of peptide, often the result of many months of hard work, are all that is available. However, the spectrum and quality of highly site-specific proteases of sequencing grade available today should allay all such fears as long as manufacturers' instructions are rigidly adhered to. The choice of protease is usually defined by examination of preliminary amino acid compositional or sequence information. By employing such mapping enzymes at 50-fold lower molar concentrations than peptide, there is little interference in subsequent structural analyses from oligopeptides resulting from protease autodigestion. Peptide digests are usually repurified by RP-HPLC and resolved oligopeptides would be separately subjected to structural analyses, the combined data from which should permit elucidation of the full primary structure of the original peptide.

2. Materials

1. All water employed in chromatographic solvent construction should be of the highest HPLC grade available (at least 18 megaohm pure), either obtained from commercial sources or produced from a purpose-designed, well-maintained HPLC water system. Periodic checks should be made on the quality of all reagents employed by running blank gradients on chromatographic columns to assess for spurious peaks owing to contaminants. Some of these peaks, however, if different in retention time from peptides of interest, can function as useful internal standards.
2. Although methanol is often employed as an HPLC solvent, there is no doubt in the author's mind that, for the purposes of peptide isolation, acetonitrile is the most efficacious solvent. This should always be of the most UV-transparent grade obtainable, designated "far UV grade" or better still, the even more UV transparent "super gradient" grade offered by some commercial companies. All solvents

for HPLC should be degassed with helium, preferably before connection to the system and ideally also periodically during use. Particle traps on helium lines should be checked and changed regularly as pernicious contaminants can be introduced into the HPLC system from this source.

3. Several different counterions can be employed in HPLC solutions, but trifluoroacetic acid (TFA) is the most commonly employed usually at a concentration of 0.1% (v/v). The TFA should be fresh, no more than 6 mo from date of manufacture, as older solutions can produce ghost peaks in chromatograms. Phosphoric acid (0.1%, v/v) can also be employed as a useful counterion for peptides whose purification is difficult to effect with TFA. This tends to promote elution of peptides at lower acetonitrile concentrations and can often significantly alter column selectivity. For acidic peptides, triethylamine (TEA) can be employed as a counterion but must be buffered to below pH 7.0 to ensure compatibility with silica-based stationary phases (*see* Note 1). All counterions should be of HPLC grade.
4. Glass and polystyrene tubes should not be used in fraction collection as their charged surfaces result in high degrees of peptide loss owing to adsorption. This effect becomes more evident as the peptide reaches purity and it is not unknown for peptides to suddenly vanish after a fractionation resulting from this process. Polypropylene tubes are the ideal for all peptidic solutions and should be used throughout peptide isolation.
5. For gel permeation chromatographic fractionations, gels with suitable exclusion and flow characteristics should be chosen. For most peptides, Sephadex G-50 or G-25 (or their Sephacryl equivalents) are ideal. The elution buffer may vary from peptide to peptide but should be free from added protein. As protein is excluded, the nature of the eluent should be such that it effects high peptide recovery and ideally should be intrinsically bacteriostatic. Acetic acid, in the range of 0.5–2*M*, has been found to be ideal for this purpose.
6. The HPLC equipment should permit gradient elution from chromatographic columns. Sophisticated systems are not necessary and indeed, in some conformations, inhibit successful isolation. Dual wavelength detection at 214 and 280 nm is highly desirable, permitting both purity assessment and aromatic amino acid residue assignment prior to sequencing. A suitable HPLC fraction collector is desirable for early fractionations. In later fractionations, as the peptide approaches purity, manual collection of absorbent peaks is more efficient and discriminatory.
7. The most commonly employed HPLC columns consist of a silica solid phase derivatized with carbon chains that may vary in chain length from C-3 (propyl), through C-8 (octyl), to C-18 (octadecyl). Diphenyl derivatives are also often employed. In initial stages of purification, high-capacity semipreparative columns are employed usually of 1-cm id and 30–60 cm in length. In later stages, analytical columns are employed of 0.46-cm id and of 15–25 cm in length.
8. Endoproteinases should be employed for digestion of peptides whose primary structures preclude complete sequencing or for those peptides of >40 residues in length. All should be of specified sequencing grade and manufacturers' instructions should be strictly adhered to.

3. Methods

3.1. Preparation of Crude Extract

1. A preparative extraction of tissues or cells containing the peptide of interest should only proceed following the evaluation of data derived from pilot experimentation. These will have determined the quantity of starting material required and the most appropriate extraction medium to be employed. Tissues should be homogenized in extraction medium, and maintaining this at a low temperature (4°C) will aid in the inhibition of endogenous protease activity. For many peptides, ethanol/0.7*M* HCl (3:1, v/v) is a highly efficient extraction medium fulfilling most, if not all, the criteria for such. This medium should be employed at a ratio of 8 vol/g tissue or 3 vol/mL of plasma, serum, culture medium, or other biological fluid. Homogenates or biological fluid extracts should be constantly stirred and kept at 4°C for 12–18 h to ensure efficient peptide solubilization.
2. The starting volume of the crude extract should be reduced by organic solvent rotary evaporation and/or lyophilization as appropriate. Peptides in the remaining solvent-free solution or reconstituted lyophilizate should be concentrated by pumping through disposable C-18 cartridges arranged in series. Low flow rates of 10–12 mL/h will ensure a high level of peptide adsorption to the octadecasilyl stationary phase. After washing the cartridges with aqueous TFA, bound peptides can be differentially eluted in a step-wise manner with increasing concentrations of acetonitrile. Once a suitable aliquot of each eluate has been screened for the peptide of interest and the peptide detected, the eluate of choice can be evaporated to near dryness (*see* Note 2).

3.2. Gel Permeation Chromatography

1. The use of 2*M* acetic acid as mobile phase permits chromatography to be carried out at room temperature with high levels of peptide recovery.
2. Prior to loading lyophilized samples onto these columns, reconstitution in mobile phase (2–3 mL) followed by a short, high-speed centrifugation to remove microparticulates, is advised as this greatly prolongs the effective life of the column. A column of 90 × 1.6 cm should be eluted at a flow rate of 10–12 mL/h to facilitate most efficient component partition, and the total volume of loaded sample should not exceed 3 mL.
3. Fractions should be collected at no more than 15-min intervals and, after the predetermined total volume (V_t) of the column has eluted, an aliquot of each fraction should be screened for the peptide of interest.

3.3. RP-HPLC

1. Gel permeation fractions containing the peptide of interest as determined by the screening assay employed, can be pooled and, if in acetic acid mobile phase, can be pumped directly onto a semipreparative RP-HPLC column that has been equilibrated in starting solvent such as 0.1% (v/v) aqueous TFA. Generally, the dimensions of the column employed at this stage should be related to the mass of

tissue employed to generate the original extract as this will reflect peptide loading, which in turn will affect component resolution. Ideally, the column should be of the highest capacity possible but the individual column dimensions may in reality be restricted by capital cost. Generally, wide-pore C-18 column chemistry should be employed at this stage as this represents the ideal compromise. For peptides greater than 4 kDa in molecular mass, wide-pore, low carbon-loaded analytical columns are recommended in subsequent fractionations. For peptides lower than 4 kDa in molecular mass, narrow-pore, high carbon-loaded columns are likewise recommended. Gradients employed in this initial semipreparative RP-HPLC fractionation should be as shallow as possible, the exact percentages being determined by the acetonitrile concentration effecting elution of the peptide from the disposable cartridges employed in an earlier stage of the purification scheme.

2. To obtain a peptide of apparent spectrophotometric homogeneity, a series of sequential analytical HPLC fractionations will usually be required. At each refractionation, fractions containing the peptide of interest should not be lyophilized. Rather, they should be diluted 1:4 with initial aqueous mobile phase and pumped directly onto the next analytical column. Dilution in start buffer is essential to remove the inherent eluting potential of the acetonitrile in these fractions.
3. After pumping the diluted fraction onto the column, one should wait until the absorbance of the effluent returns to baseline (usually 15–20 min), before initiating the elution gradient. This ensures that all the sample has adsorbed to the column and that re-equilibration has occurred. Gradients employed for elution will depend on the inherent hydrophobicity of each peptide. Generally, these should proceed rapidly to some 15% of elution solvent less than that required for elution and then progress in a shallow fashion increasing by some 0.5% of elution solvent or less/min until elution of the peptide of interest has been effected.
4. On typical analytical columns, flow rates of 1 mL/min or less favor partition of peptides with similar hydrophobic characteristics.
5. During all analytical runs, column effluents should be monitored at 214 and 280 nm. While the 214-nm detector will detect all peptidic materials, the 280 nm detector will indicate the degree of aromaticity of each peptide. With the 280-nm detector set at 10 times the sensitivity of the 214-nm detector, a single tyrosyl residue produces a similar deflection to five peptide bonds. A single tryptophanyl residue produces a similar deflection to 15 peptide bonds. Toward the end of the purification scheme, peaks of absorbing material should be collected manually. This ensures that peptide peaks can be collected into individual tubes that may not happen if two peptides elute close to one another within the same fraction collection window. The delay between detector deflection and real-time elution from the effluent port can be ascertained in any fixed volume system by calibration with a peptide standard that can be detected in fractions either immunochemically or by bioassay. This time delay should remain constant for a given flow rate if the plumbing of the system is not altered in any way.

6. For peptides that are difficult to purify to homogeneity by this standard scheme (with TFA as counterion), different counterions such as phosphoric acid or heptafluorobutyric acid (HFBA) can be employed (*see* Note 3). These two counterions render peptides more hydrophilic and hydrophobic, respectively, in relation to TFA. In addition, selectivity may be significantly altered affecting baseline resolution of components that would be impossible under the conventional scheme. Shaw et al. *(1)* is a powerful illustration of this phenomenon.
7. Inappropriate handling of pure, isolated peptides immediately prior to structural characterization can unravel previous care and attention to detail such that several months of intensive work can be negated. If peptides are to be forwarded to a core facility for structural analysis, they should be sent in a sealed polypropylene tube in the elution solvent from the final analytical HPLC fractionation. They should not be lyophilized, since this may result in major adsorptive losses. When received in the structural characterization facility, the sample can be evaporated to a volume appropriate for mass spectroscopy and microsequencing by direct application. The peptide will thus not have been subjected to lyophilization at any stage of the purification procedure except perhaps when present initially in the crude extract.

3.4. Endoproteinase Digestion

1. In those cases in which the peptide is of long chain length (usually >40 residues) or of unusual structure, direct microsequencing may not, owing to a variety of factors, result in elucidation of the entire primary structure. Mass spectroscopy data will permit estimation of the approximate number of residues in the segment not deduced by direct microsequencing. The primary structural information obtained by direct microsequencing will enable the choice of appropriate endoproteinase to be made. A considerable and ever-increasing range of site-specific endoproteinases are available commercially in highly purified and characterized sequencing grades and the choice will depend on the primary structural attributes of each individual peptide. Trypsin, chymotrypsin, Asp-N, Glu-C, Lys-C, and Arg-C are a few of those available.
2. After incubation of the peptide with an appropriate endoproteinase under the conditions specified in manufacturers' instructions, the digest is fractionated by RP-HPLC. This removes buffer salts and permits manual collection of each oligopeptide. Each fragment is then subjected to mass spectroscopy and microsequencing after which the full primary structure of the original peptide should be possible to deduce (*see* Note 4). McKay et al. *(2)* and Maule et al. *(3)* describe the use of different endoproteinases to this end.
3. In some circumstances, when the peptide produces no signal on the microsequencer, the N-terminus may be chemically-blocked. This is often owing to the presence of a pyroglutamyl group formed by the acid-mediated cyclization of an N-terminal glutaminyl residue, or by the presence of an acetylated α amino group. Both of these N-terminal modifications, which preclude initiation of Edman chemistry, can be readily and efficiently removed by highly specific commer-

cially available enzymes. Repurification by RP-HPLC can be performed as previously described. Enzymatic removal of pyroglutamyl residues is described in refs. *1* and *4*.

4. Notes

1. The silica support of most commonly employed RP-HPLC columns is damaged by mobile phases of neutral or basic pH. When employing counterions such as triethylamine, ensure that mobile-phase pH values are buffered to below pH 7.0. If pH values above this are required, then resort to polymer-based matrices such as supplied by Astec, Whippany, NJ.
2. When lyophilizing extracts, be careful not to achieve complete dryness as this may cause problems in reconstitution with resultant loss of peptide.
3. As HFBA suppresses 214 nm absorbance of peptidic material, ensure that this counterion is employed at an early stage in the purification process if required. Phosphoric acid should not be employed at the final stage of purification as this acid is not entirely volatile.
4. Sequencing grades of endoproteinases should always be employed to prevent possible aberrant cleavages by impure, nonquality-controlled preparations. The additional expense is well worthwhile when one considers that the results of several months of hard work may depend on performance.

References

1. Shaw, C., Murphy, R., Thim, L., Furness, J. B., and Buchanan, K. D. (1991) Marsupial possum neurotensin: a unique mammalian regulatory peptide exhibiting structural homology to the avian analogue. *Regul. Pept.* **35,** 49–57.
2. McKay, D. M., Shaw, C., Thim, L., Johnston, C. F., Halton, D. W., Fairweather, I., and Buchanan, K. D. (1990) The complete primary structure of pancreatic polypeptide from the European common frog, *Rana temporaria*. *Regul. Pept.* **31,** 187–198.
3. Maule, A. G., Shaw, C., Halton, D. W., Thim, L., Johnston, C. F., Fairweather, I., and Buchanan, K. D. (1991) Neuropeptide F: a novel parasitic flatworm regulatory peptide from *Moniezia expansa* (Cestoda: Cyclophyllidea). *Parasitology* **102,** 309–316.
4. Shaw, C., McKay, D. M., Halton, D. W., Thim, L., and Buchanan, K. D. (1992) Isolation and primary structure of an amphibian neurotensin. *Regul. Pept.* **38,** 23–31.

3

Amino Acid Sequencing of Neuropeptides

Ka Wan Li and Wijnand P. M. Geraerts

1. Introduction

The determination of the primary structure of peptides to-date has usually been carried out by automated amino acid sequencing involving Edman chemistry. This sensitive method is reliable, easy to perform, and the interpretation of the results is straightforward. However, other techniques such as mass spectrometry and enzymatic degradation are needed to confirm the sequence and/or to detect the posttranslational modifications. In this chapter, we will focus on the Edman degradation method, and describe the complementary methods as judged appropriate.

In the year 1950, Pehr Edman *(1)* introduced a chemical method for sequential degradation of peptides and proteins. This method became known as Edman degradation (Fig. 1). Throughout the years, the essence of Edman degradation remained largely unchanged. The sensitivity nevertheless increased tremendously owing to the improvement in automation of the degradation procedures, the implementation of gas/pulsed-liquid phase reaction vessels as containers for Edman chemistry *(2),* and the use of (micro-) high performance liquid chromatography (HPLC) for the identification of the phenylthiohydantion (PTH)-amino acids. The different forms of sequencers with different reaction vessels are described by Lottspeich et al. *(3)*. The apparatus described is an Applied Biosystems model 473A sequencer (Foster City, CA) (Fig. 2). In this model, the neuropeptide has to be loaded onto the glass fiber filter that is positioned within the reaction chamber.

Edman degradation can be divided into three steps: coupling, cleavage, and conversion (Fig. 1). The first step couples the Edman reagent phenylisothiocyanate (PITC) to the free N-terminal amino group of a peptide chain to form a phenylthiocarbamyl (PTC)-peptide. The coupling reaction has its optimum at

From: *Methods in Molecular Biology, Neuropeptide Protocols*
Edited by: G. B. Irvine and C. H. Williams Humana Press Inc., Totowa, NJ

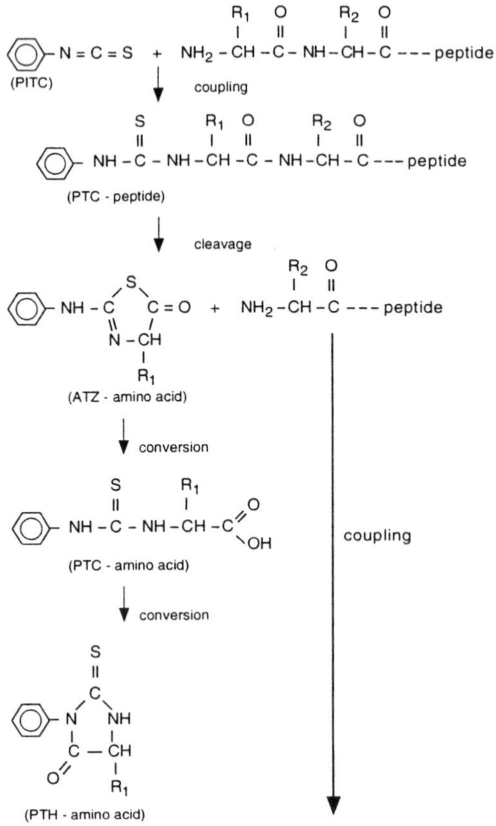

Fig. 1. The chemical reaction of the Edman cycle.

pH 8.0–8.5. In the pulsed-liquid phase sequencer, an exact volume of PITC just sufficient to wet the glass fiber filter is delivered, and a continuous stream of nitrogen transports the gas trimethylamine to the reaction chamber to create a basic milieu. Together, these procedures provide a favorable reaction environment with minimum sample wash-out. After the reaction is completed, the glass fiber filter is washed by a hydrophobic solvent, ethyl acetate, to remove the excessive PITC and its reaction byproducts.

In the second step, the cleavage reagent trifluoroacetic acid (TFA) is pulse-delivered to the PTC-peptide, and again the volume is just enough to soak the filter. The PTC-N-terminal residue is rapidly cleaved from the peptide to yield an anilinothiazolinone (ATZ) amino acid. The peptide is now one amino acid shorter with a new reactive N-terminus and can undergo another cycle of coupling and cleavage steps. The filter is washed with ethylacetate and the hydro-

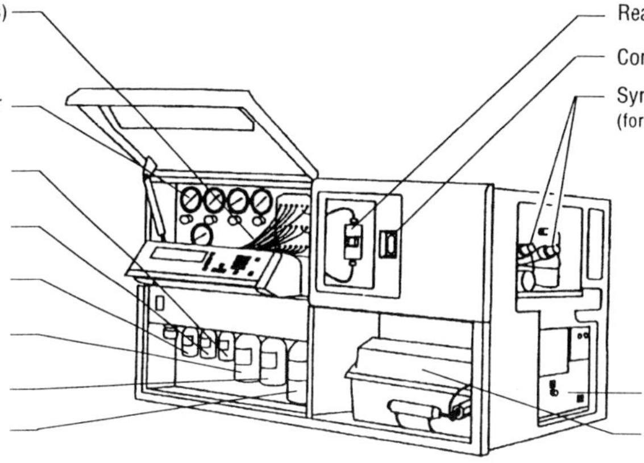

Fig. 2. The pulsed-liquid phase sequencer.

phobic ATZ-amino acid is transferred in this solvent to the conversion flask. As the ATZ-amino acid is not stable, the third step, conversion, is performed to stabilize the residue. A small amount of 25% TFA is delivered to the conversion flask, and at about 60°C, the ATZ-amino acid is hydrolyzed to a PTC-amino acid that is subsequently cyclized to a stable PTH-amino acid. After conversion is completed, the solvent is evaporated from the flask, leaving the dried residue of the PTH-amino acid behind. The residue is reconstituted in 20% acetonitrile in water and transferred to the HPLC system. The PTH-amino acid is resolved by a C18 reversed-phase column and detected by UV absorbance at 269 nm. All PTH-amino acids can be separated by gradient elution, and the sequentially released PTH-amino acids can be identified by their characteristic retention time. An example is shown in Fig. 3. Side products of the Edman degradation generated by reactions between PITC and H_2O or trimethylamine are also detected. In total, a single sequencing cycle including the Edman degradation and the PTH-amino acid detection takes about 45 min.

Edman chemistry is harsh and can destroy some amino acids partly or even completely. As a rule, approx 20% serine and 50% threonine are recovered as intact PTH-amino acids, whereas cysteine is totally destroyed.

Edman degradation requires an N-terminal reactive amino group for the reaction to proceed. Posttranslationally modified N-terminal amino acids, in the form of acetyl-, formyl-, and pyroglutamate, are blocked and cannot be sequenced. An exopeptidase, pyroglutamate aminopeptidase, can be used to remove the pyroglutamate, and the trimmed peptide is then sequencable. Other forms of blockage are generally difficult to deblock (*see* ref. *4* for chemical deblocking) and alternative methods should be considered, e.g., by tandem

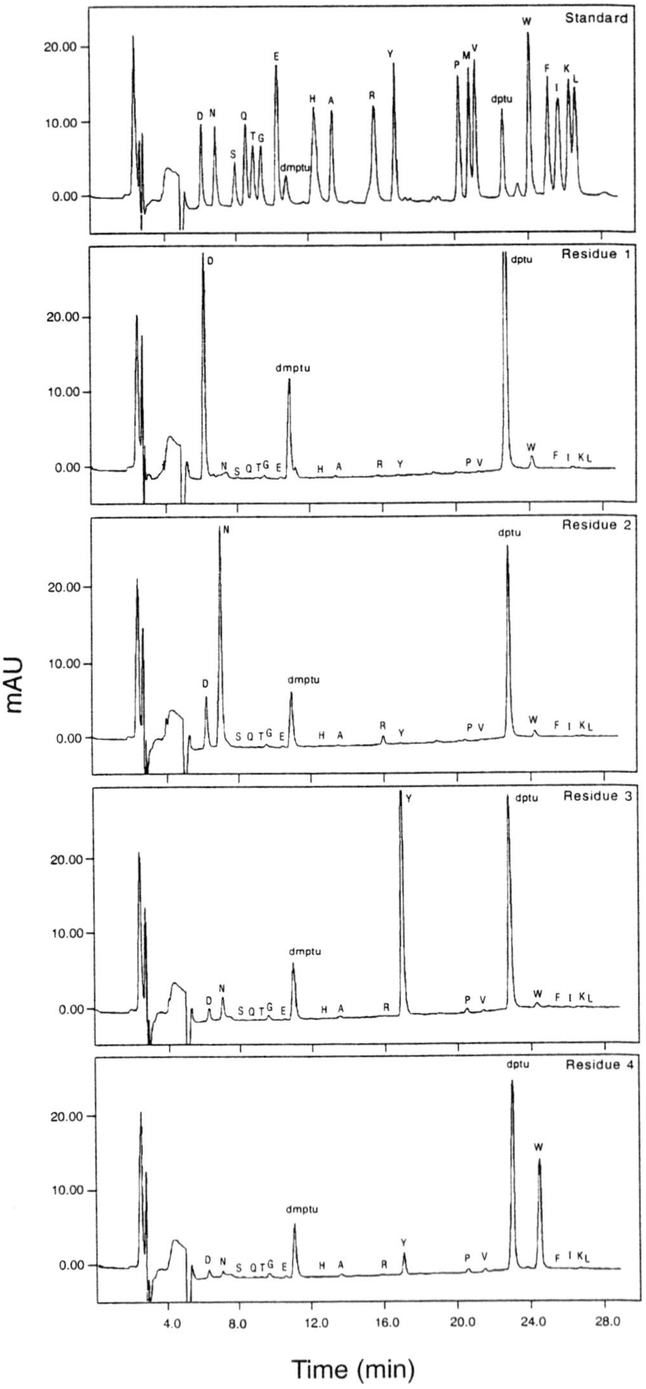
Standard
Residue 1
Residue 2
Residue 3
Residue 4
mAU
20.00
10.00
0.00
D N S Q T G E dmptu H A R Y P M V dptu W F I K L
4.0
8.0
12.0
16.0
20.0
24.0
28.0
Time (min)

mass spectrometry using fast atom bombardment or electrospray ionization source, or by postsource decay on a laser desorption mass spectrometer.

2. Materials

2.1. Apparatus

There are several commercially available automatic amino acid sequencers. The apparatus is user friendly; once the sample is loaded, the apparatus can be set to proceed with the preset number of sequencing cycles, and no further manual operation is required.

2.2. Edman Chemistry

1. 5% PITC in n-heptane.
2. 12.5% trimethylamine in water.
3. Neat TFA.
4. 25% TFA in water.
5. 20% acetonitrile in water.
6. Ethyl acetate.
7. The PTH-standard in acetonitrile.
8. Biobrene.

The HPLC-grade acetonitrile can be used to prepare the 20% acetonitrile; all other reagents should be sequencer grade and can be obtained from Perkin-Elmer Applied Biosystems (Norwalk, CT) division. The analytical-grade reagents can be used only if they are further purified *(5,6)*.

2.3. Chromatography

1. Eluent A: Add 5% HPLC-grade tetrahydrofuran to water (1L). Add further 16 mL 3*M* sodium acetate pH 3.8, and 42 mL 3*M* sodium acetate pH 4.6 to the eluent. If the eluent is not used directly, it should be purged with argon or nitrogen gas and stored air tight.
2. Eluent B: HPLC-grade acetonitrile.

2.4. Pyridylethylation of Cysteine

1. 6*M* guanidine hydrochloride-0.1*M* Tris-HCl, adjust pH to 8.5 with HCl.
2. Hydrochloric acid.

Fig. 3. *(previous page)* Sequence analysis of a model peptide. The top panel shows the separation of the standard PTH-amino acids. DMPTU and DPTU are the byproducts of the Edman sequencing. The sequence cycles 1, 2, 3, and 4 detect the PTH-amino acids aspartic acid, asparagine, tyrosine, and tryptophan, respectively. A Brownlee PTH C18 5-μm column with flow rate 210 μL/min was used. The percentage of eluent B was increased linearly from the initial level of 9% to 40% during 18 min, remained at 40% for a further 10 min, then increased to 90% during 1 min. The percentage of eluent B was returned to 9% before beginning a new sequence cycle.

3. 2-mercaptoethanol.
4. 4-vinylpyridine.

3. Methods

3.1. Pyridylethylation of Cysteine

During amino acid sequencing, the cysteine product decomposes very rapidly. Accordingly, unmodified cysteines will only show up as gaps. Therefore, to identify cysteine, it has to be derivatized to a stable form prior to sequencing. A large number of different reagents for sulfhydryl group modification have been described. Among these 4-vinylpyridine appears to be the most useful reagent, and the product can be easily identified in high yield by sequence analysis.

1. Dissolve the peptide in 100 µL of 6*M* guanidine hydrochloride-0.1*M* Tris-HCl, pH 8.5.
2. Add 7.5 µL 2-mercaptoethanol. Vortex to thoroughly mix the sample and the reagent. This reagent will cleave all the disulfide bonds.
3. Flush the vessel containing the sample for several minutes, and close the vessel tightly. Incubate it at 50°C for 2–4 h.
4. Add 15 µL 4-vinylpyridine (*see* Note 1). Vortex, flush with nitrogen, and close the vessel. Leave the sample at room temperature in the dark for 1 h.
5. Separate the derivatized peptide from the reagents by reversed-phase HPLC.

3.2. Planning the Sequence Run

1. The amount of analyte: Sequencing at 100 pmol level is straightforward, whereas at <10 pmol, the assignment of some of the amino acids such as serine, threonine, and tryptophan could be problematic. It is, therefore, useful to first estimate the amount of peptide that can be applied to the sequencer, but the exact measurement of the concentration is not essential. In most cases, peptides are purified by reversed-phase liquid chromatography (*see* Note 2), and the UV peak height (UV detector set at wavelength 214 nm to detect peptide bonds) corresponding to that of the peptide of interest will give a rough indication of the amount of peptide. In general, an amount of peptide corresponding to 0.005 absorbance units is sufficient for amino acid sequencing. However, care should be taken to minimize peptide loss prior to sequencing (*see* Note 3).
2. Purity of the sample: In case the sample contains more than one peptide, each sequencing cycle will give more than one PTH-amino acid. This will undoubtedly complicate the interpretation of the data. Purity of the sample should be ascertained before sequencing. In most cases, an indication of the purity is obtained during peptide purification. A single symmetrical UV peak usually indicates the presence of one analyte. A more reliable method is analysis of the peak scanned at multiple wavelengths simultaneously using a photodiode array detector, but this generally requires much more material.
3. Sequence cycles: Amino acid sequencing is expensive, and unnecessary sequencing cycles should be kept to a minimum. First, the sequencing cycle numbers

should be estimated prior to the actual sequencing. This can be done based on the mass of the peptide measured by a mass spectrometer. On average, for peptides above 2000 Dalton, each 110 Dalton can be taken as an amino acid residue. Second, many peptides are N-terminally blocked and therefore are not sequencable. It is desirable to observe the first four cycles for the generation of detectable PTH-amino acids before committing oneself to a long sequencing session.

3.3. Amino Acid Sequencing

All the amino acid sequencing steps should have been optimized by an experienced person. When the apparatus is constantly running in the proper way, it would give good performance for months without further adjustment of the sequencing steps.

A simple manual procedure to start neuropeptide sequencing is described. This involves loading the sample onto the glass fiber filter, placing it in the correct position, and locking fingertight inside the reaction chamber, then presetting and running the sequencing cycles.

1. Remove the upper glass cartridge block from the cartridge holder within the reaction chamber and put a teflon cartridge seal above the lower glass cartridge block.
2. Clean the cartridge block with acetonitrile on a clean cotton swab.
3. Press gently the glass fiber filter in the cartridge recess, and put the cartridge block back in the cartridge holder with the filter side faced up toward the sample drying arm that is located next to the reaction chamber.
4. Apply 30 μL Biobrene onto the glass fiber filter, and dry with a stream of nitrogen for several minutes. The gas is delivered by the sample drying arm. Biobrene is used to improve the binding of peptide to the filter.
5. Remove the upper glass cartridge, and then insert it back into the holder with the filter side faced down toward the lower block and the teflon cartridge seal.
6. Screw and hand tighten the cartridge and other fittings.
7. Run a filter conditioning cycle. This will remove many contaminants contained in the Biobrene.
8. After the filter conditioning cycle, the sample can be loaded to the filter in a way as described for Biobrene application. The maximum liquid capacity of the filter is 30 μL. If the volume of the sample is more that 30 μL, then sequential application and drying of 30 μL of the sample should be carried out.
9. Program the total sequencing cycles.
10. Run the sample cycles. Once the sequencer is started, all the sequencing steps will be performed automatically until the last cycle as predetermined by the experimenter. The PTH-amino acids are resolved by the HPLC, and the data stored in the computer.
11. After the run, peaks can be integrated, and the identities and the amount of the PTH-amino acids in each cycle are determined by comparison with the peaks given by the PTH-standard.

3.4. Improving the Data

After the sequence run, it is necessary to evaluate whether the sequence is completed/correct or not. Complementary methods are used for this purpose.

Very often, the last single amino acid to be sequenced, i.e., the C-terminal amino acid, does not bind tightly to the glass fiber filter. This could be (partially) washed off by the sequencing solvents and thus will be detected at a lower level, or not detected at all. Simple mass measurement will give an indication of whether the sequencing is completed. The measured mass should equal the calculated mass, based on the amino acid sequencing data. Alternatively, C-terminal sequencing can supplement the N-terminal sequencing. Sequential release of C-terminal amino acids at various time intervals can be performed by enzymatic degradation, e.g., by a mixture of carboxypeptidases A, B, and Y, followed by the analysis of the released amino acids by amino acid composition analysis.

Many biologically active peptides contain a C-terminal amidation instead of a free carboxyl end. Edman sequencing will not distinguish the difference. The simplest way is to perform mass spectrometry; the mass of a peptide containing a C-terminal amidation will be 1 Dalton less than that of a peptide with a free carboxyl end. Alternatively, a peptide with a C-terminal amidation cannot be digested by carboxypeptidases A and B, but can be digested by carboxypeptidase Y. Incubation of the peptide in two separate aliquots with carboxypeptidases A/B and Y, respectively, followed by amino acid composition analysis will give an indirect indication of the C-terminal amidation.

Lastly, there are many types of posttranslational modifications of amino acids, and most of them will not be detected by Edman sequencing. Here, tandem mass spectrometry will be the method of choice for the further elucidation of peptide structure.

4. Notes

1. 4-vinylpyridine should be freshly distillated prior to its use. Otherwise the (partially) oxidized 4-vinylpyridine solution may give many ghost peaks in the chromatogram, which could render the identification of the peptide problematic.
2. Edman degradation is very susceptible to contaminants. Detergents and salts should not be present in the sample. These can be easily removed by reversed-phase liquid chromatography.
3. A major cause of peptide loss prior to sequencing is the nonspecific binding of peptide to the container during its purification/storage. The problem is acute if the sample is completely dried. Whenever possible, the sample should be partially dried, or not be dried at all.

References

1. Edman, P. (1950) Method for determination of the amino acid sequence in peptides. *Acta Chem. Scand.* **4,** 283–290.
2. Hewick, R. M., Hunkapillar, M. W., Hood, L. E., and Dreyer, J. (1981) A gas-liquid-solid-phase peptide and protein sequencer. *J. Biol. Chem.* **256,** 7990–7997.
3. Lottspeich, F., Houthaeve, T., and Kellner, R. (1994) The Edman degradation, in *Microcharacterization of Proteins* (Kellner, R., Lottspeich, F., and Meyer, H. E., eds.), VCH Weinheim, pp. 117–130.
4. Hirano, H., Komatsu, S., Kajiwara, H., Takagi, Y., and Tsunasawa, S. (1993) Microsequence analysis of the N-terminally blocked proteins immobilized on polyvinylidene difluoride membrane by western blotting. *Electrophoresis* **14,** 839–846.
5. Reimann, F. and Wittmann-Liebold, B. (1989) Gas-phase sequencing of peptides and proteins, in *Advanced Methods in Protein Microsequence Analysis* (Wittmann-Liebold, B., Salnikow, J., and Erdmann, V. A., eds.), Springer-Verlag, Berlin, pp. 118–125.
6. Meinecke, L. and Tschesche, H. (1989) Water contents and quality criteria of microsequencing chemicals. Preliminary results of a reevaluation, in *Advanced Methods in Protein Microsequence Analysis* (Wittmann-Liebold, B., Salnikow, J., and Erdmann, V. A., eds.), Springer-Verlag, Berlin, pp. 126–148.

4

Neuropeptide Gene Identification Using the Polymerase Chain Reaction

Aaron G. Maule and Timothy G. Geary

1. Introduction

Polymerase chain reaction (PCR) techniques comprise some of the most powerful tools in molecular biology research. PCR is characterized primarily by the ability of polymerases to amplify specified regions of DNA both rapidly and efficiently and overcomes the lengthy procedures usually associated with in vivo DNA cloning methods. Discovery and development of the PCR technique into a viable and powerful research tool occurred in the mid-1980s *(1)*. Numerous publications give testament to the robust nature of this technique and to its flexibility, evidenced by its application in diverse experimental situations and the ever increasing range of PCR-based techniques. Many complete books are dedicated to PCR technologies and cover a broad range of aspects that are beyond the scope of this chapter (*see*, for example, *2–5*, which we have found useful). Our goal instead is to outline some basic PCR-based techniques commonly used in the identification of neuropeptide-encoding DNA and to highlight the major problems typically encountered.

PCR employs a thermostable DNA polymerase (many of which are available) and two short (typically 15–30 bases) oligonucleotide primers to selectively copy a double-stranded DNA sequence. Sequentially cycling the temperature at which the mixture is held provides successive rounds of primer annealing to the DNA template, synthesis of new DNA from the primers, and dissociation of the primer–DNA complex. Logarithmically increasing amounts of DNA synthesis thus take place without harming the enzyme; after 20 repeated cycles, the target DNA sequence can be increased a million-fold (2^{20}) in abundance.

From: *Methods in Molecular Biology, Neuropeptide Protocols*
Edited by: G. B. Irvine and C. H. Williams Humana Press Inc., Totowa, NJ

The successful application of PCR-based techniques to the identification of neuropeptide-encoding genes is primarily dependent on the quality of the oligonucleotide primers used, which in turn depends on the structural information available. This may take the form of a partial gene sequence, which enables the design of specific (with respect to nucleotide sequence) oligonucleotides to prime the PCR. However, more commonly, the only information available is in the form of an amino acid sequence with no accompanying nucleotide sequence information. The amino acid sequence may be derived from isolated, purified endogenous peptides (*see* Chapter 2) or from highly conserved regions of known peptide families, when using molecular biology techniques to search for related peptides. Since 90% of the amino acids are specified by 2 or more codons, it is unusual to find regions of no degeneracy in neuropeptide sequences. The design of degenerate oligonucleotides is therefore often a necessary first step in PCR techniques. Some guidelines for primer design can be found in Notes 1 and 2.

In the case of neuropeptide-encoding genes, standard PCR procedures typically result in the elucidation of partial sequences and do not usually allow the generation of full-length clones. It may be possible to obtain full-length sequence information using a cDNA library as the template and sense and antisense sequence specific primers coupled with vector (plasmid or phage)-specific oligonucleotide primers in amplification reactions. However, such PCRs preferentially generate truncated cDNAs; most cDNA libraries contain inserts of various lengths, and shorter sequences are preferentially amplified over longer ones. Unless sufficient nucleotide sequence is known such that sense and antisense specific primers can be designed to the 3' and 5' ends of the coding sequence, it is impossible to generate complete neuropeptide encoding genes by standard PCR protocols.

Isolation of full-length cDNAs has typically necessitated the generation of phage or plasmid cDNA libraries and the screening of millions of clones to identify full-length genes. While partial-length PCR products are useful for screening libraries, these procedures may take many months to complete. Alternatively, several modifications of the PCR technique have been developed (and been successful to varying degrees) to amplify the nucleotide sequences of 3' and 5' flanking regions and so ascertain full-length sequence information. In some situations, specific leader sequences as well as homopolymeric oligo(dT) tails characterize the mRNA such that reverse transcription (RT)-PCR enables the elucidation of the complete message, using single sequence-specific sense and antisense primers in conjunction with complementary primers to the leader sequence and the homopolymeric tail. RT-PCR amplification is carried out on mRNA by incorporating a reverse transcription step prior to the PCR. Methodologies have been developed that allow the reverse transcription step and the PCR step to be carried out in a single tube. It is also unnecessary to isolate mRNA, as RT-PCR may be successfully carried out on total RNA extracted from tissues/cells of choice. The reverse transcription step is usually primed using an antisense primer. However, if poly(A)$^+$ RNA is employed as the template, then first-strand cDNA synthesis may be primed using random hexamers or oligo(dT) primers.

Messenger RNA, which does not possess leader-sequences and/or homopolymeric tails, is not suitable for full-length gene identification by RT-PCR. Alternative strategies have been developed to enable the elucidation of full-length gene sequences using very limited sequence information. Methods available for this purpose include anchored PCR *(6)*, rapid amplification of cDNA ends (RACE) *(7)*, and one-sided PCR *(8)*. Modifications of these methods include, ligation anchored PCR *(9)* and single strand ligation of cDNA (SLIC) *(10)*. Detailed descriptions of these procedures are beyond the scope of this chapter but are reviewed elsewhere *(11)*.

DNA fragments amplified by PCR can be visualized following gel electrophoresis and isolated from the gel (*see* Section 3. for details) for further characterization. PCR products typically contain overhanging A residues, and can be conveniently cloned into vectors that contain an overhanging T residue (e.g., pCRII from Invitrogen BV, De Schelp 26, 9351 NV Leek, The Netherlands). A wide variety of DNA sequencing procedures are available for the structural analysis of subcloned PCR products and will not be discussed; for more information on this aspect, the reader is referred to standard molecular biology texts (e.g., ref. *12*). Alternatively, PCR products may be directly sequenced, though technical problems are not uncommon in this exercise (e.g., ref. *3*).

False-positive PCR products are constant and recurring problems, even if one can isolate an amplified fragment of the predicted size. It is laborious, but sometimes unavoidable, to screen for desired sequences by sequencing multiple subclones or purified fragments; Southern hybridization analysis using a homologous gene or an oligonucleotide known to be in the desired sequence is another option. The incidence of false positives may be markedly reduced by using nested primers. In this case, an amplified band is used as the template for another round of PCR, using new primers known to be internal to the desired sequence.

2. Materials

2.1. Apparatus

Equipment that enables PCR-thermal cycling procedures and electrophoresis of the DNA product is required. Thermal cycling equipment may vary from a series of temperature-regulated water baths to "state of the art" hybridization ovens for rapid PCR cycling. Most of the available PCR machines are based on programmable heating/cooling blocks and are generally very reliable. Some PCR machines offer convenient gradient facilities, e.g., the "Robocycler *Gradient* 40" (Stratagene, Cambridge, UK), which provides an annealing temperature gradient option to allow a range of annealing temperatures to be tested simultaneously. Most standard small electrophoresis units and power supplies are suitable for product analysis, e.g., a Gibco/BRL (Paisley, Scotland) Horizon mini-gel electrophoresis unit. Polypropylene microcentrifuge tubes (1.5 and 0.5 mL) are required for the preparation of master mix solutions and PCR incubations, respectively. Thin-walled PCR-tubes (0.5–0.6 mL) are recommended to increase thermocycling efficiency. Accurate adjustable micropipets with disposable sterile pipet tips are essential (positive displacement pipets are most accurate). It is also preferable to have a microcentrifuge suitable for PCR-tubes and capable of 2000*g* to enable quick pulse spins. A refrigerated microcentrifuge capable of 10,000*g* is also required.

2.2. Chemicals

2.2.1. PCR

1. Water: deionized and autoclaved (150°C).
2. *Thermus aquaticus (Taq)* DNA polymerase: 1 U/μL and stored at –20°C (*see* Note 3).
3. *Taq* DNA polymerase buffer (10X concentration): 500 m*M* KCl, 100 m*M* Tris-HCl, pH 8.8, 15 m*M* $MgCl_2$, and 1% (w/v) gelatin. Prepare fresh from autoclaved stock solutions of 1*M* Tris-HCl, pH 8.8, and 1*M* $MgCl_2$.
4. Deoxynucleotide triphosphate (dNTP) stock mix: 10 m*M* each dNTP in distilled water and stored at –20°C.
5. Template DNA: genomic (100 ng/μL) or cDNA (concentration dependent on relative abundance of target DNA, typically 0.1–100 ng/μL) stored at –20°C.
6. Control template DNA used at same concentration as experimental DNA.
7. Oligonucleotide primers (*see* Notes 1 and 2): 10 μ*M* stocks of the sense and the antisense primers in water and stored at –20°C. Sense and antisense control primers are also required in control reactions.
8. Light mineral oil (e.g., Cat. no. M3516, Sigma, Poole, Dorset, UK): autoclaved and stored at room temperature.
9. Chloroform (purity, ≥99%).

2.2.2. Reverse Transcription (RT)-PCR

1. Water: deionized and autoclaved (150°C).
2. Nuclease-free water (Promega, Madison, WI).
3. Poly(A)$^+$ mRNA (≥50 ng/μL) or total RNA from selected source (*see* Note 4).
4. Reverse transcriptase: e.g., Murine Leukemia Virus (MuLV) reverse transcriptase 50 U/μL (Applied Biosystems, Foster City, CA), stored at –20°C.
5. *Taq* DNA polymerase buffer (10X concentration) (as in Section 2.2.1., item 3).

6. dNTP stock mix: 10 m*M* each dNTP in distilled water.
7. 50 μ*M* Random hexamers or Oligo(dT)$_{12–20}$ (i.e., a mixture of polyT oligonucleotides of between 12 and 20 nucleotides in length) or 15 μ*M* specific primer (may be identical to one of the PCR primers).
8. RNase inhibitor: e.g., 20 U/μL Rnasin (Promega) stored at –20°C.
9. *Taq* DNA polymerase: 1 U/μL and stored at –20°C (*see* Note 3).
10. Oligonucleotide primers (*see* Notes 1 and 2): 10 μ*M* stocks of the sense and the antisense primers in water and stored at –20°C. Sense and antisense control primers are also required in control reactions.
11. Light mineral oil (e.g., Cat. no. M3516, Sigma): autoclaved and stored at room temperature.
12. Chloroform (purity, ≥99%).

2.2.3. Anchored PCR

1. Poly(A)$^+$ mRNA (≥50 ng μL^{-1}) or total RNA from selected source, and stored at –80°C.
2. Reverse transcriptase: e.g., Murine Leukemia Virus (MuLV) reverse transcriptase 50 U/μL (Perkin Elmer, Warrington, UK), stored at –20°C.
3. *Taq* DNA polymerase buffer (10X concentration, as in Section 2.2.1., item 3).
4. dNTP stock mix: 10 m*M* each dNTP in distilled water.
5. dATP (1 m*M*).
6. RNase inhibitor: e.g., 20 U/μL Rnasin (Promega), stored at –20°C.
7. 50 μ*M* Oligo(dT)$_{20}$ primer and sequence-specific primers (10 μ*M* each). With respect to the generation of 3' DNA ends, 2 sense primers are required. These primers should be specific to adjacent sequences in the target DNA such that the first primer (primer W) should be 5' to the second specific primer (primer X). With respect to the generation of 5' DNA ends, 2 sequence-specific antisense primers are required. These should also be specific to adjacent sequences in the target DNA such that the first antisense primer (primer Y) should be 3' to the second specific primer (primer Z).
8. TAE buffer: 40 m*M* Tris acetate, 1 m*M* EDTA (pH 7.5, adjusted with HCl at 25°C).
9. 70% Phenol/water/chloroform (Applied Biosystems).
10. 3*M* Sodium acetate, pH 5.2 (adjusted with acetic acid).
11. Absolute Ethanol, stored at –20°C.
12. 50 U/μL of Terminal deoxynucleotidyl transferase (TdT, Stratagene), stored at –20°C.
13. Terminal deoxynucleotidyl transferase (TdT) buffer (5X concentration): 500 m*M* potassium cacodylate (pH 7.2, adjusted with HCl) and 15 m*M* $CoCl_2$.
14. Dry ice.

2.2.4. Electrophoresis

1. TAE buffer: 40 m*M* Tris acetate, 1 m*M* EDTA, pH 8.3 (adjusted with HCl at 25°C). TBE buffer (100 m*M* Tris, 90 m*M* boric acid, and 1 m*M* EDTA, pH 8.3) is a suitable alternative both as the gel and running buffers.

2. Gel loading dye: 0.05% (w/v) bromphenol blue, 40% (w/v) sucrose, 0.1*M* EDTA, and 0.5% (w/v) sodium dodecyl sulfate (SDS).
3. Agarose: Agarose and nondenaturing polyacrylamide gels are suitable for the electrophoresis of PCR fragments. A range of concentrations of Nusieve GTG agarose (Flowgen, Shenstone, Lichfield, Staffordshire WS14 0EE) are recommended depending on fragment size and resolution requirements. Typically, 3–4% gels are used for DNA fragments less than 500 bp in length, 2–3% gels for DNA fragments of 500–1000 bp, and 1.5–2% gels for fragments >1000–1500 bp.
4. Ethidium bromide stock: 5 mg/μL in TAE, pH 8.3 (or TBE).

3. Methods

3.1. Standard PCR

1. Prepare a PCR master mix solution. For 6 × 50 μL PCR reactions, add: 30 μL *Taq* DNA polymerase buffer (10X), 24 μL dNTP stock, 30 μL stock (10 μ*M*) of primer A, 30 μL stock (10 μ*M*) of primer B, 3 μL (15 U) *Taq* polymerase and 177 μL H_2O (*see* Note 5). Mix thoroughly using a micropipet (*see* Note 6). Quick pulse centrifuge (2 s; 2000*g*).
2. Aliquot 49 μL of master mix to each of 6 thin-walled 0.5 mL PCR tubes.
3. Add 1 μL of DNA template to each tube.
4. Add 30 μL of sterile light mineral oil to each tube.
5. Initiate temperature cycling sequence (*see* Notes 7 and 8). Cycle consists of a denaturing step (1 min at 94°C), a primer annealing step (1–2 min at 50°C), and an extension step (1–3 min at 72°C). Annealing/melting temperatures for each primer should be similar and should guide the choice of a particular annealing temperature. Cycle segment times and temperatures need to be optimized for each PCR.
6. Repeat step 5 30–40 times (depending on template concentration and efficiency of particular PCR).
7. After the final cycle, incubate the tubes at 72°C for 10 min to ensure complete extension of PCR products and then store at 4°C prior to electrophoresis.
8. To remove the mineral oil, add 100 μL of chloroform to each tube, mix thoroughly, pulse centrifuge (5 s, 2000*g*), and recover the aqueous phase.

3.1.1. Optimization of PCR

PCR may be optimized in terms of the specificity and quantity of the product generated. Numerous factors may be systematically modified to optimize a PCR, including $MgCl_2$ concentration, pH, and thermal cycling parameters. Other modifications of the basic PCR technique that improve PCR fidelity have been described (*see* Note 9), and a number of PCR "adjuvants" can be tried to improve results (*see* Note 10).

1. A range of buffers is required to examine the optimum conditions for each particular PCR. In this respect, a series of buffers (10X concentration) with varying

concentrations of $MgCl_2$ may be employed in the reactions to test for optimal efficiency. Repeat the standard protocol (Section 3.1.), but replace the *Taq* DNA polymerase buffer (10X) with buffers containing different concentrations of $MgCl_2$ that vary from 0.5 to 5.0 m*M* in 0.5 m*M* steps. PCR optimization procedures can be simplified by using optimization kits that provide a convenient range of buffers and adjuvants for PCR, e.g., PCR Opti-Prime™ kit (Stratagene).

2. Once the optimal $MgCl_2$ concentration has been determined, other parameters may be altered to further optimize the reaction. A series of *Taq* DNA polymerase buffers with a range of pH values, e.g., 8.0, 8.2, 8.4, 8.8, and 9.0, should also be tested. Suitable pH buffers are available in PCR optimization kits.
3. Some studies have reported that the removal of K^+ from PCR reactions can reduce premature chain termination *(13)*. If this appears to be a problem, then K^+-free buffers should be tested.
4. A wide array of thermal cycling parameters may be altered to optimize particular PCR reactions (*see* Notes 7 and 8). In general terms, the shorter the cycling times the better, and these can often be altered to suit the size of the DNA product expected. The minimum denaturation time, which allows complete denaturation of the template, should be employed to restrict degradation of *Taq* DNA polymerase (though, more recently, available enzymes have considerably greater thermal stability, a feature that minimizes this problem) (*see* Note 11).

3.1.2. Electrophoresis of PCR Product

1. Prepare an agarose gel in TAE (or TBE) suitable for electrophoresis of DNA of the expected fragment size. If size is unknown, use 1–1.5% agarose (i.e., 1–1.5 g/100 mL TAE) in the first instance. Heat agarose until completely melted (temperature depends on type of agarose, Nusieve GTG agarose melts at ≥65°C), add 5 µL of ethidium bromide stock solution to 100 mL melted agarose, gently mix by swirling, and pour gel (*see* Note 12).
2. Place droplets (2 µL/reaction) of gel loading solution onto sterile parafilm. Add 5 µL of each reaction to a droplet of gel loading solution and mix with a micropipet.
3. Load the gel with 7 µL of loading solution/reaction mixture.
4. Load suitably sized DNA marker(s) to adjacent gel lane(s).
5. Electrophorese at 5 V/cm until the bromphenol blue has traveled at least 4 cm.
6. Visualize the PCR products using a short-wave UV transilluminator. Excise DNA band(s) from gel for further analysis, e.g., isolation and sequencing (*see* Note 13).

3.2. Reverse Transcriptase (RT)-PCR

In the method outlined below, primer B may be a selected antisense primer, random hexamers, or oligo$(dT)_{12–20}$. The PCR step in RT-PCR may employ oligo$(dT)_{12–20}$ with a sense primer or (where possible) a leader sequence primer with an antisense primer to generate the 3' and 5' ends of the transcripts, respectively.

1. Prepare a master mix solution. For 6 × 20 µL RT-PCR reactions, add to the first of 6 thin-walled PCR reaction tubes: 12 µL *Taq* DNA polymerase buffer (10X), 8 µL dNTP stock, 1 µL (20 U) Rnase inhibitor, 1 µL primer B (2.5 µ*M* random hexamers or oligo$(dT)_{12-20}$ or 0.75 µ*M* specific primer), 1 µL (50 U) MuLV reverse transcriptase, and 91 µL nuclease-free water.
2. Transfer 19-µL aliquots into each of the remaining 5 thin-walled PCR tubes.
3. Add 1 µL RNA sample to each tube (or 1 µg total RNA), gently mix, and pulse spin (2 s, 2000*g*). One control reaction should contain no RNA sample.
4. Cover with 40 µL light mineral oil.
5. Incubate at 25°C (10 min), 42°C (60 min), 99°C (5 min) and place on ice.
6. Remove the mineral oil (as in Section 3.1., step 8) and store the samples at –20°C.
7. Prepare master mix PCR solution as in Section 3.1. except: prepare for 12 reactions, omit the dNTPs and primer B (if this was used to prime the reverse transcription), and bring to a final volume of 480 µL with water.
8. Add 80 µL to each of the RT reaction tubes.
9. Add 50 µL light mineral oil to each reaction and repeat steps 5–8 of Section 3.1.

3.3. Anchored PCR

Anchored PCR includes a number of strategies for generating the 5' and 3' ends of mRNAs. These involve the addition of a target sequence to the end of single-stranded cDNA either by direct ligation or tailing reactions. Sample protocols for anchored PCR are outlined below (*see* Note 14).

3.3.1. Generation of 3' cDNA Ends

1. Prepare a solution (20 µL) containing: 2 µL (100 ng) poly$(A)^+$ RNA, 2 µL *Taq* DNA polymerase buffer (10X), 2 µL dNTP stock mix, 2 µL oligo$(dT)_{20}$ primer, 2 µL (40 U) RNase inhibitor, 8 µL water, and 2 µL (100 U) MuLV reverse transcriptase. Mix gently and pulse spin (2 s, 2000*g*).
2. Incubate for 60 min at 42°C and 5 min at 99°C and place on ice.
3. Prepare a PCR master mix solution as in step 1 of Section 3.1., but replace primers with an oligo$(dT)_{20}$ primer and primer W (final concentration 1 µ*M*).
4. Add 1 µL of cDNA template (from step 2).
5. Repeat steps 4–8 from Section 3.1.
6. Repeat step 3 using oligo$(dT)_{20}$ and primer X (final concentration 1 µ*M*).
7. Add 1 µL of first PCR amplification from step 3 as a template.
8. Repeat steps 4–8 from Section 3.1.
9. Analyze an aliquot by agarose gel electrophoresis.

3.3.2. Generation of 5' cDNA Ends

1. Prepare a master mix solution as in step 1 of Section 3.3.1., but use primer Y instead of oligo$(dT)_{20}$ to initiate cDNA synthesis.
2. Incubate for 60 min at 42°C and 5 min at 99°C and place on ice.
3. Add an equal volume of phenol/water/chloroform and mix using a vortex.

Centrifuge (10,000*g*, 10 min) and recover upper aqueous phase. Repeat this procedure using chloroform alone and recover upper aqueous phase.

4. Add 3*M* sodium acetate to a final concentration of 0.3*M* and 3 vol of absolute ethanol (–20°C) and mix using a vortex. Place on dry ice for 3 min, microcentrifuge (20 min, 10,000*g*).
5. Resuspend pellet in 20 µL TAE (pH 7.5).
6. Repeat step 4 and resuspend the pellet in 5 µL water.
7. Boil sample (2 min) and place on ice.
8. Prepare a solution containing: 2 µL TdT buffer (5X), 1 µL 1 m*M* dATP, 5 µL cDNA template, 1 µL H_2O, and 1 µL (50 U) TdT.
9. Incubate for 60 min at 37°C.
10. Incubate for 2 min at 65°C.
11. Ethanol precipitate DNA as in step 4 and resuspend pellet in 5 µL water.
12. Prepare a PCR master mix solution as in step 1 of Section 3.1., but replace primers with an oligo$(dT)_{20}$ primer and primer Y (final concentration 1 µ*M*).
13. Add 1 µL of cDNA template from step 11.
14. Repeat steps 4–8 from Section 3.1.
15. Repeat step 3 from Section 3.3.1. using oligo$(dT)_{20}$ and primer Z (final concentration 1 µ*M*).
16. Add 1 µL of first PCR amplification from step 14 as a template.
17. Repeat steps 4–8 from Section 3.1.
18. Analyze an aliquot by agarose gel electrophoresis.

3.4. PCR Controls

The remarkable ability of polymerases to amplify very small quantities of DNA means that contaminating DNA in any of the reactants or buffer solutions may lead to nonspecific or false-positive DNA products. To limit the amount of sequencing time required to characterize PCR products, it is essential to run a series of controls to identify and eliminate spurious amplification products. The control reactions that should be cycled with the experimental samples include (*see* Note 15):

1. Reaction in which the polymerase enzyme has been omitted.
2. Reaction in which the template has been omitted.
3. Reaction in which the test template has been replaced with another template that is known either not to contain sequences complementary to the oligonucleotide primers being used, or which is highly unlikely to contain the expected product, e.g., in the case of neuropeptide PCR, nonneural derived cDNA libraries may suffice.
4. Reactions in which only one of the primers has been added: "self-priming" occurs to an unpredictable extent in PCR, and it is wise to ensure that any amplified bands contain both primers.
5. If possible, it is advisable to prepare simultaneously a positive control, using a different set of primers that are known to amplify a different sequence template

in the DNA; in the event that the reaction of interest fails to amplify any bands, this control can rule out template quality as the cause.

4. Notes

1. General guidelines for PCR primers include the following (note that numerous computer software programs are commercially available to aid in the design of oligonucleotide primers for PCR and a number of suitable programs may be accessed on the internet).
   a. Size between 15 and 30 bases, preferably 18–25 (though there are few data to substantiate this preference).
   b. Nondegenerate primers should be used if possible, with a G + C content between 45 and 55% and an even distribution of purines and pyrimidines.
   c. G or C residues are preferred at the 3' end of the primer (termed the GC clamp); the strong hydrogen bonding of GC pairs is thought to stabilize the specific binding of the oligonucleotide to its cognate DNA sequence.
   d. Similarly, a 3' T is to be avoided as this base has the greatest tolerance for mismatch pairing.
   e. The sense and antisense primers should contain no regions of complementarity to avoid formation of primer-dimers, which will amplify preferentially in the PCR.
   f. Despite these precautions, it is well known that different primers made to the same region of DNA perform very differently in PCR; poor amplification can sometimes be remedied by simply using another primer in the same region. At present, there is no rational explanation for this observation.
   g. While considerable degeneracy can be tolerated in PCR, specificity is best served by using minimally degenerate primers. Where degenerate primers are employed, degeneracy should, if possible, be restricted to the 5' end of the oligonucleotide. Codon choices can be limited by knowledge of the codon bias in the target organism. Alternatively, an inosine can be used to replace positions of maximum degeneracy, even at the level of entire codons. It is preferable to avoid inosines in the 9 bases at the 3' end of a primer, since inosine tolerates any base paring arrangements and so can reduce specificity.
   h. Oligo(dT) primer is a mixture of 3 primers containing 12–20 T residues capped on the 3' end by A, C, or G.
   i. As noted in the Introduction, PCR products can be subcloned directly into T-overhang vectors. If a different vector is desired, restriction sites can be added to the 5' ends of the sense and antisense primers (where they will not interfere with specificity of priming) to permit restriction digestion of the amplified bands and subcloning into the same site in any vector; *see* ref. *17* for suggestions on useful restriction enzymes.
2. PCR primers should show negligible hybridization to nontarget areas in the DNA template. Also, following hybridization there should be ≤2000 base pairs between them as reaction efficiency is inversely related to size of product. Generally, primers should be added in a 10^7 molar excess with respect to the template DNA.

3. Several DNA polymerases suitable for PCR amplification procedures are currently available. Some have inherent proofreading activities that increase the fidelity of DNA synthesis; for example, *Pfu (Pyrococcus furiosus)* DNA polymerase has 12-fold higher fidelity than *Taq* DNA polymerase. Other polymerases reduce the occurrence of mismatch pausing, which can prematurely truncate DNA extension, and some are more suitable for the extension of long PCR products. Suppliers' details on the various enzymes should be analyzed to select the polymerase of choice.
4. Total RNA may be conveniently extracted using guanidine/phenol extraction procedures, e.g., Trizol (Life Technologies, Paisley, UK) or RNAzol (Biotecx Laboratories, Houston, TX). Subsequently, the poly(A)$^+$ mRNA fraction may be rapidly isolated using available isolation kits, e.g., PolyATtract Systems (Promega).
5. The amplification of DNA fragments by PCR in the latter stages of the reaction is often limited by the competition between amplification products and oligonucleotide primers for complementary sites in the DNA following denaturation. Therefore, it is essential to maintain a molar excess of reagents with respect to amplified DNA during the reaction.
6. The powerful ability of polymerase enzymes to amplify minute quantities of DNA is not only the *prima facie* benefit but also a major drawback of the PCR technique. Since only small amounts of template are required for successful amplification, it is imperative that no contaminating DNA be present in any of the reactants. To help avoid contamination of samples and buffers, it is essential that fresh micropipet tips be employed for each step in reaction preparation. It is also advisable to use different micropipets for setting up PCR and for analyzing PCR products to limit the possibility of DNA carry-over. It may also be beneficial to use a designated laboratory area for PCR experiments.

   Problems in PCR can also be caused by environmental contaminants, such as pollen (which can inhibit the reaction) or shed hair or epithelial cells, which can produce false-positive results. Performing all steps under a HEPA-filtered hood can help prevent environmental contamination.
7. The annealing temperature (T) is usually calculated as the dissociation temperature (T_d), at which 50% of the oligonucleotide primers are annealed to the template DNA. The following equation is the most commonly employed method to calculate annealing temperatures:

$$T = 4(\#G + \#C) + 2(\#A + \#T)$$

   where # = number of specified nucleotides in the primer. However, more accurate estimates of annealing temperature also make allowances for the relative positions of purines and pyrimidines; commercially available programs to assist in optimizing primer design may be useful. If the dissociation temperatures are different for the 2 primers, then the lower value should be employed in the reaction protocol.
8. Cycling parameters are often extremely critical—not only the annealing temperature but also the denaturation and extension (polymerization) temperatures.

At 72°C, *Taq* DNA polymerase extends DNA at 40–70 bases/s. Therefore, the amplification of target sequences of ≤1000 bases will require no more than 30 s at the optimal temperature. The initial denaturation step (94°C) should be the longest (up to 2 min) and may be reduced in subsequent cycles to 1 min for target sequences of ≤1000 bases. The initial step is required to ensure complete melting of long DNA templates and templates with high GC contents.

9. Modifications of the PCR reaction that can help reduce nonspecific amplifications include *Hot-start* PCR *(14)* and *Touchdown* PCR *(15)*. The former method involves the addition of a critical component for the PCR (*Taq* DNA polymerase, $MgCl_2$, or template DNA) only after the reaction mixture has reached the first denaturation step temperature. Since primer annealing and DNA synthesis begin as soon as all the components are brought together (albeit at slower rates than occur at higher temperatures), false-positive fragments can be obtained. Once a nonspecific fragment is amplified, it will continue to serve as a template even under the higher annealing temperatures used once the reaction cycling begins. Adding the critical components of the reaction only after the denaturation temperature has been reached will prevent nonspecific annealing of primers during the initial heating step. Recently, a thermally activated DNA polymerase has been used such that the complete reaction mixture can be mixed prior to the initial heating step *(16)*. Touchdown PCR involves using high initial annealing temperatures during the first few cycles and then gradually decreasing the annealing temperature such that only the most specific reactions occur in the first few cycles. This should preferentially weight subsequent amplification cycles toward the more specific product(s), thereby increasing the stringency of the PCR reaction.
10. DNAs with high G + C content provide a stern challenge to successful PCR. Several adjuvants or cosolvents have been used to minimize problems with incomplete melting, secondary structure, and so on, that accompany high G + C content. These include dimethyl sulfoxide (up to 10% by volume in the reaction mix, though this concentration inhibits *Taq* polymerase by 50%); glycerol (up to 20%); formamide (up to 10%); nonionic detergents (generally <1%); tetramethylammonium chloride (TMAC, 0.01–0.1 m*M*); and deazaGTP (3:1 ratio with GTP). Several commercially available preparations are also recommended for this purpose, e.g., PerfectMatch (Stratagene). There are few guidelines for choosing any of these, or any particular concentration; rather an empirical approach using different concentrations of several of them is most efficient.
11. In cases of low product yield, additional cycles up to 45 may be required to generate sufficient DNA for analysis (however, it has also been shown that false positive bands appear with increasing frequency as cycle number increases over 30, a precaution to bear in mind when attempting more numerous cycling events). In situations where long denaturation times are required, higher concentrations of polymerase could be added initially or more could be added during the cycling process. In situations where short DNA fragments are being amplified (<500 bp), the extension time may be omitted as there is enough

time between denaturation and annealing temperatures to allow complete product elongation.

12. Avoid getting air bubbles into the melted gel while mixing with the ethidium bromide solution as these may distort the electrophoresis of amplification products.
13. It should be noted that *Taq* DNA polymerase has an error rate of approx 2×10^{-4} nucleotides per cycle. Therefore, it is necessary to amplify and characterize PCR products at least three times to verify DNA sequence. If the PCR product is directly sequenced, however, these errors become undetectable against the background of "correct" nucleotides in any position.
14. First-strand cDNA synthesis is the most critical step in anchored PCR protocols. The fidelity of this step is pivotal to the subsequent amplification of 5' ends of cDNAs. The presence of truncated cDNA ends in the reaction mix can abolish the amplification of full-length cDNAs by PCR. In this respect, high-quality undegraded mRNA is essential. RNA with high G + C content often exhibits stable secondary structure, which will prematurely terminate reverse transcription reactions. It may be helpful to increase the temperature of the RT reaction to help eliminate such secondary structure from the RNA. It has also been found that high template (mRNA) or primer-to-enzyme ratios (>1) can inhibit the fidelity of the reverse transcription step *(11)*.
15. Control omissions from PCR reactions are replaced with an equal volume of vehicle (water or buffer).

References

1. Mullis, K. B., Faloona, F., Scharf, S. J., Saiki, R. K., Horn, G. T., and Erlich, H. A. (1986) Specific enzymatic amplification of DNA in vitro: the polymerase chain reaction. *Cold Spring Harbour Symp. Quant. Biol.* **56,** 263–273.
2. Erlich, H. A. (ed.) (1992) *PCR Technology: Principles and Applications for DNA Amplification.* W. H. Freeman and Company, New York.
3. Griffin, H. G. and Griffin, A. M. (eds.) (1994) *PCR Technology: Current Innovations.* CRC, Boca Raton, FL.
4. Innis, M. A., Gelfand, D. H., and Sninsky, J. J. (eds.) (1995) *PCR Strategies.* Academic, San Diego, CA.
5. McPherson, M. J., Hames, B. D., and Taylor, G. R. (eds.) (1995) *PCR 2: A Practical Approach.* Oxford University Press, Oxford, UK.
6. Loh, E. Y., Elliott, J. F., Cwirla, S., Lanier, L. L., and Davis, M. M. (1989) Polymerase chain reaction with single sided specificity: analysis of a T-cell receptor delta chain. *Science* **243,** 217–220.
7. Frohman, M. A., Dush, M. K., and Martin, G. R. (1988) Rapid production of full length cDNAs from rare transcripts: Amplification using a single gene-specific oligonucleotide primer. *Proc. Natl. Acad. Sci. USA* **85,** 8998–9002.
8. Ohara, O., Dorit, R. L., and Gilbert, W. (1989) One sided polymerase chain reaction: the amplification of cDNA. *Proc. Natl. Acad. Sci. USA* **86,** 5673–5677.

9. Troutt, A. B., McHeyzer-Williams, M. G., Pulendran, B., and Nossal, G. J. V. (1992) Ligation-anchored PCR—A simple amplification technique with single-sided specificity. *Proc. Natl. Acad. Sci. USA* **89,** 9823–9825.
10. Edwards, D. M., Delort, J., and Mallet, J. (1991) Oligonucleotide ligation to single-stranded cDNAs—A new tool for cloning 5' ends of messenger-RNAs and for constructing cDNA libraries by in vitro amplification. *Nucleic Acids Res.* **19,** 5227–5232.
11. Schaefer, B. C. (1995) Revolutions in rapid amplification of cDNA ends: new strategies for polymerase chain reaction cloning of full-length cDNA ends. *Anal. Biochem.* **227,** 255–273.
12. Howe, C. J. and Ward, E. S. (1990) DNA Sequencing, in *Essential Molecular Biology: A practical Approach*, vol. II (Brown, T. A., ed.), Oxford University Press, Oxford, UK, pp. 157–182.
13. Woodford, K., Weitzman, M. N., and Usdin, K. (1995) The use of K^+-free buffers eliminates a common cause of premature chain termination in PCR and PCR sequencing. *Nucleic Acids Res.* **23,** 539.
14. D'Aquila, R. T., Bechtel, L. J., Videler, J. A., Eron, J. J., Gorczyca, P., and Kaplan, J. C. (1991) Maximizing sensitivity and specificity of PCR by preamplification heating. *Nucleic Acids Res.* **19,** 3749.
15. Don, R. H., Cox, P. T., Wainwright, B. J., Baker, K., and Mattick, J. S. (1991) "Touchdown" PCR to circumvent spurious priming during gene amplification. *Nucleic Acids Res.* **19,** 4008.
16. Birch, D. E., Kolmodin, L., Wong, J., Zangenberg, G. A., and Zoccoli, M. A. (1996) Simplified hot start PCR. *Nature* **381,** 445,446.
17. Kaufman, D. L. and Evans, G. A. (1990) Restriction endonuclease cleavage at the termini of PCR products. *BioTechniques* **9,** 304–306.

5

Solid-Phase Synthesis of Neuropeptides by Fmoc Strategies

Chris Kowalczyk and Michael O'Shea

1. Introduction

The aim of this chapter is to give a detailed, step-by-step description of a procedure for obtaining a batch of a desired peptide at the required level of purity. It is assumed that the initial synthesis will be done using an automated peptide synthesizer that performs solid-phase peptide synthesis using the Fmoc strategy *(1)*. This is based on the sequential addition of amino acid residues to an insoluble polymeric support. The base-labile Fmoc group is used to protect the α-amino group of each residue. Those residues that have potentially reactive side chains are protected with acid-labile groups such as t-butyl. After removal of the Fmoc group with piperidine, the next protected amino acid is added using either a coupling reagent or preactivated amino acid derivative *(2)*. At the end of the synthesis, the peptide is cleaved from the solid support to yield a peptide acid or amide, depending on the linking agent used, and the side-chain protecting groups are removed by treating the peptide-resin with a mixture of trifluoroacetic acid and various ion scavengers *(3)*. Methyl t-butyl ether is then added to precipitate the peptide out of the cleavage mixture. The crude peptide is then dissolved and lyophilized, after which it can be purified by high performance liquid chromatography. The purified peptide can then be lyophilized for storage.

2. Materials

2.1. Apparatus

1. Automated peptide synthesizer.
2. Fume hood, for cleavage and extraction of peptide-resin.

From: *Methods in Molecular Biology, Neuropeptide Protocols*
Edited by: G. B. Irvine and C. H. Williams Humana Press Inc., Totowa, NJ

3. High performance liquid chromatography (HPLC) system, for analysis and purification of peptide. This requires two pumps, a gradient controller, a manual injector, a C18 analytical column, a C8 preparative column, a UV detector, and a chart recorder or integrator.
4. Freeze-dryer, for lyophilization of peptide.

2.2. Chemicals

1. Reagents for automated peptide synthesizer, as recommended by the instrument manufacturer.
2. Cleavage/extraction reagents: trifluoroacetic acid, ethanedithiol, thioanisole, triisopropylsilane, water, methyl t-butyl ether, 3*M* acetic acid, 30% ammonia solution.
3. HPLC reagents: acetonitrile, trifluoroacetic acid, water, formic acid. (All reagents should be HPLC grade.)

3. Methods

3.1. Synthesis

The details of the actual synthesis procedure will depend on which model of automated peptide synthesizer is to be used. The example in Sections 3.1.1. and 3.1.2. is a suggested protocol for the ABI 432A, a continuous flow synthesizer that produces 25-μmol batches *(4)*. This instrument has on-line conductivity monitoring and feedback control, which extends the length of the deprotection and subsequent coupling time when the deprotection reaction rate is slow. Other methods of on-line monitoring also exist; for example, the progress of the deprotection reaction can be followed by monitoring the UV absorbance of the released Fmoc group at 300–320 nm. Alternatively the active ester coupling reaction can be monitored by following, spectrophotometrically at 600 nm, the release of a reporter dye from the solid support *(5)*.

3.1.1. Set-Up

1. Plan the synthesis: If necessary, add extended couplings for predicted difficult sequences and add double couplings for predicted very difficult sequences (*see* Notes 1–3).
2. Wash jaw O-rings with methanol (functions 52, 53 to operate jaws).
3. Select required amino acid columns (AACs) and peptide synthesis column (PSC). Label PSC ("Pep #").
4. Load AAC wheel.
5. Check sequence.
6. Check gas supply (60–75 psi).
7. Check levels of reagent bottles. Replace, prime, flow test if necessary.
8. Run leak test.
9. Check level of waste bottle. Empty if necessary.

10. Set up run file. Print run file.
11. Insert AAC wheel: position 1 on left-hand side of jaws.
12. Insert PSC: push down, do not twist.
13. <Run Control>, set options: usually Y, Y, Y.
14. <Start>.
15. Check that the jaws close properly on the first AAC.
16. Check PSC for leaks during first cycle.

3.1.2. End of Synthesis

1. If PSC is not fully dry at end of synthesis, dry with argon for 5 min.
2. Place PSC in 50-mL polypropylene tube. (Can be stored at room temperature.)
3. Replace PSC with calibration column.
4. Discard AACs.
5. Check conductivity trace.

3.2. Cleavage and Extraction

The volumes given in Section 3.2.1. are for a 25-μmol-scale synthesis and should be increased proportionately for larger scale syntheses.

3.2.1. Cleavage

1. Select cleavage mixture:

Moieties present			
Pmc	Trt	DP	Cleavage mixture (C.M.)
–	–	–	C.M. 1: 100 μL TAn + cl
–	–	+	C.M. 1: 100 μL TAn + cl
–	+	–	C.M. 2: 50 μL TAn + 50 μL TIPS + cl
–	+	+	C.M. 2: 50 μL TAn + 50 μL TIPS + cl
+	–	–	C.M. 3: 50 μL TAn + 50 μL water + cl
+	–	+	C.M. 1: 100 μL TAn + cl
+	+	–	C.M. 4: 33 μL TAn + 33 μL TIPS + 33 μL water + cl
+	+	+	C.M. 2: 50 μL TAn + 50 μL TIPS + cl

(EDT, ethanedithiol; TFA, trifluoroacetic acid; TAn, thioanisole; TIPS, triisopropylsilane; cl = 100 μL EDT + 1.8 mL TFA.)
(Pmc, usual side-chain protecting group on Arg; Trt, usual side-chain protecting group on Cys, His, Asn, Gln; DP, Asp-Pro bond.)
2. Open PSC and transfer peptide-resin to a 5-mL polypropylene tube with pushfit cap (Elkay, Shrewsbury, MA).
3. Prepare cleavage mixture in 8-mL polypropylene fraction tube, in fume hood. Discard waste tips in bleach solution. Place cleavage mixture in dry ice until cold.
4. Pour cold cleavage mixture into peptide-resin tube, cap the tube, and mix gently. Note the time.
5. Place tube in a 50-mL polypropylene tube, cap, and place on rocking table for the required cleavage time (*see* Note 4).

3.2.2. Extraction

1. Set up a sintered-glass filter funnel (porosity 3; i.e., grade P40, pore size 16–40 µm; capacity approx 20 mL) and vacuum flask (500-mL) in fume hood. Prerinse funnel with 2 × ~10 mL of methyl t-butyl ether (MTBE).
2. At end of cleavage time, pour cleavage mixture into a 50-mL polypropylene tube. Add ~15 mL of cold MTBE and mix thoroughly. Wash out the peptide-resin tube with 3 × ~1 mL cold MTBE, and add washings to the 50-mL polypropylene tube using a Pasteur pipet. Make up volume of MTBE to 25 mL, cap, and mix thoroughly.
3. Leave the mixture to stand for approx 2 min.
4. Filter the mixture (resin plus precipitated peptide) under gentle vacuum.
5. Wash out the polypropylene tube with 2 × 25 mL cold MTBE. Filter the washings. (Use Pasteur pipet to recover resin from bottom of tube.)
6. If the amount of precipitated peptide is poor and/or the filtrate is cloudy (*see* Note 5), pour filtrate into 100-mL separating funnel. Extract with 2 × 15 mL water. Collect aqueous (lower) fraction in a 100-mL round-bottomed flask.
7. Retain ether (upper) fraction (or unextracted filtrate) in brown glass bottle. Store at room temperature (*see* Note 6).
8. Dissolve the peptide into same round-bottomed flask as in step 6 by washing the peptide plus resin on the filter funnel with 10 mL water, followed by a further 10 mL water, followed by 10 mL acetonitrile (*see* Note 7). (This solution is referred to as Pep # A.) The resin remains on the filter (*see* Note 8).
9. Take 100 µL of Pep # A for analysis. Place it in a 1.5-mL Eppendorf tube (Pep # Aa). Store at 4°C until analysis.
10. Shell freeze Pep # A in liquid nitrogen (i.e., spin the flask in the liquid nitrogen so that the contents freeze in a thin layer on the sides of the flask, thus facilitating the subsequent drying process).
11. Freeze-dry Pep # A (approx 36 h). Store at –20°C until purification.
12. After use, wash all glassware with deionized water followed by methanol. Do not use detergents.

3.3. Analysis and Purification

3.3.1. Analysis

1. Calculate predicted HPLC retention time and molecular weight of peptide, e.g., using Peptide Calculator program, available free from Finnigan MAT (San Jose, CA).
2. Dilute 20 µL of Pep # Aa to 200 µL with eluent A (*see* step 4g, Pep # Ab, acetonitrile concentration in Pep # Ab = 6–8 %).
3. Analyze Pep # Ab by HPLC.
4. HPLC Conditions:
   a. Column: Waters DeltaPak, C18, 5 µm, 100 Å, 3.9 × 150 mm. Other reverse-phase C18 or C8 columns may be used, but this one is ideal for short (5–10 residue) peptides.

b. Detector cell: analytical.
c. Flow rate: 1.0 mL/min.
d. Wavelength: 214 nm.
e. Detector range: 0.5 AUFS.
f. Loop volume: 100 µL (fill loop using 150 [± 10] µL of sample, i.e., injection volume = 100 µL).
g. Eluent A: 5.0% acetonitrile, 0.1% TFA, in water.
h. Eluent B: 60% acetonitrile, 0.085% TFA, in water.
i. Gradient: 0–60% B over 60 min.

5. Measure retention time (RTa) of main peak. Compare with calculated RT (using graph of actual vs calculated RTs).
6. If the peptide is sufficiently pure over the range (RTa ± 10% B), it may be purified by solid-phase extraction (*see* Section 3.3.2.1.), otherwise it should be purified by prep. HPLC (*see* Section 3.3.2.2.).
7. If HPLC analysis of the freeze-dried crude Pep # A is required: Pep # Ac: approx 0.5% of Pep # A dissolved in 1.0 mL eluent A; 100-µL injection, as in Section 2.3.1., step 4).

3.3.2. Purification

3.3.2.1. Solid-Phase Extraction

1. Prepare Sep-Pak C18 solid-phase extraction cartridge (Waters, Milford, MA) by washing with 5 mL of acetonitrile, followed by 5 mL of methanol, followed by 10 mL water (each at approx 5 mL/min).
2. Calculate i and j as follows: i = (RTa – RTo) – 10, where RTo = retention time of an unretained component; i.e., RT of column void volume (approx 3 min using the HPLC conditions in Section 3.3.1., step 4). j = (RTa - RTo) + 10.
3. Dissolve Pep # A in 5.0 mL ([100 – i]% eluent A: i% eluent B) (Pep # B).
4. Take 100 µL of Pep # B for analysis (Pep # Ba).
5. Pass Pep # B through Sep-Pak at approx 5 mL/min using a 5-mL syringe. Pass an additional 5 mL ([100 – i]% eluent A:i% eluent B) through Sep-Pak at approx 5 mL/min. Filtrate: Pep # C.
6. Pass 5.0 mL ([100 – j]% eluent A:j% eluent B) through Sep-Pak at approx 5 mL/min. Filtrate: Pep # D.
7. Pass 5.0 mL 100% eluent B through Sep-Pak at approx 5 mL/min. Filtrate: Pep # E.
8. Dilute Pep # Ba, D, E × 10 (10 µL diluted to 100 µL eluent A).
   Dilute Pep # C × 5 (20 µL diluted to 100 µL eluent A).
9. Analyze Pep # Ba, C, D, E by HPLC.
10. HPLC Conditions:
    a. Column: Waters DeltaPak, C18, 5 µm, 100 Å, 3.9 × 150 mm.
    b. Detector cell: analytical.
    c. Flow rate: 1.0 mL/min.
    d. Wavelength: 214 nm.
    e. Detector range: 1.0 AUFS.

f. Loop volume: 100 μL.
g. Injection volume: 20 μL.
h. Eluent A: 5.0% acetonitrile, 0.1% TFA.
i. Eluent B: 60% acetonitrile, 0.085% TFA.
j. Isocratic.
k. % B = 5.0% less than the % B at the RT of the main peak in the initial analytical HPLC = (RTa – 5)%.

11. The peptide should be present in Pep # D. Analyze Pep # D by gradient HPLC (conditions as in Section 2.3.1.), with injection volume reduced if necessary to bring main peak within chart scale.
12. Wash Pep # D into a 100-mL round-bottomed flask with approx 1 mL water.
13. Shell freeze Pep # D in liquid nitrogen.
14. Freeze-dry Pep # D (approx 24 h).
15. Weigh dried peptide. Calculate yield.
16. The molecular weight of the product should be confirmed, for example, by laser desorption mass spectrometry.

3.3.2.2. Preparative HPLC

1. Dissolve Pep # A in 4.5 mL eluent A. Try 2.0 mL eluent A first. If insoluble, try warming. If still insoluble, try 0.1–2.0 mL 100% formic acid. Make volume up to 4.5 mL with eluent A. If insoluble matter still present, centrifuge (4 Eppendorf tubes, 2 min) then inject supernatant.
2. Purify Pep # A by preparative HPLC.
3. HPLC Conditions:
   a. Column: Brownlee Prep-10, Aquapore C8, 20 μm, 10 × 250 mm.
   b. Guard column: as prep. column, except: 10 × 30 mm.
   c. Detector cell: preparative, i.e., path length = 1/10 of path length in analytical cell.
   d. Flow rate: 5.0 mL/min.
   e. Wavelength: 214 nm.
   f. Detector range: 2.0 AUFS.
   g. Loop volume: 5.0 mL.
   h. Injection volume: 4.5 mL (Pep # A 1).
   i. Eluent A: 5.0% acetonitrile, 0.1% TFA.
   j. Eluent B: 60% acetonitrile, 0.085% TFA.
   k. Gradient: 0–80% B for 80 min (or, if better resolution is required: 0–40% B for 80 min). If RTa > 40 min, prep. HPLC may be done using a gradient of (RTa – 20)% – 100% B at 1% B per min (e.g., RTa = 50 min; gradient = 30–100% B for 70 min), with Pep # A dissolved in the starting percentage of eluent B; e.g., in the above example, Pep # A 4.5 mL (70% eluent A:30% eluent B).
4. Collect all peaks in 8-mL polypropylene fraction tubes.
5. Store fractions at 4°C until analysis.

3.3.2.3. Analysis of Preparative HPLC Fractions

1. Analyze Pep # A 1 fractions by HPLC. (If the initial analytical HPLC of Pep # Ab suggests that the main peak is sufficiently pure, it may not be necessary to analyze the individual fractions.)
2. HPLC Conditions:
   a. Column: Waters DeltaPak, C18, 5 µm, 100 Å, 3.9 × 150 mm.
   b. Detector cell: analytical.
   c. Flow rate: 1.0 mL/min.
   d. Wavelength: 214 nm.
   e. Detector range: 1.0 AUFS.
   f. Loop volume: 100 µL.
   g. Injection volume: approx 20 µL.
   h. Eluent A: 5.0% acetonitrile, 0.1% TFA.
   i. Eluent B: 60% acetonitrile, 0.085% TFA.
   j. Isocratic.
   k. %B = 5.0% less than the %B at the RT of the main peak in the initial analytical HPLC = (RTa – 5)%. (The RT of the main peak should be approx 6–8 min.)

3.3.2.4. Treatment of Purified Peptide

1. Pool all the sufficiently pure fractions into a 100-mL round-bottomed flask (Pep # B).
2. Take 100 µL of Pep # B for analysis. Place sample in 1.5-mL Eppendorf tube (Pep # Ba). Dilute 20 µL (or more, if necessary) of Pep # Ba to 200 µL with eluent A (Pep # Bb). Store at 4°C until analysis.
3. Analyze Pep # Bb by gradient HPLC (conditions as in Section 3.3.1.).
4. Shell freeze Pep # B in liquid nitrogen.
5. Freeze-dry Pep # B (approx 36 h).
6. Also freeze-dry separately those fractions that contain a significant proportion of the desired peak (side fractions). These individual fractions can be freeze-dried in their fraction tubes, covered with a piece of tissue paper held in place with an elastic band.
7. Weigh dried peptide. Calculate yield.
8. The molecular weight of the product should be confirmed, for example by laser desorption mass spectrometry.

4. Notes

1. Possible problem residues:
   a. Pro-Pro.
   b. Pro at C-terminus or C-1. (Omit, or add, Glys if possible.)
   c. Long sequences of charged residues.
   d. Ile/Leu/Val coupled to Ile/Leu/Val.
   e. Phe/Gln/Arg/TrpW/Tyr—large, therefore steric hindrance.
   f. His.

2. Internal sequence peptides for antibody production: It is often useful to amidate these at the C-terminus, to make the terminal residue uncharged, i.e., like the corresponding residue within the original protein. However, do not amidate if the C-terminal -COOH is to be used for conjugation to a carrier protein.
3. N-terminal Q: Q (Gln) at the N-terminus of a synthetic peptide may cyclize to form pyroglutamic acid under acidic conditions (e.g., on an HPLC prep. column). Therefore try to avoid having Q as the N-terminal residue, e.g., by omitting Q or by adding one or more additional amino acids to the N-terminus of the peptide to be synthesized.
4. Cleavage times:

No. of Arg	Cleavage time, h
0	3.0
1	4.0
2	5.0
3+	6.0

5. If there is a good precipitate (peptide plus resin) and the filtrate is absolutely clear, it is not necessary to do an aqueous extraction of the filtrate.
6. This can be re-extracted if the yield of peptide is lower than expected. Otherwise it can be discarded at the end of the purification procedure.
7. If peptide is not soluble in water, add 100–300 μL of 3*M* acetic acid (for basic peptides) or a similar volume of 30% ammonia (for acidic peptides) to aid solution. (Acidic groups: D, E, and C-terminal carboxyl. Basic groups: R, K, H, and N-terminal amino.) Volume of Pep # A = 60 mL (or 30 mL if unextracted at step 6).
8. Allow resin to dry in filter funnel. Place resin in Eppendorf tube. Store at 4°C. This can be recleaved if the yield of peptide is lower than expected. Otherwise, it can be discarded at the end of the purification procedure.

References

1. Atherton, E. and Sheppard, R. C. (1989) *Solid Phase Peptide Synthesis, A Practical Approach,* Oxford University Press, Oxford, UK.
2. Fields, G. B. and Noble, R. L. (1990) Solid phase peptide synthesis utilizing 9-fluorenylmethoxycarbonyl amino acids. *Int. J. Peptide Protein Res.* **35,** 161–214.
3. Applied Biosystems Inc., Foster City, CA (1990) *Introduction to Cleavage Techniques.*
4. Applied Biosystems Inc., Foster City, CA (1993) *Synergy User Manual.*
5. Calbiochem-Novabiochem, Nottingham, UK (1994) *Catalog and Peptide Synthesis Handbook.*

6

Incorporation of Stable Pseudopeptide Bonds

Methylene Amino, Thioether, and Hydroxyethylene Derivatives

Graeme J. Anderson

1. Introduction

The introduction of pseudopeptide bonds (amide bond surrogates) into the peptide backbone during synthesis is now a common technique in peptide chemistry *(1)*. These pseudo-peptide bonds are introduced in order to satisfy criteria such as stability to enzymatic degradation, transition state analogs/ enzyme inhibition, alteration in peptide backbone conformation (with corresponding changes in flexibility and hydrogen-bonding character), increased receptor specificity, increased potency, and biological responses *(2)*.

Since the early 1980s, the number of pseudopeptide bonds reported in the literature has increased markedly, most notably with the research effort into the development of small, stable renin inhibitors *(3)*. Included among the moieties that have been used are the methylene amino (reduced amide), hydroxyethylene, ketone, alkene, ether, and thioether isosteres (*see* Fig. 1).

This chapter describes in detail the preparation of pseudodipeptides resulting from the replacement of the amide bond with the methylene amino, thioether, and hydroxyethylene moieties (Note 1). Incorporation of these units into the peptide as a whole is normally achieved by a combination of solution and solid phase techniques and is outside the scope of this chapter. Details may be found within the references in each section.

The use of pseudopeptide inserts is a rapidly evolving field, and one that allows neuropeptide scientists great scope and versatility in the design and synthesis of novel analogs. These dipeptide fragments can be incorporated into the peptide sequence, generating a specific mutation. This may lead to changes in a number of biological properties of the parent peptide (such as enhanced

From: *Methods in Molecular Biology, Neuropeptide Protocols*
Edited by: G. B. Irvine and C. H. Williams Humana Press Inc., Totowa, NJ

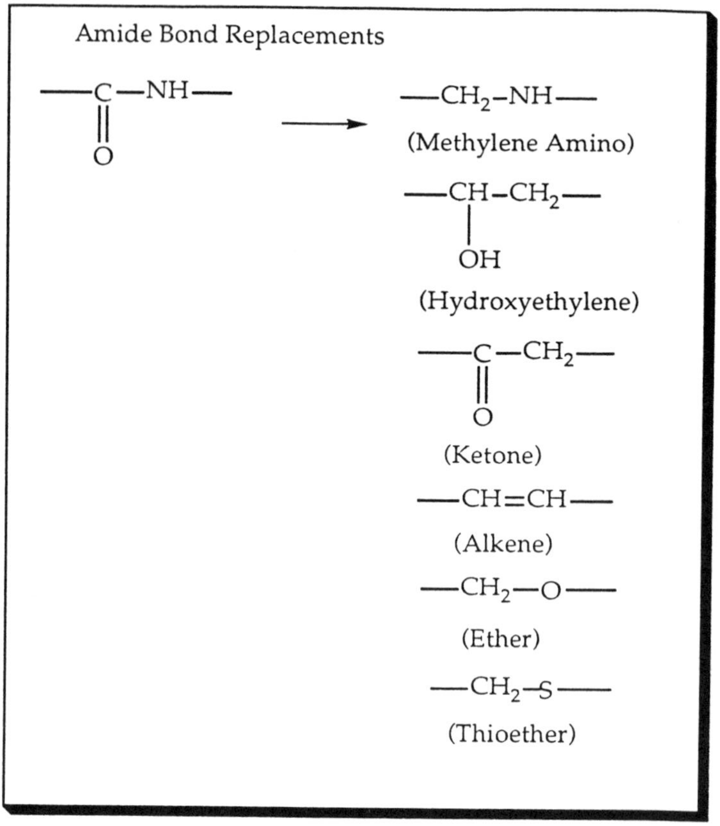

Fig. 1. Examples of the types of functional groups that can be incorporated in place of the peptide bond.

potency, increased in vivo stability, oral activity, enzyme inhibition, and so on) and enable the discovery of new "designer" drugs.

2. Materials

Little specialized equipment is required during these preparations, aside from that which is available in most organic chemistry laboratories, including techniques for structure confirmation such as NMR, mass spectroscopy, and so on.

In general, most solvents should be distilled prior to use, and reagents should be of the highest purity available. Specialized hydrogenation equipment is needed for Section 3.3.1., during preparation of lactone H, but many synthetic laboratories will be equipped with this.

Peptides and pseudopeptides are normally quite hygroscopic in nature and are stable if stored below –20°C.

2.1. Preparation of Methyleneamino Dipeptides

1. Solvents: dichloromethane, ethyl acetate, ether.
2. Boc-amino acids (*see* Note 1).
3. Amino acid esters (*see* Note 1).
4. Triethylamine.
5. Benzotriazol-1-yloxytris [dimethylamino]-phosphonium hexafluorophosphate (BOP).
6. O:*N*-dimethyl hydroxylamine hydrochloride.
7. 3*M* HCl.
8. Saturated $NaHCO_3$.
9. Saturated NaCl.
10. Anhydrous $MgSO_4$.
11. Lithium aluminium hydride.

2.2. Preparation of Thioether Dipeptides

1. Amino acids (*see* Note 1).
2. Boc-aminoalcohols (*see* Note 1).
3. Solvents: ether.
4. Dimethylformamide.
5. Conc. H_2SO_4.
6. KBr.
7. $NaNO_2$.
8. Anhydrous $MgSO_4$.
9. Cesium thiobenzoate ($CsSCOC_6H_5$).
10. 1*M* solution of ammonia.
11. Pyridine.
12. Toluene *p*-sulfonyl chloride.
13. KOH.
14. 5% solution of $Na_2S_2O_3$.
15. A supply of Argon gas.

2.3. Preparation of Hydroxyethylene Dipeptides

1. Solvents: Tetrahydrofuran, ether, 200 mL of 2% acetic acid in toluene (v/v).
2. Diisopropylamine.
3. 1.6*M* n-butyl lithium in hexane (Aldrich, Milwaukee, WI).
4. Ethyl propiolate (distilled).
5. 5% Pd/$BaSO_4$ (hydrogenation catalyst).
6. Hexamethyldisilazane.
7. Requisite alkyl bromide (*see* Note 1).
8. Saturated NH_4Cl solution.
9. Saturated $NaHCO_3$ solution.
10. 10% citric acid solution.
11. n-Butylamine.
12. Silica gel for chromatography.
13. A supply of solid CO_2 (dry ice).

3. Methods

3.1. Synthesis of Methylene Amino Dipeptides

The methylene amino bond has been introduced into a wide variety of peptides as a means of investigating hydrogen-bonding sites and has led to analogs that have been used as enzyme inhibitors and competitive antagonists at receptor sites, or are stable to enzymatic degradation *(4,5)*.

The synthetic protocol involves (Fig. 2) the reaction of a protected amino aldehyde (prepared from the corresponding amino acid) with an excess of amino acid ester in the presence of sodium cyanoborohydride.

3.1.1. Preparation of Amino Aldehydes (A)

Protected amino aldehydes (A) are prepared (Fig. 3) by a two-step procedure *(6)*, via formation of the N-methoxy N-methyl-Boc-carboxamide and subsequent reduction to the aldehyde with lithium aluminium hydride.

3.1.1.1. N-Methoxy N-Methyl-Boc-Carboxamides

1. Dissolve 10 mmol of the Boc amino acid in dichloromethane and stir the solution.
2. Add 1.012 g (10 mmol) of triethylamine (TEA), then 3.483 g (10 mmol) of benzotriazol-1-yloxytris [dimethylamino]-phosphonium hexafluorophosphate (BOP), followed by 1.117 g (11 mmol) of *O,N*-dimethylhydroxylamine hydrochloride, and 1.113 g (11 mmol) of TEA. Continue stirring until reaction is complete (*see* Note 2).
3. Dilute the reaction mixture with dichloromethane (250 mL), transfer the mixture to a separating funnel, and wash successively with 3*M* HCl (3 × 30 mL), saturated $NaHCO_3$ (3 × 30 mL), and saturated NaCl (3 × 30 mL).
4. Combine the organic phases and dry them over anhydrous $MgSO_4$. Filter to remove the $MgSO_4$. Evaporate the solvent and purify the product by flash chromatography (silica gel) (*see* Note 3) or by recrystallization (*see* Note 4). An oil is normally formed in 70–95% yield.

3.1.1.2. Boc Amino Aldehydes

1. Dissolve 2 mmol of the N-methoxy N-methyl-Boc-carboxamide so formed in 20 mL of anhydrous ether (Note 5), stir the solution, and add 95 mg (2.5 mmol) of $LiAlH_4$ (*see* Note 6 for reaction time).
2. Add a solution of 477 mg (93.5 mmol) of $KHSO_4$ in 10 mL of water to hydrolyze the reaction mixture (**Care!** Heat will be evolved).
3. Allow the mixture to cool to room temperature. Transfer the mixture to a separating funnel. Extract the aqueous phase 3 times with 50-mL portions of ether, and combine these extracts. Wash the ether consecutively with 3*M* HCl (3 × 20 mL), saturated $NaHCO_3$ (3 × 20 mL), and saturated NaCl (3 × 20 mL), and dry with $MgSO_4$. Filter off the $MgSO_4$. Evaporate the solvent to leave an almost pure aldehyde (as an oil or low melting point solid) that can be used in the next step unpurified (*see* Note 7).

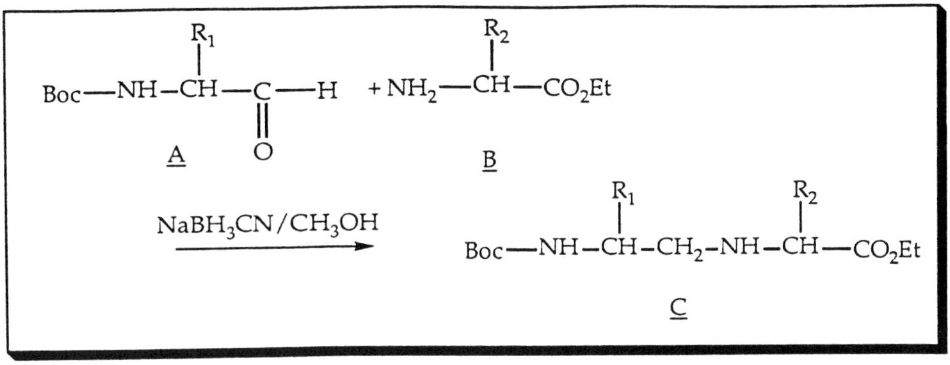

Fig. 2. Synthetic scheme for the preparation of methylene amino dipeptide analogs.

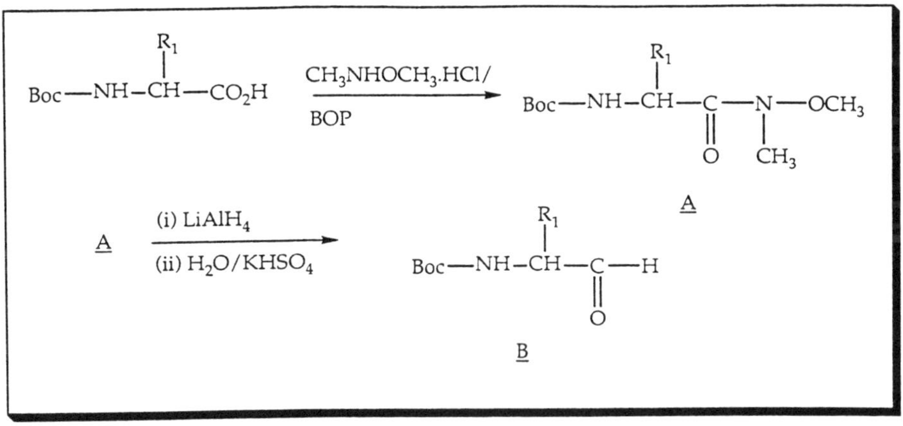

Fig. 3. Preparation of protected amino aldehydes from protected amino acid.

3.1.2. Preparation of Methylene Amino Pseudopeptides (C) (7)

1. Dissolve 1.5 mmol of the amino aldehyde (A) (Section 3.1.1.) in 10 mL of methanol containing 1% (0.1 mL) of acetic acid. Stir the solution and add 1 mmol of the appropriate amino acid ethyl ester (B) followed by addition of 0.189 g (3 mmol) of sodium cyanoborohydride in portions over 45 min (*see* Note 8).
2. Cool the reaction in an ice bath and add 100 mL of saturated $NaHCO_3$, followed by 150 mL of ethyl acetate.
3. Transfer the mixture to a separating funnel and run off the lower aqueous layer. Wash the organic layer with 20 mL of water and dry it over $MgSO_4$. Filter off the $MgSO_4$ and evaporate the ethyl acetate.

4. Further purify the residue by silica gel chromatography (*see* Note 3). Evaporate the eluate to dryness and triturate with diethyl ether to give a white product. Collect this by filtration. Yields are generally between 65 and 80%.

3.2. Synthesis of Thioether Dipeptides

The thioether moiety was one of the first amide bond surrogates to be reported *(8)*. Interest has focused on its hydrophobic nature and its resistance to enzymatic degradation; it has been introduced in a number of different peptides, giving rise to numerous potent analogs of, for example, collagen and LHRH *(9)*.

The generally adopted methodology (Fig. 4) involves the reaction of a thioacid (prepared from an amino acid) with the toluenesulfonate derivative of an amino alcohol, to give the required thioether adduct.

3.2.1. Preparation of Thioacids from Amino Acids (D)

These are prepared (Fig. 5) *(10,11)* via conversion of the α-amino acid to the α-bromo acid, followed by nucleophilic displacement with a thiol moiety, which results in an overall inversion of configuration at the chiral center.

3.2.1.1. Preparation of α-Bromo Acids *(10)*

1. Dissolve 5 mmol of the α-amino acid and 2.38 g (20 mmol) of KBr in 2.45 g (25 mmol) of concentrated sulfuric acid and cool in an ice bath.
2. Dissolve 0.69 g (10 mmol) of $NaNO_2$ in 10 mL of water. Add this to the sulfuric acid and stir the reaction at room temperature for 1 h.
3. Transfer the mixture to a separating funnel. Extract the mixture three times with 20-mL portions of ether (*see* Note 9). Combine the extracts and wash them with 5% sodium thiosulfate (3 × 20 mL) and dry over $MgSO_4$.
4. Filter to remove the $MgSO_4$. Evaporate the solvent *in vacuo* and purify the residue using silica gel chromatography (*see* Notes 3 and 10). Products are obtained in good yield as oils or as solids of low melting point. This procedure results in ~95% optical purity from either R or S amino acid.

3.2.1.2. Preparation of α-Thioacids *(11)*

1. Stir 10 mmol of the α-bromo acid and 2.97 g (10.5 mmol) of cesium thiobenzoate together in 15 mL of dimethylformamide. After 1 h, add ether (45 mL). Wash the mixture two or three times with 10-mL volumes of water. Dry the ether over $MgSO_4$. Filter off the $MgSO_4$. Evaporate off the ether.
2. Recrystallize the product (benzoylthio acid) obtained in step 1 above (*see* Note 11). Yields are 70–80%.
3. Stir the benzoylthio acid (5 mmol) obtained in the previous step with 20 mL of 1*M* NH_3 for 3.5 h. This ammonolysis produces the thiol.
4. Purify the crude product by flash chromatography on silica gel (Note 12). Yield of pure product is ~50%. During the ammonolysis procedure, (S)-bromo acids

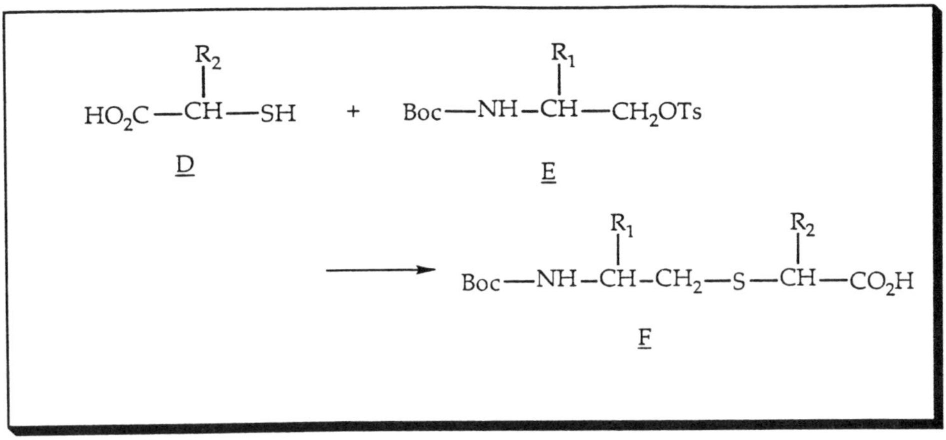

Fig. 4. General methodology used for the preparation of thioether dipeptide analogs.

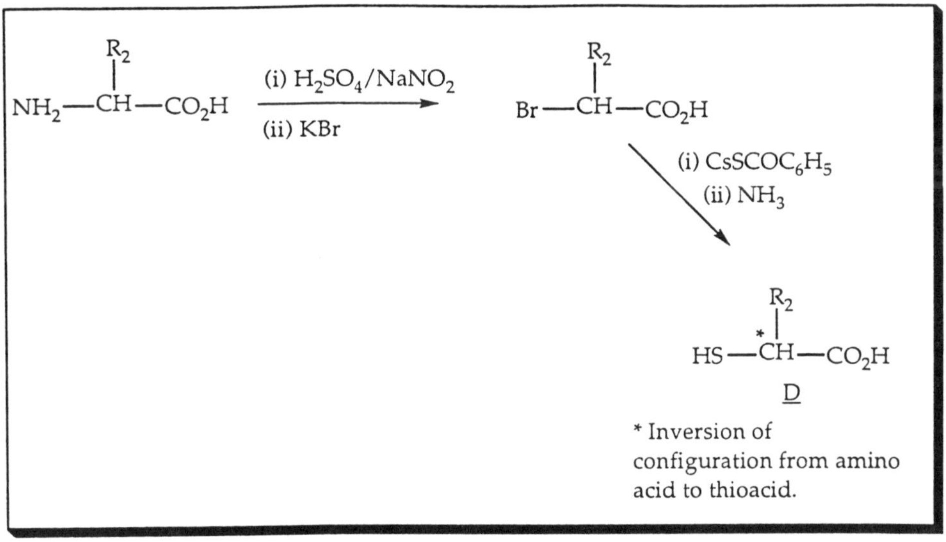

Fig. 5. Synthesis of α-thio acids, via α-bromo acid intermediates.

are converted to the (R)-thioacid and vice versa, with negligible racemization occurring (enantiomeric excess ~80–95%).

3.2.2. Preparation of Amino Toluenesulfonates (E) (12)

1. Dissolve the requisite Boc-amino alcohol (9 mmol) in pyridine (15 mL), and cool in an ice bath.

2. Add toluenesulfonyl chloride (3.5 g, 184 mmol) in three equal amounts at 10-min intervals and stir the solution at –20°C for 12 h and then at 4°C for 24 h.
3. Pour the solution into an ice slurry (150 mL), adjust the pH to ~2.5 with 2*M* HCl, and extract the mixture with ethyl acetate (3 × 100 mL).
4. Combine the organic phases and wash with water (3 × 100 mL), saturated NaCl (3 × 100 mL), and dry over $MgSO_4$. Remove the $MgSO_4$ by filtering.
5. Evaporate the solvent to leave an oil, and purify by silica gel chromatography (generally CH_2Cl_2/hexane, 1:1). The yield of the amino toluenesulfonate is normally 80–90%.

3.2.3. Preparation of Thioether Pseudopeptide (F) (12)

1. Add the α-thioacid (3.75 mmol) (D) (Section 3.2.1.) in dimethylformamide (2 mL) to an aqueous solution of KOH (0.63 g, 11 mmol) followed by the amino toluenesulfonate (E) (4.75 mmol) (Section 3.2.2.).
2. Stir the solution under argon for ~4 d, then pour into cold water (100 mL) and wash with ether (2 × 50 mL).
3. Saturate the aqueous phase with citric acid and extract with ethyl acetate (3 × 50 mL). Wash the organic phases with water (3 × 50 mL), then saturated NaCl solution (3 × 50 mL), and dry over $MgSO_4$.
4. Filter to remove the $MgSO_4$. Evaporate the solvent, leaving an oily residue, and purify by column chromatography (silica gel, $CHCl_3$/MeOH [95:5] as eluent).

3.3. Synthesis of Hydroxyethylene Dipeptides

Hydroxyethylene-containing peptides were first synthesized in the early 1980s, as inhibitors of aspartyl proteinases such as renin *(13)* and HIV-1 protease *(14)*, in which the hydroxyethylene isosteres function as transition state mimics for amide bond hydrolysis. This has led to analogs that are stable toward enzymatic cleavage in vivo.

The method developed for synthesis of hydroxyethylene-containing pseudodipeptides involves the synthesis of a γ-lactone precursor from a variety of sources *(15–17)*, including the amino aldehyde method, which is presented here *(18)*. This lactone is then ring-opened with alkali or amines to yield the desired hydroxyethylene derivative (Fig. 6).

Stereochemical control during conversion of G to H and H to I is achieved by steric effects and gives the predominant isomer indicated (2R,4S).

3.3.1. Preparation of γ-Lactone Derivatives (I) (18)

1. Dissolve diisopropylamine (0.88 g, 8.69 mmol) in anhydrous tetrahydrofuran (5 mL) and cool to –50°C using a stirred mixture of dry ice/chlorobenzene in a Dewar flask.
2. Add 4.5 mL of a 1.6*M* n-butyl lithium solution (7.25 mmol) in hexane dropwise and further cool the solution to –78°C using a stirred mixture of dry ice/acetone in a Dewar flask.

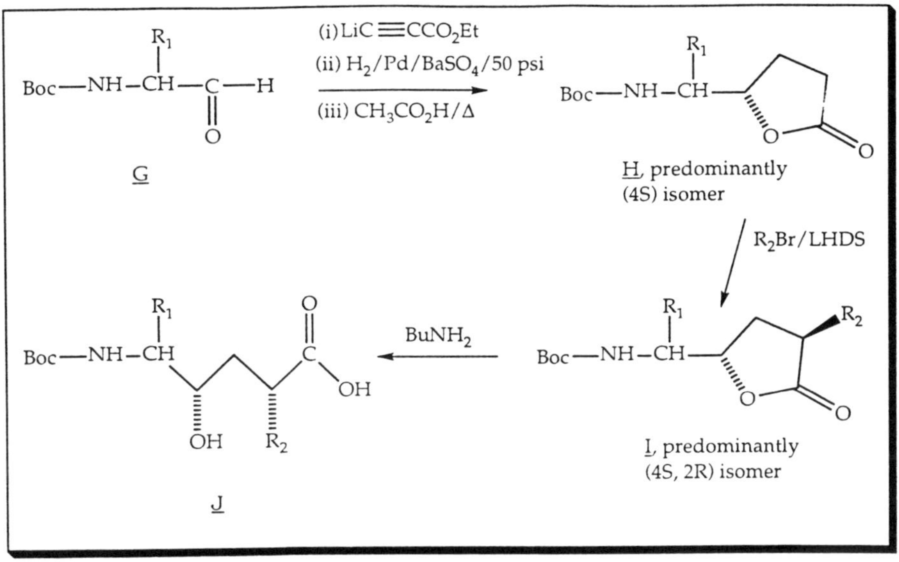

Fig. 6. Hydroxyethylene dipeptide preparation, via γ-lactone precursor.

3. Add freshly distilled ethyl propiolate (0.71 g, 7.25 mmol) dropwise and stir the solution for 30 min.
4. Add the amino aldehyde (Section 2.1.1.) (4.83 mmol) in 5 mL of tetrahydrofuran over a 30-min period (*see* Note 13).
5. Add a mixture of 0.4 mL of acetic acid and 1.6 mL of tetrahydrofuran to quench the reaction and warm the mixture to room temperature.
6. Add 50 mL of ether, transfer the mixture to a separating funnel, and wash it twice with 30-mL portions of 10% citric acid and 30-mL portions of saturated $NaHCO_3$. Dry the ether over $MgSO_4$. Filter to remove the $MgSO_4$. Evaporate the ether to yield the crude epimeric hydroxyacetylenic ester intermediate as an oil.
7. Purify the product by silica gel chromatography (*see* Note 14) to give a yellow oil of low yield (generally 35–40%).

3.3.1.1. Hydrogenation and Lactonization (to give H)

1. Dissolve 14.7 mmol of acetylenic ester (as prepared in steps 1–7 above) in ethyl acetate (50 mL) and hydrogenate at 50 psi for 90 min in the presence of 2.76 g of 5% $Pd/BaSO_4$ catalyst.
2. Remove the catalyst by filtration and evaporate the solvent *in vacuo*.
3. Dissolve the residue in 200 mL of toluene:acetic acid mixture (98:2) and reflux for 3 h. Remove the solvent to yield crude H (both isomers, predominantly [4S] as shown).
4. Separate the isomers on a silica gel column (*see* Note 3) to give ~80% yield of the (4S) isomer indicated (*see* Fig. 6). For R_1 = 2-methylpropyl (i.e., corresponding to leucine side-chain) elute with ether:hexane (4.6:10 by volume) to separate isomers. For other cases, *see* Note 15.

3.3.1.2. Alkylation of Lactone H (to Give Lactone I)

1. Dissolve hexamethyldisilazane (1.39 g, 8.49 mmol) in 3.5 mL of tetrahydrofuran and cool to 0°C in a bath of ice water.
2. Add 5.3 mL (8.11 mmol) of a 1.6*M* solution of n-butyl lithium in hexane and cool the solution further to –78°C in a stirred bath of dry ice/acetone in a Dewar flask.
3. Add the resulting lithium hexamethyldisilazane suspension to the lactone H (3.69 mmol) in 3 mL of tetrahydrofuran and allow to stir at –78°C for another 15 min.
4. Slowly add freshly distilled alkyl bromide (4.06 mmol) in 2 mL of tetrahydrofuran, and warm the mixture to –40°C for 2 h (Note 16).
5. Add 2 mL of aqueous saturated NH_4Cl to quench the reaction.
6. Add ether (30 mL) and 10% aqueous citric acid, wash the organic layer with 10% citric acid (3 × 30 mL) and saturated $NaHCO_3$, and dry over $MgSO_4$. Filter to remove the $MgSO_4$. Evaporate the solvent to yield crude I (both isomers, predominantly the *trans* lactone) (*see* Notes 17 and 18).

3.3.2. Preparation of Hydroxyethylene Dipeptides (J) (17)

1. Dissolve the resulting alkylated γ-lactone derivative I (3 mmol) in n-butylamine (80 mL) and stir for 24 h at 40°C in a thermostatically controlled water bath.
2. Evaporate the solvent and purify the residue by flash chromatography (Note 3) using a mixture of hexane and ethyl acetate (2:1 v/v) as eluent to obtain a pure white solid in 70–90% yield ([2R,4S]-isomer).

4. Notes

1. The methods described are quite general for preparation of modified dipeptides. Thus the choice of reactant will be governed by the dipeptide sequence desired (*see also* Note 16).
2. During preparation of the N-methoxy N-methyl-Boc-carboxamides, follow the course of the reaction by TLC (ethyl acetate/hexane, 1:1 or 1:2 as eluents). Reaction is normally complete within 30–60 min.
3. For a typical flash chromatography column of 2-cm diameter, approx 100 mL of swollen silica gel are required to give a bed height of 30 cm. On this column, 2–3 g of material can be purified. Dissolve the material to be purified in the eluting solvent and allow to run into the column. Then place more eluting solvent onto the bed of silica gel. Fill the solvent reservoir and pass the solvent through the column by use of a peristaltic pump. Twenty-five to thirty fractions (5 mL each) are collected. Follow the elution of the desired compound by TLC examination of the fractions. Pool the fractions containing the required compound and remove the solvent by evaporation.
4. Recrystallize the crude N-methoxy N-methyl-Boc-carboxamides from ethyl acetate.
5. Alternatively dissolve the N-methoxy N-methyl-Boc-carboxamides in anhydrous tetrahydrofuran during the preparation of the aldehyde derivative.

6. Follow the reduction to the amino aldehyde by $LiAlH_4$ by TLC (ethyl acetate/hexane, 1:2). This is normally complete within 20 min.
7. Normally prepare aldehydes for immediate use. However, these can be stable for up to 2 wk if stored under argon.
8. Follow the coupling to form the methylene amino dipeptide by TLC (ethyl acetate; ethyl acetate /pyridine/acetic acid/water, 80:20:3:3 as eluents). This is normally complete within 1 h. Carry out silica gel chromatography with ethyl acetate as eluent.
9. For the more polar amino acids, a better procedure is to extract the α-bromo acids from the reaction mixture using ethyl acetate.
10. During α-bromo acid preparation, elute fractions from the silica column using hexane, followed by ethyl acetate, as eluents.
11. Recrystallize the benzoylthio-acid intermediate from petroleum ether.
12. Use dichloromethane as eluent for chromatographic separation of the α-thioacids.
13. During preparation of the hydroxyacetylenic esters (step i), the solution becomes clear after all of the aldehyde is added. The reaction is generally finished after a further 1 h stirring. Its course can be followed using TLC (ethyl acetate/hexane, 3:7 as eluents).
14. Change the eluent (ethyl acetate/hexane) composition from 15:85 to 25:75 during the purification to aid in the hydroxyacetylenic esters separation.
15. The precise conditions for separation of isomers will vary with the nature of the compounds synthesized and may need to be determined empirically.
16. During the alkylation procedure to form I, after addition of lactone the solution again becomes clear. The choice of bromo compound to be added will obviously depend on the nature of the amino acid side-chain in the native peptide, e.g., Br-CH_2-Ph would be added for a mimetic of phenylalanine and so forth.
17. The ratio of *trans* lactone:cis lactone is of the order of 16:1 after the alkylation procedure.
18. For an alternative method of γ-lactone ring-opening, *see* ref. *15*.

References

1. Morgan, B. A. and Gainor, J. A. (1989) Approaches to the discovery of non-peptide ligands for peptide receptors and peptidases. *Ann. Rep. Med. Chem.* **24,** 243–252.
2. Spatola, A. F. (1993) Synthesis of pseudopeptides. *Meth. Neurosci.* **13,** 19–42.
3. Kleinert, H. D., Baker, W. R., and Stein, H. H. (1991) Renin inhibitors. *Adv. Pharmacol.* **22,** 207–250.
4. Szelke, M., Leckie, B., Hallett, A., Jones, D. M., Sueiras, J., Atrash, B., and Lever, A. F. (1982) Potent new inhibitors of human renin. *Nature* **299,** 555–557.
5. Wyvratt, M. J. and Patchett, A. A. (1985) Recent developments in the design of angiotensin-converting enzyme inhibitors. *Med. Chem. Rev.* **5,** 483–531.
6. Fehrentz, J.-A. and Castro, B. (1983) An efficient synthesis of optically active α-(butoxycarbonylamino)- aldehydes from α-amino acids. *Synthesis* 676–678.

7. Martinez, J., Bali, J. P., Rodriguez, M., Castro, B., Magous, R., Laur, J., and Lignon, M.-F. (1985) Synthesis and biological activity of some pseudo-peptide analogues of tetragastrin: importance of peptide backbone. *J. Med. Chem.* **28,** 1874–1879.
8. Yankeelov, J. A., Jr., Fok, K.-F., and Carothers, D. J. (1978) Peptide-gap inhibitors: stereoselective synthesis of enantiomeric dipeptide analogues of glycylleucine which contain methylene thioether groups substituted for peptide linkages. *J. Org. Chem.* **43,** 1623,1624.
9. Spatola, A. F., Agarwal, N. S., Betag, A. L., and Yankeelov, J. A., Jr. (1980) Synthesis and biological activity of pseudo-peptide analogues of LH-RH. *Biochem. Biophys. Res. Comm.* **97,** 1014–1023.
10. Chang, S.-C., Gil-Av, E., and Charles, R. (1984) Extension of the gas chromatographic separation of enantiomers on chiral phases: resolution of α-halogenocarboxylic acids. *J. Chromat.* **289,** 53–63.
11. Strijtveen, B. and Kellogg, R. M. (1986) Synthesis of (racemisation prone) optically active thiols by S_N2 substitution using cesium thiocarboxylates. *J. Org. Chem.* **51,** 3664–3671.
12. Smith, C. W., Saneii, H. H., Sawyer, T. K., Pals, D. T., Scahill, T. A., Kamdar, B. V., and Lawson, J. A. (1988) Synthesis and renin inhibitory activity of angiotensinogen analogues having dehydrostatine, Leu [CH_2S]Val or Leu [CH_2SO]Val at the P_1-P_1' cleavage site. *J. Med. Chem.* **31,** 1377–1382.
13. Boyd, S. A., Fung, A. K. L., Baker, W. R., Mantei, R. A., Armiger, Y.-L., Stein, H. H., Cohen, J., Egan, D. A., Barlow, J. L., Klinghofer, V., Verburg, K. M., Martin, D. L., Young, G. A., Polakowski, J. S., Hoffman, D. J., Garren, K. W., Perun, T. J., and Kleinert, H. D. (1992) C-terminal modifications of non-peptide renin inhibitors: improved oral bioavailability via modification of physiochemical properties. *J. Med. Chem.* **35,** 1735–1746.
14. Lyle, T. A., Wiscount, C. M., Guare, J. P., Thompson, W. J., Anderson, P. S., Darke, P. L., Zugay, J. A., Emini, E. A., Schlief, W. A., Quintero, J. C., Dixon, R. A. F., Sigal, I. S., and Huff, J. R. (1991) Benzocycloalkyl amines as novel C-termini for HIV protease inhibitors. *J. Med. Chem.* **34,** 1228–1230.
15. Evans, B. E., Rittle, K. E., Homnick, C. F., Springer, J. P., Hirshfield, J., and Veber, D. F. (1985) A stereocontrolled synthesis of hydroxyethylene dipeptide isosteres using novel, chiral aminoalkyl epoxides and γ-(aminoalkyl) γ-lactones. *J. Org. Chem.* **50,** 4615–4625.
16. Baker, W. R. and Pratt, J. K. (1993) Dipeptide isosteres. 2. Synthesis of hydroxyethylene dipeptide isostere diastereoisomers from a common γ-lactone intermediate. Preparation of renin and HIV-1 protease inhibitor transition state mimics. *Tetrahedron* **49,** 8739–8756.
17. Herold, P., Duthaler, R., Rihs, G., and Angst, C. (1989) A versatile and stereocontrolled synthesis of hydroxyethylene dipeptide isosteres. *J. Org. Chem.* **54,** 1178–1185.
18. Fray, A. H., Kaye, R. L., and Kleinman, E. F. (1986) A short, stereoselective synthesis of the lactone precursor to 2R, 4S, 5S hydroxyethylene dipeptide isosteres. *J. Org. Chem.* **51,** 4828–4833.

7

Synthesis of Conformationally Restricted Peptides

Annette G. Beck-Sickinger

1. Introduction

Peptides are very flexible molecules, in contrast to proteins, which are stabilized by disulfide bridges and salt bridges in their tertiary structure. Peptides can adopt several conformations, at least in aqueous solutions. Some limitations, however, are imposed caused by their primary sequence. Two torsion angles characterize the free rotation of each peptide unit: the rotation about the Cα-CO-bond is called ψ-angle, the rotation about the Cα-NH-bond the ϕ-angle (Fig. 1 *[1]*). The torsion angle of the peptide bond, CO-NH, is about 180° and almost fixed owing to its double-bond character. Only certain combinations of ψ- and ϕ-angles are possible because of the steric hindering of carbonyl oxygen, amide hydrogen, and side-chain atoms of the amino acids.

Although small peptides are very flexible in solution, they can adopt a very specific conformation at their receptors *(2)*. Different receptor subtypes, however, may recognize different conformations of the same peptide. In order to characterize these subtypes, but also to find smaller selective peptides or finally nonpeptide drugs, knowledge of the bioactive conformation of a neuropeptide agonist or antagonist is the main concern in structure–activity studies *(3,4)*. In this chapter, three ways to constrain the conformation of a small peptide are described that include the incorporation of nonprotein amino acids, spacer templates, and the synthesis of cyclopeptides. Whereas for the incorporation of nonprotein amino acids and the synthesis of cyclopeptides, protocols are included, only general remarks are given for the use of spacer templates as the synthetic procedures are very complex. It is assumed that the reader is familiar with the usual methods of solid-phase peptide synthesis.

From: *Methods in Molecular Biology, Neuropeptide Protocols*
Edited by: G. B. Irvine and C. H. Williams Humana Press Inc., Totowa, NJ

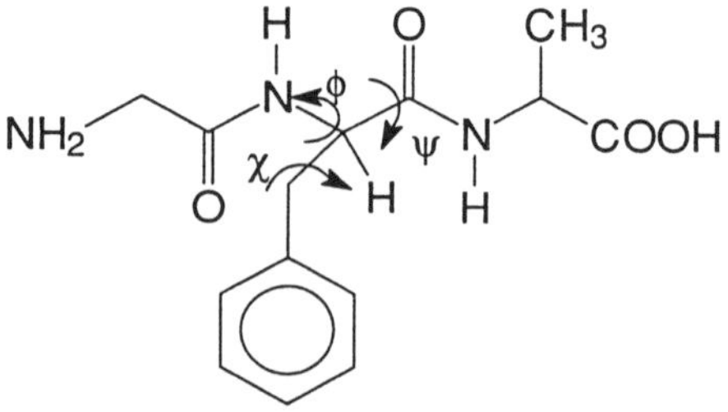

Gly-Phe-Ala

Fig. 1. Conventional notations for the various angles of rotations about bonds in a peptide chain.

1.1. Incorporation of Nonprotein Amino Acids

Whereas the natural amino acids have a broad spectrum of possible combinations of ψ- and ϕ-angles, these can be reduced by a number of nonprotein amino acids. The most frequently used amino acid is aminoisobutyric acid (Aib), the smallest C^{α}-di-alkylated amino acid. Other symmetric or unsymmetric dialkylated residues are reported (Fig. 2A) *(5)*. Modified proline residues, such as Aoc or Oic (Fig. 2B) have also been used in the synthesis of conformationally constrained peptides. Smaller or bigger ring sizes (Pip, Tic, Ctp, Acp, Pac) and heteroatom analogs (Thi) are also known (Fig. 2B). Further modifications, that limit the numbers of allowed conformations are N-alkylation, local backbone constraints (tetrazole analogs, ψ [CN_4]; olefinic analogs, ψ [C=C]), and further modifications of the amide bond, which are described in Chapter 6. Alkylation of C_{β}-atoms, dehydroamino acids, and cyclopropyl amino acid substitution (Fig. 2C) can further reduce the number of conformations and lead to highly active peptides.

Whereas for all of these amino acids N^{α}-protection is similar to proteinogenic amino acids (Chapter 5), carboxyl activation has to be significantly improved. Especially for amino acids with alkylated backbone atoms, the coupling yields obtained with conventional activation (1-hydroxybenzotriazol [HOBt], diisopropylcarbodiimide [DIC]) fail or are insufficient. Improved activation and coupling is reported with the reagents listed in Table 1 and used for *in situ* activation. The first class of reagents is based on phosphonium or uronium salts, which, in the presence of a tertiary base, can smoothly convert the protected amino acids to a variety of activated species. Also, HBTU (hydroxybenzotriazol-1-yl-oxy-tetra-[dimethylamino]-uronium tetrafluoroborate and TBTU (benzotriazol-1-yl-oxy-tetra-[dimethylamino]-uronium tetrafluoroborate) generate HOBt-esters and are widely used in solid-phase

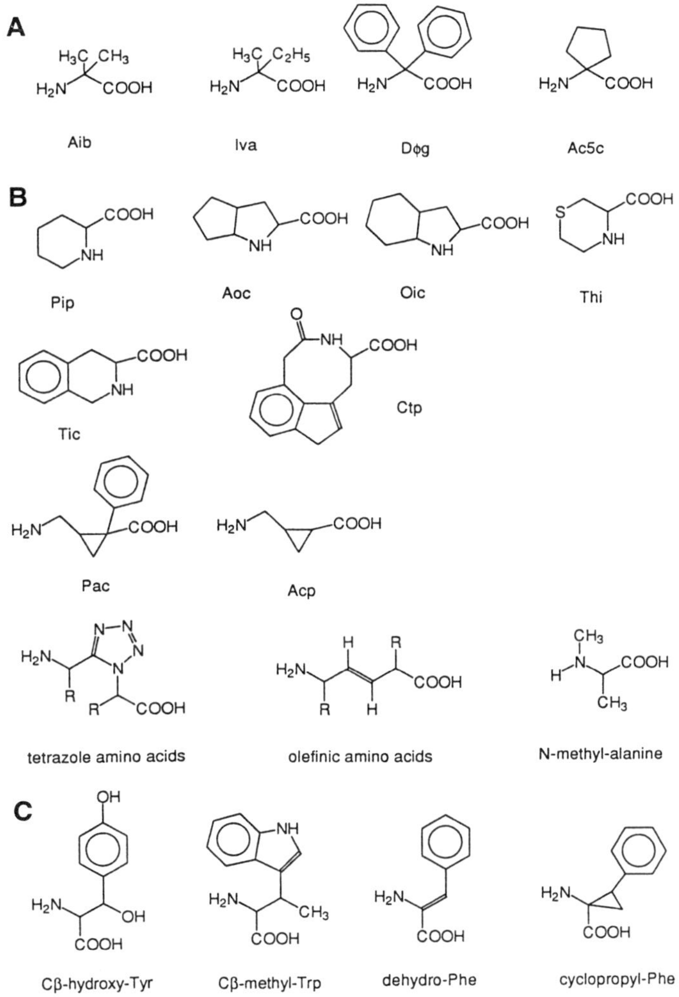

Fig. 2. Non proteinogenic amino acids, which can be used for the synthesis of constrained peptides. **(A)** Cα,α-dialkylated amino acids restrict ψ- and ϕ- torsion angles. **(B)** amino acids with secondary amide groups, cyclopropyl amino acids, olefinic, and tetrazole analogs give a high degree of conformational rigidity. **(C)** Amino acids, which restrict the χ-angle of peptides.

peptide synthesis. PyBOP (benzotriazol-1-yl-oxy-tris-pyrrolidino-phosphonium hexafluorophosphate) should be used instead of BOP (benzotriazol-1-yl-oxy-tris-dimethylamino-phosphonium hexafluorophosphate), which forms a carcinogenic byproduct during handling. PyBro (bromo-tris-pyrrolidino-phosphonium hexafluorophosphate) is reported to be excellent for the coupling of

Table 1
Reagents Used for Effective Activation of Sterically Hindered Amino Acids

Formula	Abbreviation	Application
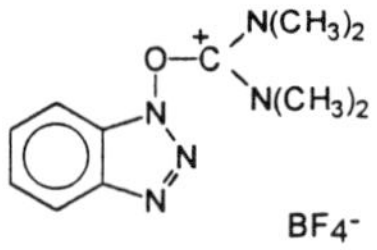 BF_4^-	TBTU	solid phase peptide synthesis
PF_6^-	HBTU	
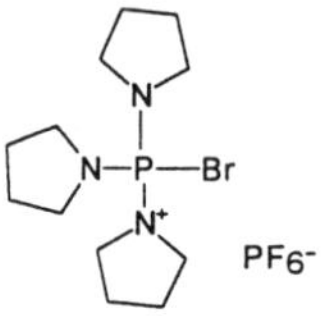	PyBroP	coupling to N-methyl amino acids
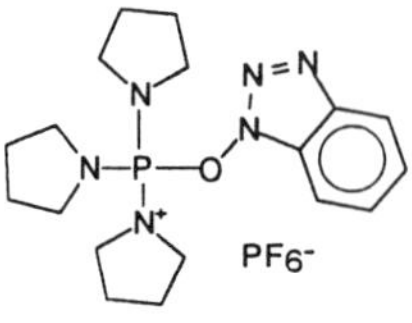	PyBOP	coupling of sterically hindered amino acids
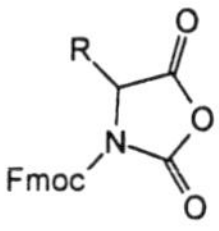	Fmoc-NCA	has to be specially produced for each amino acid
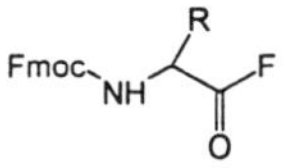	Fmoc-fluoride	has to be specially produced for each amino acid

N^{α}-alkylated amino acids. Successful couplings are also reported with the activated amino acid N-carboxyanhydrides (NCAs) and amino acid fluorides. Because they have to be specially prepared, these derivatives are most frequently used to couple proteinogenic amino acids to sterically hindered residues.

1.2. Templates that Induce Secondary Structure

Whereas a number of amino acids can reduce the conformational space and constrain flexibility, several templates and amino acid linkers have been specially designed to initiate a desired conformation.

Building blocks that initiate or stabilize α-helices, are quite complex molecules and are difficult to synthesize. Kemps' tricyclic prolyl-based structure is one notable exception. A further example is shown in Fig. 3. The synthesis of these molecules, however, requires several steps of organic synthesis.

Whereas helix-inducing building blocks usually require three hydrogen bonds for stabilization, one stabilizing hydrogen bond frequently is sufficient to stabilize or induce a turn conformation. Azabicycloalkanes, α-aminomethyl-phenylacetic acid, α-hydrazino residues, lactam constraints, and spirolactam analogs are a few examples of turn-inducing building blocks (Fig. 3).

The incorporation of these building blocks is usually not as difficult as the incorporation of sterically hindered amino acids. However, the chemical sensitivity of each new template has to be characterized and tested to see whether it is stable under the conditions of solid-phase peptide synthesis (Fmoc strategy: piperidine in dimethylformamide for 20 min, cleavage with trifluoroacetic acid and scavenger *[6–8]*). The synthesis itself most frequently contains several steps of organic synthesis and should be elaborated in collaboration with an expert laboratory.

Including templates, which induce secondary structure, into neuropeptides may not always have the desired effect owing to other amino acids within the sequence. Sometimes it is more helpful to start with flexible linkers, such as 6-aminohexanoic acid or other ω-amino alkanoic acids, in order to identify the distance between two segments that are to be brought together. Characterization of the resulting conformation can be achieved by nuclear magnetic resonance spectroscopy (Chapter 16) and circular dichroism spectroscopy (Chapter 15).

1.3. Cyclopeptides

In addition to the use of single residues or building blocks, the conformation of peptides can be significantly constrained by cyclization. Three methods are most frequently applied:

1. Cyclization by disulfide formation between two Cys residues.
2. Cyclization by lactamization of N- and/or C-terminus or by the amino- and carboxy group-containing side chains Lys, Orn, Dab (diaminobutyric acid), Asp, and Glu.
3. Backbone to side-chain cyclization.

The position of the ring, the number of amino acids that are bridged, the configuration of the bridge residues, and the length can be varied and used to optimize structure-affinity/activity studies. Several examples are shown in Fig. 4. In addition, the orientation of the lactam bond is frequently important.

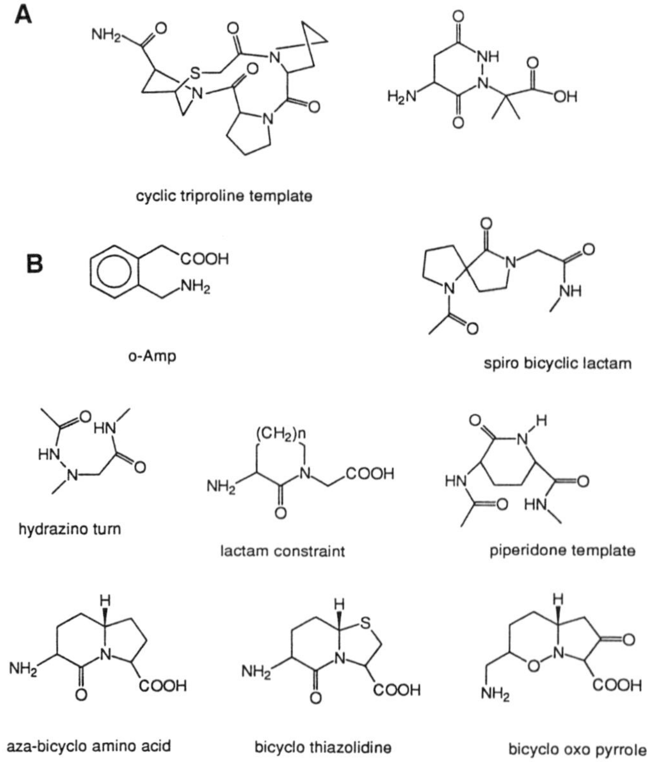

Fig. 3. Templates to induce a specific secondary structure of peptides. **(A)** Building blocks for the generation of α-helices. **(B)** β- or γ-turn mimetics.

2. Materials

2.1. Incorporation of Nonprotein Amino Acids

1. N^{α}-Fmoc-protected, nonprotein amino acids.
2. Dimethylformamide (highest purity, free from amines).
3. 1-hydroxy-benzotriazol (HOBt).
4. Benzotriazol-1-yl-oxy-tetra-(dimethylamino)-uronium tetrafluoroborate (TBTU; or other uronium-based reagents) or benzotriazol-1-yl-oxy-tris-pyrrolidino-phosphonium hexafluorophosphate (PyBOP; or other phosphonium-based reagents).
5. Diisopropylethylamine.
6. A 20% solution (v/v) of piperidine in dimethylformamide.
7. Trifluoroacetic acid.
8. Scavenger (a 1:1 mixture of thioanisole and ethanedithiol or reagent K *[6]*).
9. Diethyl ether (free from peroxides).

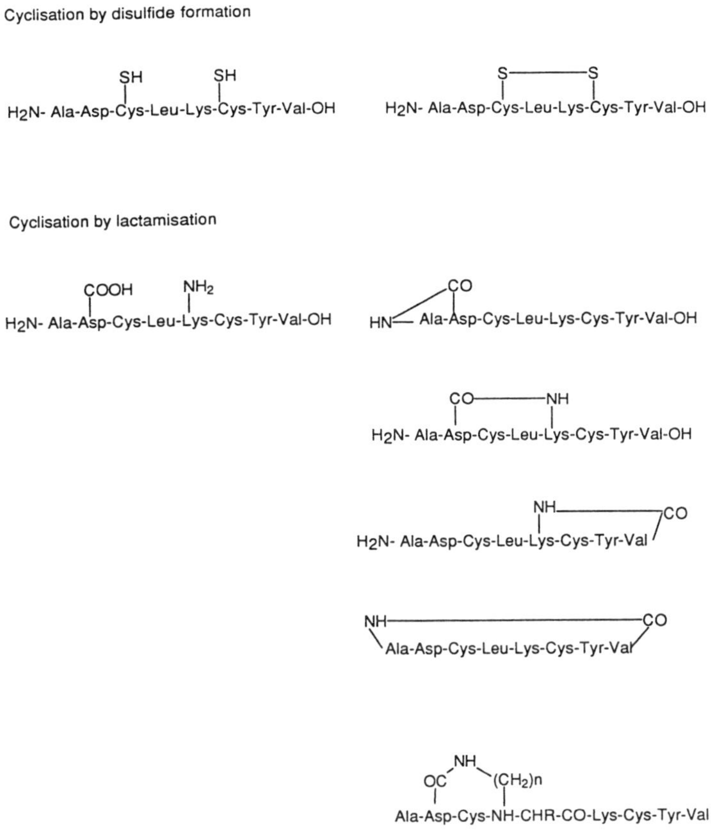

Fig. 4. Cyclic constraints that are used for rigid peptides.

10. Reagents to prepare the three solutions for the Kaiser test are as follows:
    a. Solution I: Dissolve 1.0 g ninhydrin in 20 mL ethanol.
    b. Solution II: Dissolve 80 g phenol in 20 mL ethanol.
    c. Solution III: Dissolve 6.5 mg of potassium cyanide (**Care!** severe poison) in 100 mL of water (1 m*M* solution). Add 0.4 mL of this to 20 mL of distilled pyridine.

2.2. Synthesis of Cyclopeptides by Cys-Oxidation

1. Linear peptide, obtained by solid-phase peptide synthesis.
2. Ellman's reagent prepared as follows: dissolve 25 mg of 5,5-dithiobis-(2-nitrobenzoic) acid in 25 mL of ethanol, plus 25 mL of 125 m*M* Tris-HCl buffer, pH 8.2.
3. 10 m*M* ammonium acetate buffer, pH 7.6.
4. Reaction flask with a gas inlet.

2.3. Synthesis of Cyclopeptides by Lactamization

1. Fully protected peptide fragment, in which selective deprotection of one amino and one carboxy group has been carried out.
2. Methylene chloride.
3. Dimethylformamide (highest purity, free from amines).
4. Benzotriazol-1-yl-oxy-tetra-(dimethylamino)-uronium tetrafluoroborate (TBTU).
5. Diisopropylethylamine.
6. A 20% solution (v/v) of piperidine in dimethylformamide.
7. Trifluoroacetic acid.
8. Scavenger (for example: a 1:1 mixture of thioanisole and ethanedithiol or reagent K *[6]*).
9. Diethyl ether (free from peroxides).

3. Methods (*see* Note 1 before proceeding)

3.1. Incorporation of Nonprotein Amino Acids

1. Wash the resin-bound peptide with 5 × 15 mL portions of dimethylformamide (3 min/wash). Filter to remove the dimethylformamide.
2. Deprotect by adding 20% piperidine in dimethylformamide (15 mL for 20 min). Filter to remove the dimethylformamide.
3. Repeat step 1.
4. Dissolve 3 equivalents (*see* Note 1) of the required Fmoc-protected nonprotein amino acid in dimethylformamide (5 mL/g) and add this to the resin.
5. Add 3 equivalents (*see* Note 1) of PyBOP (or TBTU, HBTU, PyBro), 4–6 equivalents of diisopropylethylamine (*see* Note 1), and gently shake the reaction vessel.
6. After 4 h, remove the solution by filtering the resin and wash the resin four times with dimethylformamide as in step 1.
7. Test for free NH_2 groups as follows. Take a small aliquot of the resin and add one drop of each of the Kaiser solutions I, II, and III. Heat to 110°C for 5 min. A negative test (lack of blue color) indicates that no free amino groups are present, and that the reaction is complete. A blue color of the resin (positive test) requires the repeat of steps 4–7 (*see* Note 2).
8. Continue peptide synthesis in the conventional manner, but check the coupling yield of the next amino acid after the nonprotein amino acid as well by using the Kaiser test as in step 7 (*see* Note 2).

3.2. Synthesis of Cyclopeptides (*see* Note 3)

3.2.1. Synthesis of the Linear Precursor Peptides

The design of the precursor is one of the most important steps in the synthesis of cyclopeptides: The anchor that is used for peptide attachment to the resin has to be compatible with the normal strategy of peptide synthesis *(6, 7)* and also with the cyclization process. The amino acid residues that are planned to be involved in the bridge have to be specifically protected and their side-chain protection groups selectively removed. For Cys side-chain protection, Cys(Trityl) has been found to be suitable (*see* Note 4) as it leads to the free SH group, if ethanedithiol (5%; *see* Note 5) is included in the scavenger mixture.

3.2.2. Oxidation of Cysteine Residues (Air Oxidation)

1. Synthesize the linear precursor by Fmoc-strategy, incorporating the Cys(Trt) residues as required.
2. Cleave the peptide from the resin by treating with trifluoroacetic acid (90%) and scavenger (10%).
3. Add diethyl ether to precipitate the peptide and collect the precipitate by centrifugation.
4. Wash the precipitated peptide twice by resuspension and centrifugation with cold diethyl ether.
5. Dissolve the peptide in water or a mixture of water and t-butanol and lyophilize.
6. Dissolve the peptide in 10 m*M* ammonium acetate buffer (pH 7.6) to give a peptide concentration of 1 m*M*.
7. Supply the reaction vessel with a gas inlet and bubble air (2–3 bubbles/s) through the liquid overnight. Completion of the reaction should be tested by Ellman's reagent, which yields a yellow spot on TLC in the presence of free SH groups (Notes 6 and 7).
8. Lyophilize the oxidized peptide and purify by preparative reversed-phase HPLC (*see* Note 8).

3.2.3. Synthesis of Cyclopeptides by Lactamization

In this case, the linear precursor of the cyclic peptide has to be carefully designed. Several examples for successful strategies are shown, using Fmoc chemistry for synthesis of the linear peptide (*see* Note 9).

3.2.3.1. Coupling of N- to C-Terminus

1. Synthesize the linear peptide in the normal way by Fmoc-strategy using a super acid sensitive resin (SASRIN, Bachem, Bubendorf, Switzerland or Chlorotrityl-Resin, NovaBiochem, Läufelfinger, Switzerland). Side-chains of Lys, Orn, or Dab should be protected with Boc, and Asp and Glu with t-butylester (*see* Note 10).
2. Remove the N-terminal Fmoc-group with 20% piperidine in dimethylformamide as described in Section 3.1. (step 2).
3. Cleave the peptide from the resin by washing the resin 10 times with 0.5% trifluoroacetic acid in methylene chloride and combine the filtrates. This will lead to the fully side-chain protected peptide with free N- and C-terminus.
4. Neutralize the filtrate with pyridine and remove the solvent by evaporation *in vacuo*.
5. Dissolve the peptide in sufficient dimethylformamide to give a concentration of 1 m*M* and add 4 equivalents (*see* Note 1) of TBTU and 8 equivalents (Note 1)

of diisopropylethylamine. After 4 h, reaction should be complete (Note 11). Remove the solvent by evaporation *in vacuo*.
6. Remove excess reagents by gel chromatography or preparative HPLC (*see* Note 8) and recover the cyclopeptide by lyophilization.
7. Deprotect the cyclopeptide by treatment with trifluoroacetic acid (90%) and scavenger (10%) for 1 h.
8. Add sufficient diethyl ether to precipitate the peptide completely and collect the precipitate by centrifugation.
9. Wash the precipitate twice by resuspension and centrifugation with cold diethyl ether.
10. Dissolve the peptide in water or a mixture of water and t-butanol and lyophilize.

3.2.3.2. Coupling of Amino Group Side-Chain (e.g., Lys, Orn, or Dab residue) to C-Terminus

1. Synthesize the linear peptide by Fmoc-strategy (the N-terminal amino acid should be incorporated with N^{α}-Boc-protection) attached to a super acid sensitive resin (SASRIN, Chlorotrityl-Resin). The side-chain of the Lys (or Orn, Dab) residue, which is to be part of the bridge, should have methyltrityl (Mtt) protection, which is subsequently cleaved with 1% trifluoroacetic acid. (Fmoc-Lys[Mtt]-OH is obtainable from NovaBiochem.) Other Lys, Orn, or Dab residues within the sequence should be side-chain protected with Boc, and Asp and Glu with t-butyl.
2. Cleave the peptide by washing the resin 10 times with 1% trifluoroacetic acid in methylene chloride. Filter off the solution after each wash and combine the filtrates. This will lead to the fully protected peptide with one free amino group and free C-terminus.
3. Neutralize the solution with pyridine and remove the solvent by lyophilization.
4. Dissolve the peptide in sufficient dimethylformamide to give a 1 m*M* solution. Add 4 equivalents (*see* Note 1) of TBTU and 8 equivalents (*see* Note 1) of diisopropylethylamine. After 4 h evaporate the solvent *in vacuo*.
5. Purify the cyclic peptide by gel chromatography or preparative HPLC (Note 8) and lyophilize.
6. Add trifluoroacetic acid (90%) and scavenger (10%) to deprotect the peptide.
7. Add diethyl ether to precipitate the peptide and collect the precipitate by centrifugation.
8. Wash the precipitate twice by resuspension and centrifugation with cold diethyl ether.
9. Dissolve the peptide in water or a mixture of water and t-butanol and lyophilize.

3.2.3.3. Coupling of Amino Group (Side-Chain or N-Terminus) to Carboxy Group of the Side-Chain (*see* Note 12)

1. Synthesize the linear peptide by Fmoc-strategy using benzyl side-chain protection and a more acid-stable resin (PAM-resin). Residues that are involved in the cyclization, are protected with Boc/t-butyl (*see* Notes 13 and 14).

2. Add a solution of 20% trifluoroacetic acid in methylene chloride to remove side-chain protection.
3. Add 4 equivalents (*see* Note 1) of TBTU and 8 equivalents (*see* Note 1) of diisopropylethylamine to bring about cyclization on the polymer.
4. After 4 h, remove excess reagents by washing the resin five times with dimethylformamide and methylene chloride and finally with diethyl ether.
5. Dry the resin.
6. Cleave the peptide from the resin with 1*M* trifluoromethanesulfonic acid in trifluoracetic acid/scavenger for 3 h at 0°C.
7. Add diethyl ether to precipitate the peptide and collect the precipitate by centrifugation.
8. Wash the precipitate twice by resuspension and centrifugation with cold diethyl ether.
9. Dissolve the peptide in water or a mixture of water and t-butanol and lyophilize.

4. Notes

1. Exact amounts of reagents used will depend on the capacity of the resin (i.e., of available sites per unit weight of resin) and the amount of resin used. When weights and volumes are given, these apply to 1 g of resin with a loading of approx 0.5 mmol/g.
2. In order to be sure that the Kaiser test works properly, it is advisable to include a negative (for example a resin, loaded with a hydroxyl groups containing anchor) and a positive control (e.g., deprotected resin). Since coupling efficiency is influenced by the surrounding amino acids, the following modifications should be applied if double coupling is required. Change the solvent (methylene chloride, N-methylpyrrolidone, dimethylacetamide, or 25% dimethylsulfoxide) and wash the resin before the double coupling with chaotropic salts such as 0.8*M* LiCl.
3. The first indication of a successful cyclization can be obtained by HPLC (*see* Note 8). Retention time usually shifts and indicates the completeness of the cyclization. Matrix-assisted laser desorption mass spectrometry (Chapter 14) or electrospray mass spectrometry (Chapter 13) can be used to confirm the structure. The loss of 18 amu represents lactam cyclization; loss of 2 amu represents disulfide formation. The latter can also be effectively characterized using Ellman's reagent (*see* Section 2.2.). Gas-phase sequencing (Chapter 3) can be used to determine the correct position of the bridge, which usually is characterized by two gaps within the sequence. Circular dichroism and nuclear magnetic resonance will indicate the change of conformation compared to the linear peptide.
4. If more than one disulfide bridge is to be formed, the corresponding Cys residues have to be side-chain protected in different ways, e.g., using a combination of Trt (trityl) and Acm (acetamido) protecting groups.
5. The scavenger ethanedithiol, used to obtain free SH groups from Cys (Trt) residues, has a very unpleasant smell. In order to avoid isolation in the labora-

tory, prepare a bath containing H_2O_2 (10%) and KOH (1–5%) before using ethanedithiol. Everything that has been in contact with ethanedithiol (glassware, disposable tips, and so on) should be directly transferred to this bath after use. After overnight soaking, disposable materials can be put to waste and glassware can be cleaned.

6. Oxidation can take several days, and the reaction should be continued until no educt is found by HPLC (*see* Note 8) or Ellman's test. If the reaction is incomplete, check pH. If dimers are found, reduce the concentration of peptide and increase the concentration of ammonium acetate.
7. Alternative methods for the formation of the disulfide bridge are oxidation with trichloromethylsilane (100 equivalents in trifluoroacetic acid) in the presence of diphenylsulfoxide (reaction time, 10 min) *(9)* and oxidation with $K_3Fe(CN)_6$. A solution of $K_3Fe(CN)_6$ (0.2 m*M*) in water is added dropwise to the peptide solution (0.2 m*M*) in ammonium acetate buffer (1*M*, pH 7.0) until a pale yellow color is visible, which is stable for at least 1 h. Lyophilize the oxidized peptide and purify it by preparative HPLC (*see* Note 8).
8. Conditions for HPLC will depend on the sequence of the particular peptide and must be determined empirically.
9. Alternatively, by applying Boc-strategy for Nα-protection, the side-chains used for cyclization can be introduced with Fmoc-/OFm-protecting groups. These can be selectively cleaved on the polymer. The deprotected side-chains are then cyclized on the polymer and the cyclic peptide can then be cleaved from the resin with HF or trifluoro-methansulfonic acid. In addition, less stable linkers can be used and cyclization can be performed in solution as described.
10. An alternative suitable side-chain protection group is trityl.
11. If the cyclization is incomplete, the solvent can be changed (mixtures of dimethylformamide/methylene chloride), the reaction time can be prolonged, and the concentration and the excess of coupling reagents can be increased. If dimers are found, the concentration of the linear peptides should be diluted and solvents that do not favor aggregation should be used.
12. Alternative cyclization can include the backbone utilizing side-chains linked to the amide group of the backbone. Frequently N-(ω-amino-alkylidene) residues are incorporated into the sequence and cyclization is performed by lactamization to the C-terminus, a side-chain carboxy group or to a N-(ω-carboxy-alkylidene) residue. Selective protection strategy is required *(10)*.
13. A different protocol for the coupling of amino group (side-chain or N-terminus) to carboxy group of the side-chain can be performed with the allyl-linker (Hycram, Orpegen, Heidelberg) and allyl-based side-chain protecting groups. Cleavage is performed with the exclusion of oxygen by hydrogenolysis over a Pd^0 catalyst.
14. For the coupling of amino group (side-chain or N-terminus) to carboxy group of the side-chain of peptide amides, different linkers have to be chosen. MBHA-resin can be applied and protocols can be performed as discussed above in Section 3.2.3.3.

References

1. Fasman, G. D. (1989) *Prediction of protein structure and the principles of protein conformation*. Plenum, New York.
2. Fauchere, J.-L. (1986) Elements for the rational design of peptide drugs. *Adv. Drug Res.* **15,** 29–69.
3. Hruby, V. (1992) Strategies in the development of peptide antagonists. *Progr. Brain Res.* **92,** 215–224.
4. Rizo, L. and Gierasch, L. (1992) Constrained peptides: models of bioactive peptides and their protein structures. *Annu. Rev. Biochem.* **61,** 387–418.
5. Toniolo, C. (1990) Conformationally restrained peptides through short range cyclisation. *Int. J. Pept. Protein Res.* **35,** 287–299.
6. Fields, G. and Noble, R. L. (1990) Solid phase peptide synthesis utilizing 9-fluorenylmethoxycarbonyl amino acids. *Int. J. Pept. Protein Res.* **35,** 161–214.
7. Atherton, E. and Sheppard, R. C. (1989) *Solid Phase Peptide Synthesis: A Practical Approach.* IRL, Oxford, UK.
8. Jung, G. and Beck-Sickinger, A. G. (1992) Multiple peptide synthesis. *Angew. Chem. Int. Ed. Engl.* **31,** 367–383.
9. Akaji, K., Tatsumi, T., Yoshida, M., Kimura, T., Fuiwara, Y., and Kiso, Y. (1992) Disulfide bond formation using the silyl chloride-sulfoxide system for the synthesis of cystine peptide. *J. Am. Chem. Soc.* **114,** 4137–4143.
10. Gilon, C., Halle, D., Chorev, M., Selinger, Z., and Byk, G. (1991) Backbone cyclization: A new method for conferring conformational constraint on peptides. *Biopolymers* **31,** 745–750.

8

Purification of Synthetic Peptides by High Performance Liquid Chromatography

D. David Smith and Ann M. Hanly

1. Introduction

Synthesis of peptides on a solid support is described in detail in Chapter 5 of this book. Contributing to the ongoing success of Merrifield's solid-phase peptide synthesis methodology *(1)* was the use of high performance liquid chromatography (HPLC) for the purification of the desired peptide from the byproducts generated by this technique.

Impurities found with the desired peptide are derived from three sources: namely, coupling of amino acid derivatives to the growing peptide chain, cleavage of the peptide from the solid support, and deprotection of side-chains of the assembled sequence. Whereas highly optimized chemistries keep side reactions to a minimum, they have not been completely eliminated *(2)*. Impurities often have small differences in structure such as the deletion of one amino acid residue resulting from a slow coupling reaction or a rearranged/derivatized side-chain group formed during the cleavage of the peptide from the solid support *(3)*. As a result, impurities often have similar physical and chromatographic properties to those of the desired peptide, which can result in a challenging purification.

Over the past 15 yr, reversed-phase (RP)-HPLC has been the method of choice to assess the purity of synthetic peptides. The high resolution of this chromatography is ideally suited for the separation of peptides that differ in structure by as little as the configuration of one asymmetric carbon *(4)*. A wide variety of conditions, solvents, and columns have been used. The packing material is spherical or irregular-shaped silica particles, 5–10 µm in diameter, derivatized with a hydrophobic functional group. Eluents are usually mixtures of water and water-miscible organic solvents, such as methanol and

From: *Methods in Molecular Biology, Neuropeptide Protocols*
Edited by: G. B. Irvine and C. H. Williams Humana Press Inc., Totowa, NJ

acetonitrile. Whereas the addition of buffers and/or additives such as phosphoric acid *(5),* phosphate buffers *(6),* ammonium acetate *(7),* pyridinium acetate buffers *(8),* formic acid *(9),* hydrochloric acid *(10),* or heptafluorobutyric acid *(11)* improves peak shape and resolution, trifluoroacetic acid (TFA) *(12)* is the most widely adopted additive for RP-HPLC of peptides *(13)*. TFA has the advantage of being UV transparent. When used with acetonitrile, peptides can be detected at wavelengths as low as 208 nm. In addition, TFA is volatile, eliminating the need for desalting.

Unfortunately, use of TFA for preparative RP-HPLC of peptides results in poor peak shape with exaggerated tailing, and low recovery of product *(14)*. The same is also true when ammonium acetate is employed as a buffer with the added disadvantage that peptides cannot be detected below 280 nm. To overcome these problems, Rivier et al. introduced triethylammonium phosphate buffers (TEAP) as an alternative *(15,16)*. Peptides can be detected at wavelengths as low as 210 nm in these buffers with greater than 90% recovery after passage through the chromatography column. A pH of 2.5–3.0 suffices for the chromatography of most peptides. However, for peptides containing an excess of acidic amino acid residues, greater success may be achieved at higher pH of 6.5–7.0 *(17)*. Unfortunately, the TEAP buffer is nonvolatile and, therefore, a desalting step is required prior to removing the solvent by lyophilization.

The use of ion-exchange HPLC for analysis or purification of synthetic peptides is less well documented *(18)*. Disadvantages of early columns were low resolution of peaks accompanied by poor peak shape and low recovery of peptide product *(19)*. Additionally, silica-based packing materials cannot be used at pH >7.0. Recently, the Poly(SULFOETHYL) Aspartamide ion-exchange column was introduced as a strong cation exchange column for the purification of peptides *(20)*. The silica packing material is coated with a hydrophilic aspartamide polymer that minimizes nonspecific binding, resulting in high recovery of product. The polymer is derivatized with ethyl sulfonic acid groups whose low pK_a allows cation-exchange chromatography at pH 3.0. Peptides with a net positive charge as low as 1, at this pH, will bind to the column, making this a very useful tool for purification of peptides.

Ion-exchange and reversed-phase chromatography have complementary modes of separation (charge vs hydrophobicity), making these techniques ideal for the purification of a desired peptide from the crude material obtained from the cleavage reaction. Although numerous examples exist using low-pressure ion-exchange chromatography coupled with RP-HPLC *(21,22)*, few exist using ion-exchange HPLC and RP-HPLC *(23)*. This chapter will describe a general protocol that has been applied to the purification of over 20 peptides ranging in length from 17 to 37 residues. The purification of calcitonin gene-related peptide *(8–37)* is presented as an example.

2. Materials

At all times, the highest quality of reagents should be used. The following lists suitable suppliers. Glass-distilled acetonitrile and methanol is obtained from Burdick & Jackson (Muskegon, MI) and filtered through a 0.45-µm membrane. Water can be obtained from a Barnstead Nanopure system filtered through a 0.22-µm membrane. Sequanal grade TFA and triethylamine are from Pierce (Rockford, IL) and HPLC grade phosphoric acid (85%) and reagent grade potassium hydroxide and potassium chloride are from Fisher Scientific (Pittsburgh, PA).

2.1. Buffers for Ion-Exchange HPLC

1. Buffer A: 0.34 mL of 85% phosphoric acid is added to 800 mL of water and the mixture is diluted to 1 L with acetonitrile (*see* Note 1). The pH is adjusted to 3.0 with 1*M* potassium hydroxide solution, and the buffer is filtered through a 0.45-µm membrane prior to use.
2. Buffer B: 18.72 g of potassium chloride is dissolved in 500 mL of Buffer A and the resulting solution is filtered through a 0.45-µm membrane prior to use.

2.2. Buffers for RP-HPLC

1. Buffer A: 6.8 mL of 85% phosphoric acid is added to 800 mL of water and the pH is adjusted to 2.25 with triethylamine. The mixture is then diluted with water to 1 L and filtered through a 0.45-µm membrane prior to use.
2. Buffer B: 300 mL of acetonitrile is added, with stirring, to 200 mL of Buffer A, and the resulting solution is filtered through a 0.45-µm membrane prior to use.
3. Solvent C: 500 µL of TFA is dissolved in 500 mL of water with stirring.
4. Solvent D: To 200 mL of water, with stirring, is added 450 µL of TFA followed by 300 mL of acetonitrile (*see* Note 2).

2.2. Apparatus

All chromatography can be performed on the same biocompatible, gradient HPLC system capable of handling the high concentrations of chloride ions used in ion-exchange HPLC. Alternatively, the RP-HPLC may be performed on a standard stainless steel gradient HPLC instrument. A biocompatible system from the Waters (Milford, MA) consists of a 625LC controller and fluid handling unit (pump), a Rheodyne 9125 manual injector, and 486 UV variable wavelength detector. The UV detector is connected to a Linear 1201 chart recorder obtained from Isco (Lincoln, NE). For preparative chromatography, 4-mL fractions are collected using a Cygnet fraction collector from Isco equipped with 16 × 100 mm glass tubes. Preparative ion-exchange HPLC is performed using a column (10 × 200 mm) packed with Poly(SULFOETHYL) Aspartamide silica (5 µm, 300 Å), at a flow rate of 4 mL/min. Analytical RP-HPLC is performed utilizing a column (4.6 × 150 mm) packed with Vydac C_{18} silica (5 µm, 300 Å), at a flow rate of 1 mL/min. Preparative RP-HPLC utilizes a column (10 × 250 mm) packed with Vydac C_{18} silica (10–15 µm, 300 Å), at a flow rate of 4 mL/min. All columns are available from the NEST Group (Marlborough, MA).

3. Methods

3.1. Preparative Ion-Exchange HPLC

1. Dissolve the lyophilized material (<150 mg), obtained from the cleavage reaction (*see* Note 3), in Buffer A (<50 mL), and filter the solution through a 0.45-µm membrane syringe filter (*see* Note 4).
2. Wash the Poly(SULFOETHYL) Aspartamide column for 15 min with water, 5 min with Buffer B, and equilibrate the column for 30 min with Buffer A. Set detector wavelength to 230 nm (*see* Note 5).
3. Add the sample to an empty 50-mL syringe barrel previously connected to the RUN–INJECT–DRAW valve of the fluid handling unit, and switch the valve to INJECT to load the sample onto the column. Alternatively, this sample may be pumped onto the column through the Buffer A solvent line. Often, the UV absorption of the effluent will rise as unretained material is eluted from the column. Collect the effluent in a separate flask (*see* Note 6).
4. As the level of the sample solution reaches the bottom of the syringe barrel, wash with Buffer A (2 × 5 mL) and then switch the valve back to RUN.
5. Wash the column with Buffer A until all unretained material is eluted from the column. This is conveniently accomplished by monitoring the UV absorption of the effluent until it returns to zero (*see* Note 7).
6. Connect the effluent tube to the fraction collector and start a linear solvent gradient from 0–100% Buffer B over 40 min (*see* Note 8).
7. Wash the column with water for 15 min and a mixture of water and methanol (1/1, v/v) for 15 min.
8. Analyze fractions by analytical RP-HPLC (*see* Note 5) and pool those containing the desired product (*see* Note 9).

3.2. Preparative RP-HPLC (see Note 10)

1. Using the same Vydac analytical column, determine the percentage of Buffer B (X%) that elutes the desired product with a capacity factor (k') of 4, under

isocratic conditions, at a flow rate of 1 mL/min (*see* Note 11). The k' is calculated according to the following formula:

$$k' = \frac{t_r - t_o}{t_o}$$

where t_r = retention time of desired product; t_o = retention time of unretained material under same elution conditions. Sodium nitrate is used to determine t_o. Wash column with a mixture of water and methanol (2/8, v/v) and remove from instrument.

2. Connect the preparative Vydac column to the instrument and equilibrate with a solvent composition of (X – 10)% Buffer B for 30 min. Set detector wavelength to 230 nm (*see* Note 5).
3. Load pooled fractions from previous ion-exchange HPLC onto the column through the RUN–INJECT–DRAW valve as before using Buffer A for washing.
4. Wash column, with the above solvent composition used for equilibration, until all unretained material is removed from the column. Once again, this can be done by monitoring the UV absorption of the effluent.
5. Connect the effluent tube to the fraction collector and start a linear solvent gradient from (X – 10)% to (X + 10)% B over 50 min.
6. Pool all fractions containing only the desired product, as determined by analytical RP-HPLC (*see* Note 12).
7. Wash preparative column with methanol for 30 min and equilibrate with solvent C for 30 min (*see* Note 13).
8. Dilute pooled fractions with an equal volume of water and load onto preparative column through the RUN–INJECT–DRAW valve as before.
9. Wash with Solvent C until all triethylamine phosphate salt is removed from the column (*see* Note 14).
10. Connect the effluent tube to the fraction collector and start a linear gradient from 10 to 100% Solvent D over 50 min (*see* Note 15).
11. Analyze fractions by RP-HPLC (*see* Note 12). Pool and lyophilize those that contain only the desired product (*see* Note 16).

4. Notes

1. Up to 25% acetonitrile in the ion-exchange buffers further minimizes secondary nonspecific interactions, resulting in narrower peak widths and increased resolution. In our laboratory, 20% acetonitrile is used routinely.
2. The lower amount of TFA (0.09%) in solvent D is to offset a rising baseline derived from elution of TFA-based impurities with increasing amounts of acetonitrile in the eluent.
3. Lyophilized material obtained directly from the cleavage reaction usually does not contain enough salts to prevent the product from binding to the ion-exchange column. However, if the crude cleaved peptide requires a postcleavage modification that generates salts, such as disulfide bond formation, a gel filtration or dialysis step is required prior to ion-exchange chromatography.

4. There is no limit on the volume of the sample solution, although larger volumes will require longer than 12 min to load the sample onto the column at a flow rate of 4 mL/min.
5. For preparative chromatography, the lower absorbance of peptide bonds at 230 nm is used to obtain a chromatographic trace below 2.0 absorbance units full scale (AUFS), the maximum absorbance of the 486 UV detector. A wavelength of 220 nm is used for analytical RP-HPLC.
6. This is a precautionary step in the event that the product is not retained on the column. For strong cation-exchange chromatography at pH 3.0, this is usually because of too much salt in the sample (*see* Note 3).
7. If the absorption does not fall back exactly to zero, the solvent gradient is started when the absorption is <0.05 AUFS and unchanged over 5 min.
8. The primary role of the ion-exchange HPLC step is to remove impurities that coelute with the desired product on RP-HPLC. Peptides with similar hydrophobic properties but different net charges would be expected to have grossly different retentive characteristics on an ion-exchange column. This step is considered to be a clean up of the crude mixture. The same gradient of 0–0.5*M* K^+ ions is used for all peptides. Peptides with net charges as high as +7 are eluted under these conditions. In the event that a peptide could not be obtained homogeneous after the RP-HPLC, the ion-exchange step could be optimized using a shallower gradient. This has not been necessary for the purification of over 20 peptides in our laboratory.
9. Gradient analytical RP-HPLC is preferred to ensure complete removal of highly hydrophobic impurities after each injection.
10. Preparative RP-HPLC is carried out under highly optimized conditions using the general protocol of Hoeger et al. *(17)*, using a slightly higher TEAP concentration of 100 m*M*. Briefly, the percentage of Buffer B (X%) is determined that elutes the product, isocratically, from an analytical column with a k' value of 4. The product is purified using a shallow gradient from (X – 10)% to (X + 10)% Buffer B. The elution time is 2 min/cm of column, which corresponds to 50 min for a 25-cm column.
11. Determine the value of X% for each fresh batch of TEAP buffer, since small differences in buffer composition (especially Buffer B) can cause shifts of retention times of up to 3 min.
12. This is conveniently done using the isocratic mode under solvent conditions that elute the desired product within 10 min. Impurities found in these fractions have similar chromatographic properties to those of the product eliminating the need for a wash step in between injections.
13. The rapid, convenient method of Rivier *(15)* is used to desalt the peptide from the TEAP buffer. Using solvent mixtures containing TFA, the peptide is retained on the same preparative RP-HPLC column while the TEAP salts are removed. A linear solvent gradient of increasing acetonitrile concentrations elutes the peptide from the column.
14. 100 m*M* TEAP has a lower extinction coefficient than 0.1% TFA at 220 nm; therefore, look for the absorbance of the effluent to drop as TEAP is eluted from the column.

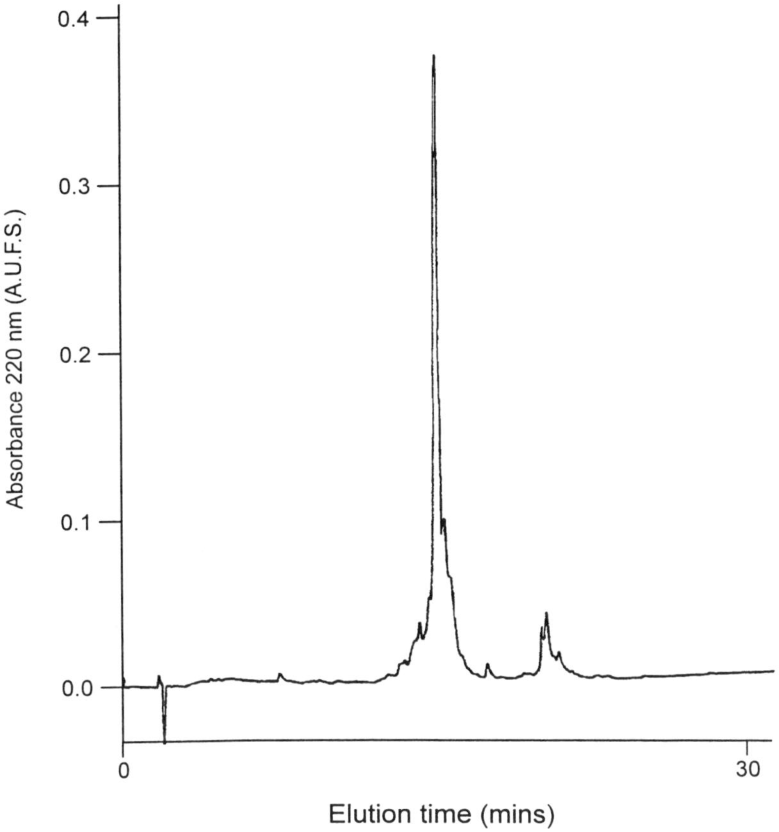

Fig. 1. Analytical RP-HPLC of crude material from TFA cleavage. Column: Vydac C_{18}, 4.6 × 150 mm. Buffers: 0.1% TFA in water = A; 0.09% TFA in acetonitrile:water (60:40) = B. Gradient: 10–100% B (over 30 min). Flow rate, 1 mL/min. 1050 psi back pressure. Chart speed: 20 cm/h.

15. A steep solvent gradient is employed to ensure the highest recovery of product from the column. In addition, the run time may also be shortened to 30 min.
16. This protocol was used for the purification of calcitonin gene-related peptide fragment (8–37) {CGRP(8–37)}, a CGRP competitive antagonist. A Perkin Elmer (Foster City, CA) 432A peptide synthesizer was used to synthesize the peptide on a 0.025-mmol scale employing Fmoc amino acid derivatives and HBTU-mediated coupling reactions. Cleavage from the resin with TFA in the presence of thiol scavengers yielded the crude product as a white lyophilized powder. Analytical RP-HPLC showed the crude product to contain one major component and several minor components (Fig. 1). The retention times of most of the minor components were within 2 min of that of the major

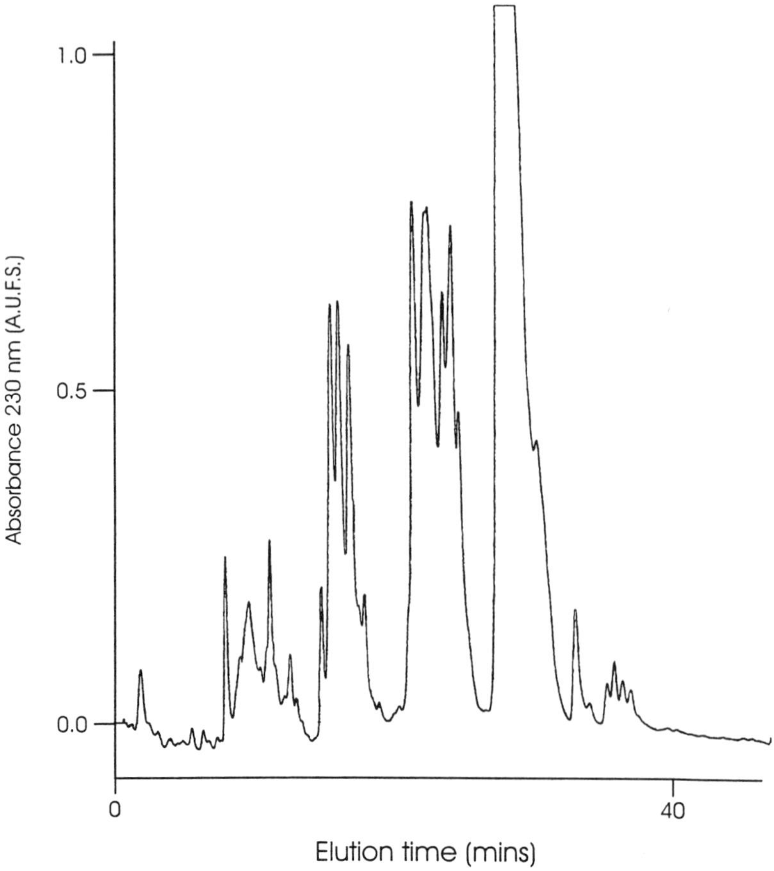

Fig. 2. Preparative ion-exchange HPLC of crude material from TFA cleavage. Column: Poly(SULFOETHYL) Aspartamide, 10 × 200 mm. Buffers: potassium phosphate 5 m*M* pH 3.0, 20% CH_3CN = A; 0.5*M* KCl in A = B. Gradient: 10–100% B (over 40 min). Flow rate, 4 mL/min. 450 psi back pressure. Chart speed: 20 cm/h.

component, resulting in a poor fractionation of the mixture. Preparative ion-exchange HPLC of the crude product, however, resulted in baseline-to-baseline separation of the major component from the majority of minor components (Fig. 2). Fractions containing the major component were pooled and subjected to isocratic analytical RP-HPLC using the TEAP buffer-solvent system. The major component eluted with a k' of 4.2 using 65:35 (vol:vol) mixture of Buffer A and Buffer B. Using the same TEAP buffers, the pooled fractions from the ion-exchange HPLC were loaded on to the preparative Vydac C_{18}

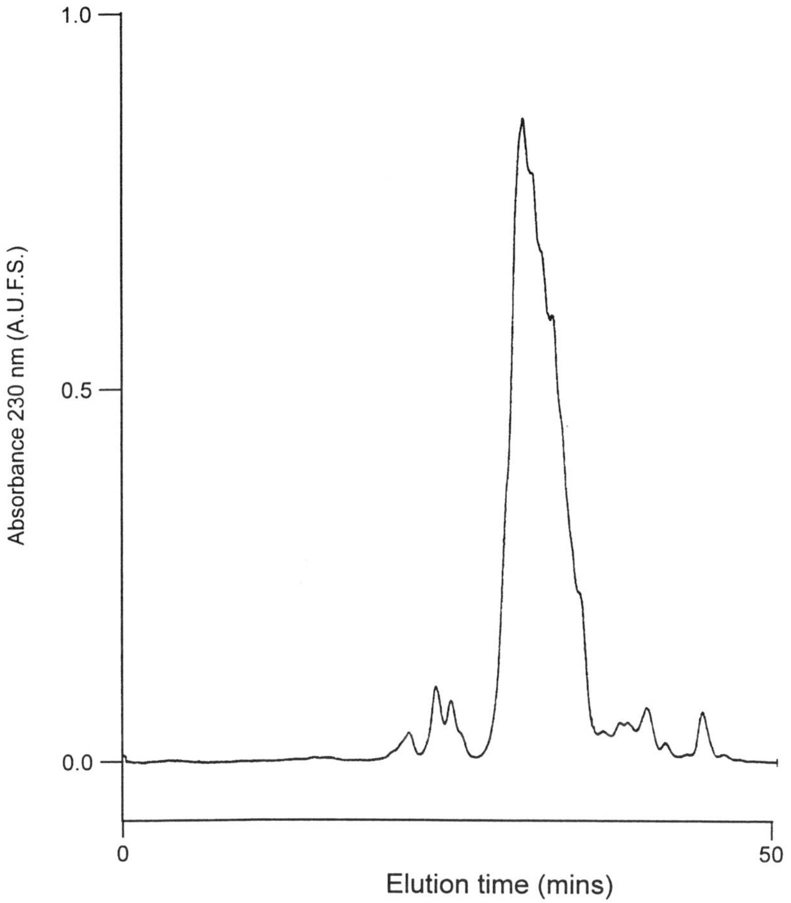

Fig. 3. Preparative RP-HPLC of major component from Fig. 2. Column: Vydac C_{18}, 10 × 250 mm. Buffers: 100 m*M* TEAP pH 2.25 = A; acetonitrile:A (60:40) = B. Gradient: 25–45% B (over 50 min). Flow rate, 4 mL/min. 450 psi back pressure. Chart speed: 20 cm/h.

RP-HPLC column that had been previously equilibrated with a 75:25 (vol:vol) mixture of Buffer A and Buffer B. As seen in Fig. 3, the major component was well resolved from the remaining minor components using the linear solvent gradient of 25–45% Buffer B over 50 min. Fractions containing only the major component were pooled and desalted, on the same column, using the TFA-based solvent described in Fig. 4. Fractions containing the desalted major component were pooled and lyophilized to yield 10 mg of a fluffy white powder. Analytical RP-HPLC showed the major component to have been purified to apparent homo-

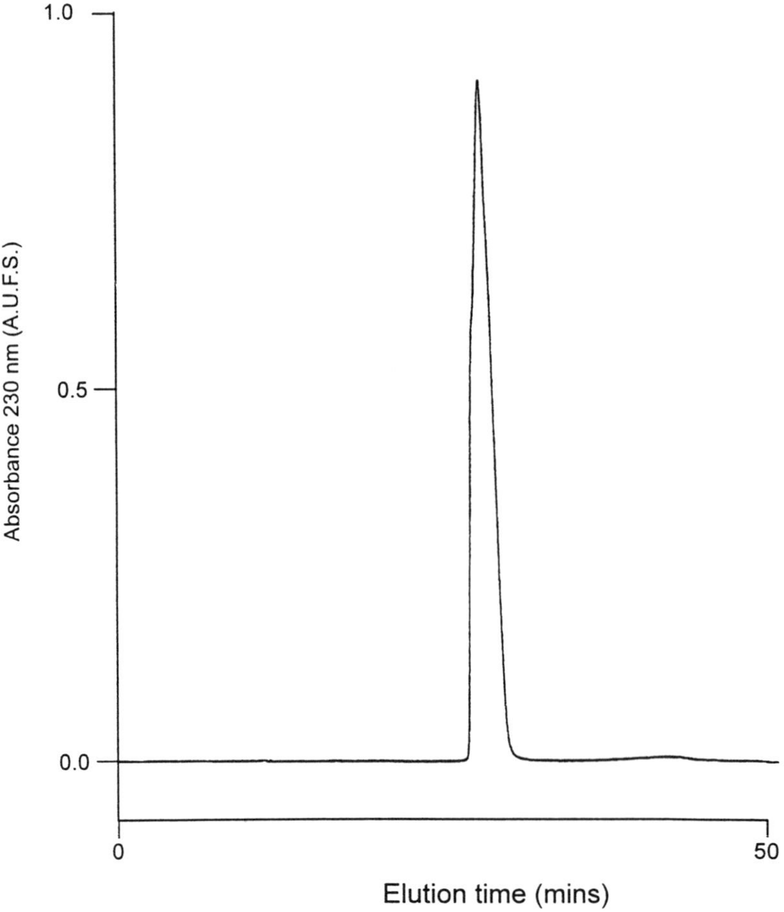

Fig. 4. Preparative desalt by RP-HPLC of major component from Fig. 3. Column: Vydac C_{18}, 10 × 250 mm. Buffers: 0.1% TFA in water = A; 0.09% TFA in acetonitrile:water (60:40) = B. Gradient: 10–100% B (over 30 min). Flow rate, 4 mL/min. 450 psi back pressure. Chart speed: 20 cm/h.

geneity (Fig. 5) and amino acid analysis and mass spectrometry confirmed its structure to be that of CGRP (8–37).

Acknowledgments

This work was supported in part by U.S. Public Health Service Grant HL51131 and the State of Nebraska, Department of Health Smoking & Cancer Research Grant, LB595.

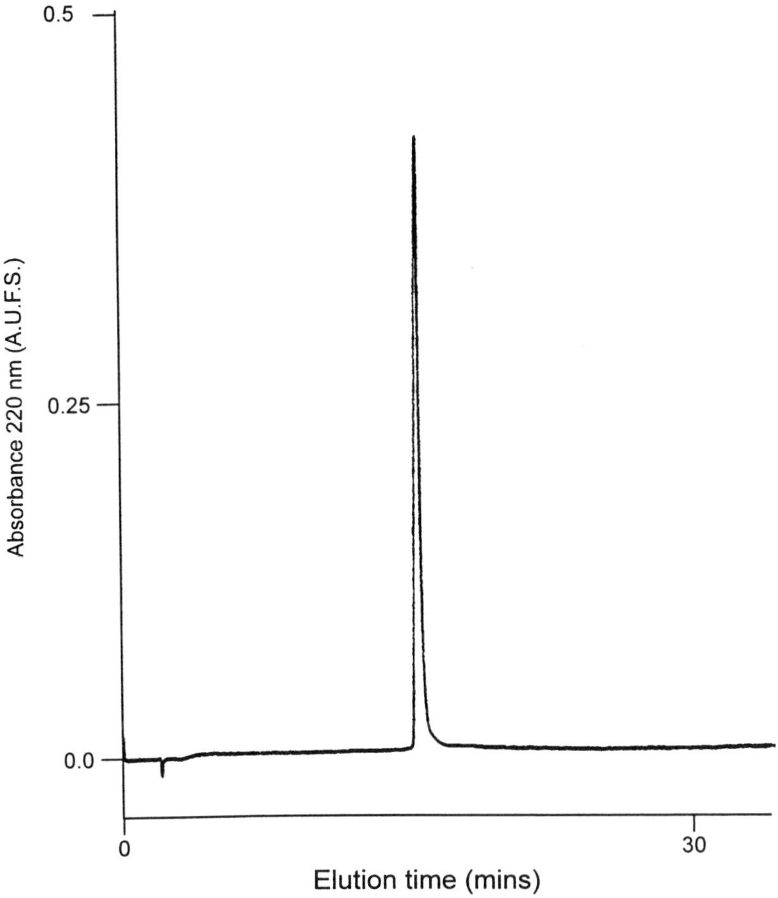

Fig. 5. Analytical RP-HPLC of lyophilized pool of fractions of major component from Fig. 4. Column: Vydac C_{18}, 4.6 × 150 mm. Buffers: 0.1% TFA in water = A; 0.09% TFA in acetonitrile:water (60:40) = B. Gradient: 10–100% B (over 30 min). Flow rate, 1 mL/min. 450 psi back pressure. Chart speed: 20 cm/h.

References

1. Merrifield, R. B. (1963) Solid phase peptide synthesis. I. The synthesis of a tetrapeptide. *J. Am. Chem. Soc.* **85,** 2149–2154.
2. Schnolzer, M., Jones, A., Alewood, P. F., and Kent, S. B. H. (1992) Ion-spray tandem mass spectrometry in peptide synthesis: structural characterization of minor by-products in the synthesis of ACP (65-74). *Anal. Biochem.* **204,** 335–343.
3. Giralt, E., Andreu, D., Miro, P., and Pedroso, E. (1983) Solid phase synthesis of tyrosine-containing histone fragments. *Tetrahedron* **39,** 3185–3188.

4. Moore, G. J. (1982) III. Reversed phase high pressure liquid chromatography for the identification and purification of neuropeptides. *Life Sci.* **30,** 995–1002.
5. Hancock, W. S., Bishop, C. A., Prestidge, R. L., Harding, D. R. K., and Hearn, M. T. W. (1978) High-pressure liquid chromatography of peptides and proteins II. The use of phosphoric acid in the analysis of underivatised peptides by reversed-phase high-pressure liquid chromatography. *J. Chromatogr.* **153,** 391–398.
6. Molnar, I. and Horvath, C. (1977) Separation of amino acids and peptides on non-polar stationary phases by high-performance liquid chromatography. *J. Chromatogr.* **142,** 623–640.
7. Rivier, J. E., Lazarus, L. H., Perrin, M. H., and Brown, M. R. (1977) Neurotensin analogues. Structure–activity relationships. *J. Med. Chem.* **20,** 1409–1412.
8. Rubenstein, M., Stein, S., Gerber, L. D., and Udenfriend, S. (1977) Isolation and characterization of the opioid peptides from rat pituitary: β-Lipotropin. *Proc. Natl. Acad. Sci. USA* **74,** 3052–3055.
9. Takagaki, Y., Gerber, G. E., Nihei, K., and Khorana, H. G. (1980) Amino acid sequence of the membranous segment of rabbit liver cytochrome b_5. *J. Biol. Chem.* **255,** 1536–1541.
10. O'Hare, M. J. and Nice, E. C. (1979) Hydrophobic high-performance liquid chromatography of hormonal polypeptides and proteins on alkylsilane-bonded silica. *J. Chromatogr.* **171,** 209–226.
11. Harding, D. R. K., Bishop, C. A., Tartellin, M. F., and Hancock, W. S. (1981) Use of perfluoroalkanoic acids as volatile ion-pairing reagents in preparative HPLC. *Int. J. Peptide Protein Res.* **18,** 214–220.
12. Bennett, H. P. J., Browne, C. A., and Solomon, S. (1980) The use of perfluorinated carboxylic acids in the reversed phase HPLC of peptides. *J. Liq. Chromatogr.* **3,** 1353–1365.
13. Bennett, H. P. J., Browne, C. A., and Solomon, S. (1981) Purification of the two major forms of rat pituitary corticotropin using only RP-HPLC. *Biochemistry* **20,** 4530–4538.
14. Linde, S. and Welinder, B. S. (1991) Non-ideal behaviour of silica-based stationary phases in trifluoroacetic acid-acetonitrile-based reversed-phase high-performance liquid chromatographic separations of insulins and proinsulins. *J. Chromatogr.* **536,** 43–55.
15. Rivier, J. E. (1978) Use of trialkyl ammonium phosphate (TAAP) buffers in reversed phase HPLC for high resolution and high recovery of peptides and proteins. *J. Liq. Chromatogr.* **1,** 343–366.
16. Rivier, J. E. (1986) Preparative purification of synthetic peptides. *J. Chromatogr.* **288,** 303–328.
17. Hoeger, C., Galyean, R., Boublik, J., McClintock, R., and Rivier, J. (1987) Preparative reversed phase high performance liquid chromatography: effects of buffer pH on the purification of synthetic peptides. *BioChromatography* **2,** 134–142.
18. Mant, C. T. and Hodges, R. S. (1989) Optimization of peptide separations in HPLC. *J. Liq. Chromatogr.* **12,** 139–172.

19. Burke, T. W., Mant, C. T., Black, J. A., and Hodges, R. S. (1989) Strong cation-exchange high-performance liquid chromatography of peptides. Effect of non-specific hydrophobic interactions and linearization of peptide retention behaviour. *J. Chromatogr.* **476,** 377–389.
20. Alpert, A. J. and Andrews, P. C. (1988) Cation exchange chromatography of peptides on Poly(SULFOETHYL) Aspartamide silica. *J. Chromatogr.* **443,** 85–96.
21. Smith, D. D., Li, J., Wang, Q., Murphy, R. F., Adrian, T. E., Elias, Y., Bockman, C. S., and Abel, P. W. (1993) Synthesis and Biological Activity of C-terminally truncated fragments of human-α-calcitonin gene-related peptide. *J. Med. Chem.* **36,** 2536–2541.
22. Smith, D. D., Conlon, J. M., Petzel, J., Chen, L., Murphy, R. F., and Morley, B. J. (1994) Solid-phase peptide synthesis and biological activity of bovine thymopoietin II (bTP-II). *Int. J. Peptide Protein Res.* **44,** 183–191.
23. Andrews, P. C. (1988) Ion exchange HPLC for peptide purification. *Peptide Res.* **1,** 93–99.

9

Molecular Weight Estimation for Neuropeptides Using Size-Exclusion High Performance Liquid Chromatography

G. Brent Irvine

1. Introduction

Size-exclusion high performance liquid chromatography (HPLC) has enabled purification of peptides and proteins to be carried out 10–100 times faster than conventional gel filtration chromatography on soft gels. The prepacked columns (about 1 × 30 cm) contain small particles (13 µm or less), resulting in tens of thousands of theoretical plates per meter. They can be operated at flow rates of about 1 mL/min, giving run times of <20 min. The improved peak sharpness and speed have led to a resurgence of interest in the technique. As well as being a standard chromatographic mode for the purification of peptides and proteins, size-exclusion chromatography can be used for estimation of molecular weights. Polyacrylamide gel electrophoresis in sodium dodecyl sulfate-containing buffers is widely used for determining the molecular weight of protein subunits, but is operating near the limits of resolution of the technique in the molecular weight range below a few thousand Daltons (*see* Chapter 10).

For polymers of the same shape, plots of log molecular weight against the distribution coefficient (K_d) (*see* Note 1) give straight lines within the range $0.1 < K_d < 0.8$ *(1)*. This is true only for ideal size-exclusion chromatography, in which the support does not interact with solute molecules (*see* Note 2). In any case, it must be kept in mind that it is the size, rather than the molecular weight, of a solute molecule that determines its elution volume. Hence calibration curves prepared with globular proteins as standards cannot be used for the assignment of molecular weights to proteins with different shapes, such as

From: *Methods in Molecular Biology, Neuropeptide Protocols*
Edited by: G. B. Irvine and C. H. Williams Humana Press Inc., Totowa, NJ

the rod-like protein, myosin. It has been found that the most reliable measurements of molecular weight by size-exclusion HPLC are obtained under denaturing conditions, when all proteins have the same random coil structure. Disulfide bonds must be reduced, usually with dithiothreitol, in a buffer that destroys secondary and tertiary structure. Buffers containing guanidine hydrochloride *(2,3)* or sodium dodecyl sulfate (SDS) *(4)* have been used for this purpose.

The use of denaturants has many drawbacks when working with proteins, but these properties may actually be advantageous when studying peptides (*see* Note 3). The method described in Section 3. is for the estimation of molecular weights of peptides in 6*M* guanidine hydrochloride. Under these conditions the recently developed Superdex Peptide medium from Pharmacia (Uppsala, Sweden) can be used for determination of molecular weights in the range of 500–5000, whereas TSK G2000SW covers the range of 3000–30,000. The use of the former column to separate neuropeptides in a volatile mobile phase prior to mass spectrometry is described in Chapter 13.

2. Materials

2.1. Apparatus

An HPLC system, or fast protein liquid chromatography (FPLC) system, for isocratic elution is required. This comprises a pump, an injector, a size-exclusion column, a UV detector, and a data recorder. The results described in Fig. 1 were obtained using two columns, spanning different size-exclusion ranges. These were a TSK G2000SW column (Toyo Soda, Tonda, Japan) and a Superdex Peptide HR 10/30 column (Pharmacia).

2.2. Chemicals

1. 0.1*M* disodium hydrogen orthophosphate (Na_2HPO_4).
2. 0.1*M* sodium dihydrogen orthophosphate (NaH_2PO_4).
3. 0.1*M* sodium phosphate, pH 6.0: Adjust the pH of 500 mL of 0.1*M* NaH_2PO_4 to a value of 6.0 by addition of 0.1*M* Na_2HPO_4 (approx 70 mL).
4. 6*M* guanidine hydrochloride/0.05*M* sodium phosphate: Dissolve 573.2 g guanidine hydrochloride (*see* Note 4) in 0.1*M* sodium phosphate, pH 6.0 (500 mL). Make up to 1 L with water. Filter through a 0.22-µm filter (Millipore Bedford, MA, *see* Note 5).
5. Solutions of standard peptides and proteins: Dissolve each peptide in filtered 6*M* guanidine hydrochloride/0.05*M* sodium phosphate at a concentration of about 1 mg/mL (*see* Note 5). Peptides and proteins suitable for use as standards are listed in Table 1.

Table 1
Polypeptide Standards

Polypeptide	Mol wt
Immunoglobulin G	160,000
Carbonic anhydrase	29,000
Myoglobin	16,900
Cytochrome c	12,400
Ubiquitin	8565
ACTH	4500
Glucagon	3550
Dynorphin A	2147
Neurotensin	1673
Tyr^0 Bradykinin	1223
LHRH	1182
Angiotensin II	1046
YGGFMRFamide	875
YGGFM	588
YGGF	442

All polypeptides listed were obtained from Sigma, Poole, UK, except for YGGFMR-Famide from Peninsula, Belmont, CA, and YGGF from Bachem, Bubendorf, Switzerland.

3. Method

1. Allow the column to equilibrate in the 6*M* guanidine hydrochloride/0.05*M* sodium phosphate until the absorbance at 280 nm is constant. The flow rate during equilibration should be only half that used during subsequent runs (*see* Note 6).
2. Inject a solution (20 µL) of a very large molecule, such as immunoglobulin G, and determine V_o from the absorbance profile (*see* Note 6).
3. Inject a solution (20 µL) of one of the standard peptides. Record its elution volume, V_e. Repeat this procedure until all the standards have been injected. Calculate K_{av} (*see* Note 1) for each peptide and plot K_{av} against log molecular weight. Typical plots for both columns are shown in Fig. 1.
4. Inject a solution (20 µL) of the peptide of unknown molecular weight in 6*M* guanidine hydrochloride/0.05*M* sodium phosphate. If the peptide contains disulfide bonds, then they must be reduced before injecting the sample (*see* Note 7). Record the elution volume, V_e.
5. If the sample contains more than one peptide, and the peaks cannot be assigned with certainty, collect fractions and assay each fraction for the relevant activity.
6. Calculate K_{av} for the unknown peptide and use the calibration plot to obtain an estimate of its molecular weight.
7. Store the column in an appropriate solvent and wash out the pumps and injector with water (*see* Note 8).

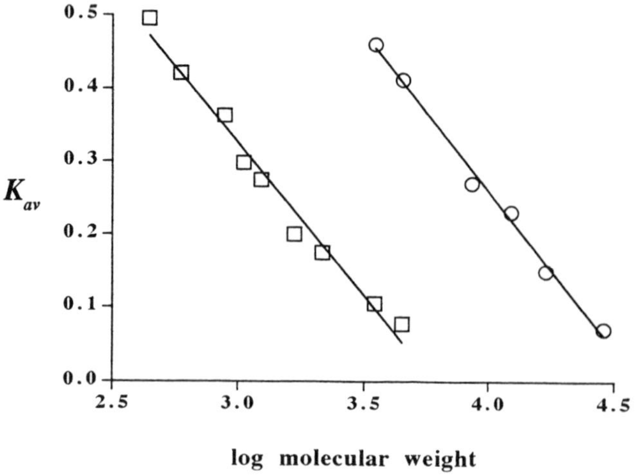

Fig. 1. Plots of K_{av} vs log molecular weight for the polypeptides listed in Table 1. 20 μL of a solution containing about 20 μg of each polypeptide was injected. The equipment was a Model 501 Pump, a 441 Absorbance Detector operating at 280 nm, a 746 Data Module (all from Waters), and a Rheodyne model 7125 injector (Rheodyne, Cotati, CA) with a 20-μL loop. The flow rate was 1 mL/min for the TSK G2000SW column and 0.7 mL/min for the Superdex Peptide HR 10/30 column. For the Superdex Peptide HR 10/30 column, V_o was determined to be 7.36 mL from the elution peak of immunoglobulin G, whereas V_c was calculated from the column dimensions (1.0 × 30.6 cm) to be 24.0 mL. The regression line $y = -0.416x + 1.572$, $r^2 = 0.982$ was computed using all the points shown as □. For the TSK G2000SW column, V_o was determined to be 5.23 mL from the elution peak of immunoglobulin G, whereas V_c was calculated from the column dimensions (0.75 × 30 cm) to be 13.25 mL. The regression line $y = -0.426x + 1.964$, $r^2 = 0.994$ was computed using all the points shown as ○.

4. Notes

1. The support used in size-exclusion chromatography consists of particles containing pores. The molecular size of a solute molecule determines the degree to which it can penetrate these pores. Molecules that are wholly excluded from the packing emerge from the column first, at the void volume, V_o. This represents the volume in the interstitial space (outside the support particles) and is determined by chromatography of very large molecules. Under the conditions described in this chapter, immunoglobulin G is sufficiently large for estimation of V_o for both columns.

   Molecules that can enter the pores freely have full access to an additional space, the internal pore volume, V_i. Such molecules emerge at V_t, the total volume available to the mobile phase, which can be determined from the elution volume of molecules that are so small that they do not experience exclusion. Hence $V_t = V_o + V_i$. A solute molecule that is partially restricted from the pores will emerge with elution volume, V_e, between the two extremes, V_o and V_t. The distribution coefficient, K_d, for such a molecule represents the fraction of V_i available to it for diffusion. Hence:

$$V_e = V_o + K_d . V_i .$$

and

$$K_d = \frac{V_e - V_o}{V_i} = \frac{V_e - V_o}{V_t - V_o}$$

Measurement of V_t is not trivial, however, because small molecules may also experience nonideal size-exclusion (*see* Note 2). In addition, in the case of the Superdex Peptide HR 10/30 column, size-exclusion effects may still operate even in the molecular weight range near 100 Dalton. Thus, it is simpler to measure the total internal volume of the packed column, V_c. This is the sum of V_t and the solid support volume, V_s, that is not accessible to solvent. V_c (mL) can be calculated from the equation for the volume of a cylinder, $\pi r^2 h$, where r is half the internal diameter of the column and h is the column length, both expressed in cm. The parameter K_{av} can then be used instead of K_d.
where:

$$K_{av} = \frac{V_e - V_o}{V_c - V_o}$$

The ratio of K_{av}:K_d is constant for an individual packing *(5)*.

2. Silica to which a hydrophilic phase such as a diol has been bonded still contains underivatized silanol groups. Above pH 3, these are largely anionic and will interact with ionic solutes, leading to nonideal size-exclusion chromatography. Depending on the value of its isoelectric point, a protein can be cationic or anionic at pH 7. Proteins that are positively charged will undergo ion exchange, causing them to be retarded. Conversely, anionic proteins will experience electrostatic repulsion from the pores, referred to as ion exclusion, and will be eluted earlier than expected on the basis of size alone. When size-exclusion chromatography is carried out at a low pH, the opposite behavior is found, with highly cationic proteins being eluted early and anionic ones being retarded. To explain this behavior, it has been suggested that, at pH 2, the column may have a net positive charge *(6)*. In order to reduce ionic interactions, it is necessary to use a mobile phase of high ionic strength. On the other hand, as ionic strength increases, this promotes the formation of hydrophobic interactions. To minimize both ionic and hydrophobic interactions, the mobile phase should have an ionic strength between 0.2 and 0.5*M* *(7)*.
3. For a particular column, the mol-wt range in which separation occurs is reduced in denaturing solvents. This is because the radius of gyration, and hence the hydrodynamic size, of a molecule increases when it changes from a sphere to a random coil. For example, the separation range of a TSK G2000SW column

operating with denatured proteins is 3000–30,000 (Fig. 1) compared to 5000–100,000 for native proteins in 0.1*M* sodium phosphate buffer, pH 7.0 containing 0.25–0.3*M* NaCl *(8,9)*. However, when working with peptides, this lower molecular weight range is actually more suitable.

4. A high-quality grade of guanidine hydrochloride, such as Sigma G-4505, must be used to give low background absorbance at 280 nm and to avoid contamination of the column. Solutions of guanidine hydrochloride absorb light in the far UV range, so that monitoring the absorbance in the most sensitive region for peptides (200–220 nm) is no longer possible and a longer wavelength, such as 280 nm, must be used. Only peptides that contain tyrosine or tryptophan will absorb light at 280 nm.
5. All samples injected onto the column must be clean. If there is any uncertainty about presence of insoluble material or dust particles, centrifuge the samples in a microfuge or filter through a 0.45-μm filter (Millipore type HV).
6. The TSK column can withstand back pressures up to 20 Bar (300 psi) and can be run at flow rates up to 1.2 mL/min. Back pressure for the Superdex Peptide column should not exceed 15 Bar (217 psi), giving maximum flow rate of 1.2 mL/min. Flow rates less than half the maximum value should be used when switching mobile phases. Flow rates must also be reduced when mobile phase of high viscosity, such as 6*M* guanidine hydrochloride, is used. Flow rates of 0.7 mL/min and 1.0 mL/min, respectively, were used in the experiments with the Superdex Peptide column and the TSK G2000SW column.
7. Disulfide bonds can be reduced by incubation of the peptide in 6*M* guanidine hydrochloride, containing 10 m*M* dithiothreitol, pH 8.5, at 37°C for 2 h.
8. It is advisable to flush out the column with water at the end of each day and to store it in an appropriate antimicrobial solvent, such as 0.02% sodium azide or 20% ethanol if it is not being used for several days. It is also necessary to wash out the chromatography system with water, since high concentrations of salts, especially those containing halide ions, can adversely affect pumps and stainless steel.

References

1. Gooding, K. M. and Regnier, F. E. (1990) Size exclusion chromatography, in *HPLC of Biological Macromolecules* (Gooding, K. M. and Regnier, F. E., eds.), Marcel Dekker, New York, pp. 47–75.
2. Ui, N. (1979) Rapid estimation of molecular weights of protein polypeptide chains using high-pressure liquid chromatography in 6 M guanidine hydrochloride. *Anal. Biochem.* **97,** 65–71.
3. Kato, Y., Komiya, K., Sasaki, H., and Hashimoto, T. (1980) High-speed gel filtration of proteins in 6 M guanidine hydrochloride on TSK-GEL SW columns. *J. Chromatogr.* **193,** 458–463.
4. Josic, D., Baumann, H., and Reutter, W. (1984) Size-exclusion high-performance liquid chromatography and sodium dodecyl sulphate-polyacrylamide gel electrophoresis of proteins: a comparison. *Anal. Biochem.* **142,** 473–479.

5. *Gel Filtration Principles and Methods,* 6 ed. (1993) Pharmacia Publication 18-1022-18; ISBN91-97-0490-2-6.
6. Irvine, G. B. (1987) High-performance size-exclusion chromatography of polypeptides on a TSK G2000SW column in acidic mobile phases. *J. Chromatogr.* **404,** 215–222.
7. Regnier, F. E. (1983) High performance liquid chromatography of proteins. *Methods Enzymol.* **91,** 137–192.
8. Kato, Y., Komiya, K., Sasaki, H., and Hashimoto, T. (1980) Separation range and separation efficiency in high-speed gel filtration on TSK-GEL SW columns. *J. Chromatogr.* **190,** 297–303.
9. Irvine, G. B. and Shaw, C. (1986) High-performance gel permeation chromatography of proteins and peptides on columns of TSK-G2000-SW and TSK-G3000-SW: a volatile solvent giving separation based on charge and size of polypeptides. *Anal. Biochem.* **155,** 141–148.

10

Molecular Weight Determinations Using Polyacrylamide Gel Electrophoresis with Tris-Tricine Buffers

G. Brian Wisdom

1. Introduction

Polyacrylamide gel electrophoresis (PAGE) in buffers containing the anionic detergent sodium dodecylsulfate (SDS) is a very powerful technique for small-scale separation of polypeptides and for assigning molecular weights to these molecules. However, the majority of systems used (e.g., the one described by Laemmli *[1]*) cannot separate polypeptides with masses below about 15 kDa. Various methods have been described to extend the range of SDS-PAGE; these have included the use of high concentration gels and the incorporation of materials such as urea to resolve the polymers of low molecular weight. The most generally used technique is the one developed by Schagger and von Jagow *(2)*. This technique employs a discontinuous gel system containing SDS. However, the interference of SDS with the stacking and separation of small polypeptides is diminished by changing the trailing ion (in the cathode buffer) from glycine to the more mobile Tricine (N-tris[hydroxymethyl]-methylglycine) and by lowering the pH of the separating gel.

Schagger and von Jagow *(2)* examined four types of gel system; this chapter describes, with minor modifications, the two most generally useful ones. System A is more simple and resolves polypeptides in the range of 5–100 kDa. System B has a spacer gel in addition to the stacking gel; this modification is necessary for the resolution of polypeptides and oligopeptides below 5 kDa and the system has a range of 1–60 kDa.

From: *Methods in Molecular Biology, Neuropeptide Protocols*
Edited by: G. B. Irvine and C. H. Williams Humana Press Inc., Totowa, NJ

2. Materials

2.1. Equipment

1. Electrophoresis apparatus (for vertical gels).
2. Power pack.

2.2. Gel Components

1. Acrylamide monomers solution (49.5%T, 3%C): 48 g of acrylamide and 1.5 g of bis-acrylamide are dissolved in water and made up to 100 mL (stored at 4°C).
2. Gel buffer: 3*M* Tris-HCl buffer, pH 8.25, containing 0.1% SDS (stored at 4°C).
3. Glycerol.
4. Ammonium persulfate: 10% (w/v) in water (prepared fresh each day).
5. TEMED (N,N,N',N'-tetramethylethylenediamine).

2.3. Sample Buffer

Double strength sample buffer: 0.1*M* Tris-HCl buffer, pH 6.8, containing 4% SDS, 5% 2-mercaptoethanol, 20% glycerol, and 0.01% Coomassie brilliant blue G250 or equivalent (*see* Note 1).

2.4. Electrode Buffers

1. Anode buffer: 0.2*M* Tris-HCl buffer, pH 8.9 (stored at 4°C).
2. Cathode buffer: 0.1*M* Tris/0.1*M* Tricine buffer, pH 8.25, containing 0.1% SDS (stored at 4°C).

2.5. Staining and Destaining Solutions (see Note 2)

1. Fixer: 10% acetic acid, 50% methanol.
2. Stain: 10% acetic acid containing 0.025% Coomassie brilliant blue G250.
3. Destain: 10% acetic acid.

2.6. Molecular Weight Standards

Various companies sell molecular weight standards suitable for the calibration of gels of this type, for example, Novex (San Diego, CA), Pharmacia Biotech (Uppsala, Sweden), and Sigma (St. Louis, MO).

3. Methods

3.1. Gel Preparation

Gels are cast in vertical cassettes. The dimensions of these can be varied, but it is important to have sufficient path length in the separating gel to give adequate separation of the standards and the poly/oligopeptides of interest. The stacking gel and the spacer gel (if used) should each be about 10% of the total length of the gel but the optimal arrangement for a particular application must be determined empirically. The volumes given in Table 1 are for use with a

Table 1
Composition of Gels

Component	Separating gels A (10%T/3%C)	Separating gels B (16.5%T/3%C)	Spacer gel (10%T/3%)	Stacking gel (4%T/3%C)
Glycerol, g	2.0	2.0	—	—
Gel buffer, mL	5.0	5.0	1.67	1.24
Monomer solution, mL	3.03	5.0	1.01	0.40
Deionized water, mL	to 15.0	to 15.0	2.32	3.36
Ammonium persulfate, µL	75.0	50.0	20.0	40.0
TEMED, µL	7.5	5.0	2.0	4.0

Hoefer Tall Mighty Small apparatus (SE 280) and are ample for the preparation of two, 8 × 11 cm gels with a thickness of 0.75 mm.

System A (10%T/3%C) comprises a separating gel and a stacking gel, whereas System B (16.5%T/3%C) has a spacer gel between the separating and stacking gels.

1. Combine the components of the separating gel in the order given in Table 1, mix by gentle swirling, add to the cassette, overlay with water, and leave to polymerize for 0.5 h. (Degassing of the gel solution is not necessary.)
2. Remove the overlay using a sheet of filter paper, but avoid touching the gel surface.
3. Prepare the spacer gel, if required, and the stacking gel in the same manner. Place a comb in the stacking gel to form sample wells.

3.2. Sample Preparation

Dissolve the samples in an equal volume of sample buffer and place them in a boiling water bath for 5 min.

3.3. Electrophoresis

1. Attach the gel to the electrophoresis apparatus and place the cathode and anode buffers in the upper and lower reservoirs respectively.
2. Flush the wells in the stacking gel with cathode buffer using a Pasteur pipet or syringe to remove unpolymerized components.
3. Load the samples and standards with a syringe (through the cathode buffer) into the wells.
4. Apply a constant voltage of 30 V until the samples have moved out of the stacking gel (about 1 h) and then apply an increased voltage until the dye front is at the bottom of the gel. For System A, 150 V are used (takes about 2 h), whereas for System B, 90 V are applied (for about 4 h).
5. Stain the gel by removing it from the cassette and immediately placing it in fixer for 30 min. Transfer it to stain solution for 1 h, and finally destain the gel with

four changes of destaining solution over 1 h (*see* Note 3). All these steps are carried out on an orbital shaker.

6. Measure the migration distances of the standard polypeptides, plot these as a function of the log of their molecular weights (*see* Note 4), and determine the unknowns.

4. Notes

1. The dye bromophenol blue (0.01%) may also be used to monitor the progress of gel electrophoresis runs. However, in the systems described above, it migrates behind some of the smaller oligopeptides so Coomassie brilliant blue G250, although it forms a broad band, is preferred.
2. Other staining methods can be applied, but it is essential to ensure that oligopeptides and small polypeptides are fixed in the gel and remain fixed during destaining. Some molecules bind Coomassie blue poorly; silver staining may help in these cases.
3. The times given are for 0.75-mm gels. Polypeptides in thicker gels will require longer times for fixing and staining.
4. The calibration curve has an inflection at about 6 kDa. Bradykinin (1.1 kDa) or bacitracin (1.4 kDa) can be used as a convenient reference for expressing migration distances or ratios.

Acknowledgment

I am grateful to George Allen for excellent technical assistance and advice.

References

1. Laemmli, U. K. (1970) Cleavage of structural proteins during the assembly of the head of bacteriophage T4. *Nature* **227,** 680–685.
2. Schagger, H. and von Jagow, G. (1987) Tricine-sodium dodecyl sulfate-polyacrylamide gel electrophoresis for the separation of proteins in the range from 1 to 100 kDa. *Anal. Biochem.* **166,** 368–379.

11

Determination of Neuropeptides by Capillary Electrophoresis

SungAe S. Park, Wei-Lun Hung, Daniel E. Schaufelberger, Norberto A. Guzman, and Juan P. Advis

1. Introduction

In today's world, we face a significant degree of miniaturization in many aspects of technology. Wonders of miniaturization are also occurring in the areas of physical and biological sciences. The development of a 21st-century technology has enabled investigations and processes to be performed at concentration levels that were previously unimaginable. Unfortunately, as the concentration levels and quantities of samples involved become progressively smaller, chemical-scale analytical technology rapidly approaches its limit of usefulness.

Frequently, many applications require analytical systems capable of handling nanoliter volumes of sample containing subnanomolar quantities of materials. In the clinical environment, in which the early detection of the onset of a disease process may be essential to a patient's survival, determinations of analytes at concentrations approaching even the single molecule level may soon be essential. As a result of this new awareness, microscale will eventually be defined by atto-, zepto-, and eventually, yoctomolar concentration ranges.

Of the methods to arise in this 21st-century technology, modern capillary electrophoresis (CE) has evolved as a powerful technique yielding remarkable information in a variety of applications. The method offers unique characteristics that include high resolution and efficiency, the potential for high speed, great mass sensitivity, and extremely low sample consumption. It has been applied successfully to the separation and analysis of a variety of simple and complex molecules, ranging from ions, amino acids, and nucleotides to pep-

From: *Methods in Molecular Biology, Neuropeptide Protocols*
Edited by: G. B. Irvine and C. H. Williams Humana Press Inc., Totowa, NJ

tides, oligonucleotides, and glycoproteins, including viral and subcellular particles *(1–14)*.

Undoubtedly, CE is gradually becoming an all-purpose method, solving analytical problems in many biological and chemical disciplines. By definition, capillary electrophoresis is the electrophoretic separation of a substance within a narrow tube. Contrary to high performance liquid chromatography (HPLC), which separates on the basis of a mechanically driven movement of a mobile phase across a stationary phase, capillary electrophoresis functions by generating an electrically driven motion of fluids and ions within a relatively narrow capillary tube. Because narrow capillary tubes have very small internal volumes, they can dissipate heat efficiently through their thin walls. This prevents disruption of separations by induced thermal convection currents. Capillary electrophoresis can make use of relatively large electrical fields to induce rapid and efficient separations within very small amounts of sample.

Currently, the most common methods to determine neuropeptides are HPLC, radioimmunoassays (RIA), and other immunoassays such as enzyme-linked immunosorbent assay (ELISA). However, there are many limitations in these assays and a need for more sensitive and practical techniques is essential. An emerging technique is capillary electrophoresis coupled to laser-induced fluorescence (CE-LIF) detection as a very sensitive method to determine a variety of analytes *(15–18)*.

This chapter deals with the use of CE-LIF detection as an assay for the determination of in vivo release of multiple neuropeptides in perfusate samples from the brain hypothalamic median eminence (ME). Determination and quantification of in vivo neuropeptide from the ME is essential for understanding the neuroendocrine control of reproduction. Among the multiple brain peptides involved in triggering a preovulatory surge of luteinizing hormone (LH) from the anterior pituitary (an obligatory event for reproduction to occur) are luteinizing-hormone releasing hormone (LHRH), neuropeptide Y (NPY), and β-endorphin (βEND). Previous data from our laboratories indicate that multiple in vivo neuropeptide release preceding the preovulatory LH surge can be monitored in ME perfusate samples using CE-LIF *(19)*. Furthermore, it has been suggested that enzymatic degradation of LHRH neuropeptide by endopeptidase 24.15 might play a role in the genesis of this preovulatory surge *(20,21)*. Thus, in an attempt to generate endogenous samples to be tested in a CE-LIF-based assay, we have studied the possible effect that inhibition of ME endopeptidase-24.15 (EP-24.15) might have on preovulatory in vivo ME-neuropeptide release. These studies were carried out by a perfusion technique that uses a push–pull cannula (PPC). The PPC consists of a rigid probe inserted into specific regions of the brain containing an inlet and outlet where a fluid is mechanically pumped. In contrast with a microdialysis probe, the push–pull probe does not contain a microdialysis membrane at the end of the probe. Normally, the PPC is perfused with an artificial cerebrospinal fluid solution, and then chemicals in the vicinity of the end of the probe mix, and continue through the outlet to be further collected and analyzed *(22–24)*.

PPC perfusate samples *(22–24)* were obtained both under basal conditions and during ME perfusion of an inhibitor of EP-24.15. This inhibitor prevents the activity of the endopeptidase, which degrades LHRH and other neuropeptides *(21)*. The amount of neuropeptide was determined by specific radioimmunoassay and by CE-LIF assay in aliquots of each PPC sample *(19)*. A good correlation was observed between RIA and CE-LIF results. In vivo ME perfusion of the EP-24.15 inhibitor to early follicular ewes increased LHRH release to the extent that it advances the onset of the preovulatory LH surge by 4–6 h. This is the first evidence that inhibition of ME endopeptidase 24.15 increases in vivo ME-LHRH release and causes a premature onset of the preovulatory LH surge. Furthermore, these data indicate that a CE-LIF-based assay can be used to simultaneously detect ME endogenous release of multiple neuropeptides under defined physiologically relevant conditions. A description of the CE-LIF assay is presented.

2. Materials

2.1. CE Instrumentation

The method described in this chapter was carried out utilizing a Beckman P/ACE 2100 capillary electrophoresis system equipped with System Gold software for data analysis, purchased from Beckman (Fullerton, CA). A laser-induced fluorescence detector was employed for monitoring all separations performed in this study. An argon-ion laser was used (excitation 488 nm and emission 520 nm) as the detection system. Fused-silica capillary (75 μm id × 57 cm, 50 cm to detector) was purchased from PolyMicro Technologies (Phoenix, AZ). Filters of 0.45-μm porosity (Nalgene Filterware), screw-cap polyethylene conical vials (1.5 mL) (used for sample derivatization), and Fisher Vortex Genie 2 were obtained from Fisher Scientific (Fair Lawn, NJ). Regular glass vials (4.5-mL) (used as buffer container and for sample vial holder) and 100-μL total volume polyethylene conical microvial (used as sample container) were purchased from Beckman.

2.2. Chemicals

1. Luteinizing-hormone releasing hormone (0.5 mg), β-endorphin (0.5 mg), and neuropeptide Y (0.2 mg) were obtained as ready-to-use vials (containing either 0.2 or 0.5 mg/vial) from Peninsula Laboratories (Belmont, CA).

2. 5-Carboxyfluorescein succinimidyl ester (FSE) (Beckman LIF kit) was purchased as a ready-to-use vial (containing 0.5 mg/vial of chromophore) from Beckman or as a solid reagent from Molecular Probes (Eugene, OR).
3. Sodium tetraborate (Borax) was obtained from Sigma (St. Louis, MO). Concentrated hydrochloric acid, sodium phosphate dibasic heptahydrate, and 50% sodium hydroxide solution were purchased from Fisher. Deionized water was obtained from HYDRA ultra pure system (Research Triangle Park, NC).
4. 100 m*M* sodium tetraborate/50 m*M* sodium phosphate. Weigh 3.81 g of sodium tetraborate and 0.71 g of sodium phosphate, and then transfer into a 100-mL volumetric flask. Dissolve and dilute the solids to the mark with deionized water and divide into two portions.
5. Derivatization buffer: 100 m*M* sodium tetraborate/50 m*M* sodium phosphate, pH 8.0. Adjust the pH of one portion (50 mL) of 100 m*M* sodium tetraborate/50 m*M* sodium phosphate to pH 8.0 with concentrated HCl. Filter this solution through a 0.45-µm filter.
6. Separation buffer: 50 m*M* sodium borate/25 m*M* sodium phosphate, pH 10.0. Mix the remaining portion (50 mL) of 100 m*M* sodium borate/50 m*M* sodium phosphate buffer with 50 mL of deionized water. Adjust the pH to 10.0 with 50% sodium hydroxide. Filter this solution through a 0.45-µm filter and degas well using vacuum while it is mechanically stirred using a stir bar (*see* Note 1).

2.3. Sample Preparation

Samples were obtained from an ewe during the early follicular phase before and after median eminence perfusion of an inhibitor of endopeptidase 24.15 (EP-24.15) (cpp-Ala-Ala-Phe: Nova Biochem, San Diego, CA) through the PPC probe.

2.4. PPC Sampling

Our laboratory developed and characterized a multiple guide cannula assembly (MGCA) and removable PPC probes to repetitively sample in vivo ME-LHRH release *(22)*. A summary of this technique is as follows:

1. The multiple guide cannula assembly has three main components, a platform, a grid plug, and a slug plug, all made of Daldrin plastic. The platform (height, 2 mm; length, 68 mm; width, 35 mm) provides support for the grid or slug plugs. It is attached to the skull using four stainless steel screws (length, 18 mm; diameter, 3.5 mm) and cranioplastic cement (Plastic One, Roanoke, VA). The cranial port is located in its center. The grid plug (diameter, 2 cm; height, 1.5 cm) has parallel guide holes positioned 1.5 mm apart (48 holes in a 6 × 8 array) through which removable probes are positioned into their target areas. The slug plug is an exact replica of the grid plug but without the parallel guide holes. This plug closes the cranial port at all times except during sampling, when it is replaced by the grid plug. Both plugs fit tightly into the cranial port and are fixed by screws to the

platform. Each MGCA was attached to an experimental ewe's skull at least 2 wk prior to starting PPC sampling.

2. General health checkups were done a day before MGCA surgery. Feed and water were withheld for 24 h before surgery. At this time, precautionary antibiotic treatment was started with procaine penicillin G and dihydrostreptomycin (Burns Vet Supply, Rockville Centre, NY; 1 million IU, for intramuscular administration). Also, the antiinflammatory drug, dexamethasone (Sigma; 20 mg, im) was given. The head of the animal was cleanly shorn and disinfected with iodine solution.
3. Anesthesia was induced with 5% halothane and maintained with 1–1.5% of the gas. The head of the animal was fixed in a stereotaxic system (David Kopf, Tujunga, CA) with the ear bars tightly screwed against the supermastoid foramina on each side of the head. The correct positioning of the head in the stereotaxic apparatus was assessed by a dorsal roengentogram. After skin incision, a circular piece of dorsal cranium between the occipital and frontal bones was removed. The exact location where the bone was to be removed, so that the ME could be reached through the opening, was determined with the help of a lateral roengentogram. The platform was attached to the cranium with the help of screws and cranioplastic cement. A probe was inserted into the third cerebral ventricle and a radio-opaque dye (100 μL, Renographin) was injected before a lateral roengentogram was taken with the infusion probe *in situ*. The position of the tip of this probe was used to determine which guide cannula of the multiple array must be used, and the depth below the top of the multiple guide cannula required to reach the ME. The MGCA assembly was replaced with the plug before finishing the surgical procedure. Each ewe recovered in a padded pen illuminated with infrared lamps. Antibiotic treatment and disinfection of the head's top was continued for the next week.
4. Each ewe's jugular vein was cannulated prior to PPC perfusion, thus allowing us to obtain simultaneous PPC perfusate samples and jugular blood samples. The PPC probe consists of two concentric stainless steel cannulae, with the inner cannula projecting 1 mm beyond the outer one. The PPC probe was inserted into the brain through the guide cannula in the multiple array that will be defined as being on top of the ME, based on the lateral roengentogram obtained during surgery. During 6 h sampling, artificial cerebrospinal fluid (CSF) with or without inhibitors was infused through the inner concentric tube of the PPC probe (2 h CSF, 2 h CSF + inhibitors, 2 h CSF), using a peristaltic pump (supply pump, Rabbit-Plus, Rainin Instruments, Woburn, MA). Simultaneously, fluid was removed from the space between the two concentric tubes of the PPC probe using another peristaltic pump (collection pump), calibrated at the same rate as that of the supply pump (10 μL/min). Thus, artificial CSF bathed the surrounding area of the probe's tip (theoretically a sphere 1.5 mm in diameter) and was pulled through the outer cannula by the outlet peristaltic pump. The CSF flow was calibrated at 100 μL/10 min. The perfusate was received by a fraction collector into a different acidified borosilicate tube every 10 min, which was then frozen until assayed for LHRH. The composition of the artificial CSF was 127.6 m*M* NaCl,

2.5 m*M* KCl, 1.4 m*M* $CaCl_2$, 1.0 m*M* $MgSO_4$, and 12.0 m*M* Na_3P0_4. This solution (pH 7.4) was stored frozen until used for PPC sampling.

3. Methods

3.1. CE Instrument Set-Up

1. A Beckman P/ACE system 2100 coupled to laser-induced fluorescence detector can be utilized. The instrument should be connected to 80 psi nitrogen gas (*see* Note 2).
2. For fluorescence detection, use a cartridge-cassette containing a 75 µm id × 57 cm (50 cm to detector) (*see* Notes 3 and 4).
3. Install an emission band pass filter (520 nm) in the LIF detector as indicated in the Beckman manual.

3.2. Preparation of Reagents Used for Derivatization of Peptides

1. 1 m*M* FSE solution: Take a ready-to-use vial containing 0.5 mg of FSE (as described in the Beckman LIF kit) or weigh 0.5 mg of FSE and dissolve with 1 mL of derivatization buffer. Prepare the FSE solution freshly prior to derivatization of the sample.
2. Standard solutions: Add water to each standard of LHRH (0.5 mg/vial), β-END (0.5 mg/vial), and NPY (0.2 mg/vial) to make 0.5 mg/mL final concentration. These standard solutions should be diluted serially to generate a calibration curve necessary for the quantitation of samples present in brain perfusates.
3. Derivatized sample and standards: Pipet 50 µL of a sample or standard into a 1.5-mL screw cap polyethylene vial (derivatization vial). Then add 50 µL of FSE solution while the sample is vortexed at speed 3 of the Vortex instrument.

3.3. Capillary Electrophoresis Procedure

1. Pipet 30–50 µL of samples derivatized with FSE into a 100-µL total volume conical microvial that fits into a spring or adaptor within the vial. Place the spring into an empty vial; then place the plastic vial to be inserted into the top of the spring (vial #20 or higher number of the carousel). Fill three vials with separation buffer (one for rinse; vial #34 and two for separation; vial #11 and #1), one vial with 0.1*N* sodium hydroxide (vial #32), and one vial with deionized water (vial #33). An empty vial is placed at position #10 of the outlet tray for disposal of waste (*see* Note 5).
2. The sample to be analyzed is introduced into the capillary by pressure injection (0.5 psi for 5 s). The capillary is rinsed between runs with 0.1*N* sodium hydroxide using a pressure of 20 psi.
3. A summary of the method is described below:
   a. Instrument: Beckman PACE with LIF detector.
   b. Column: 57 cm (50 cm) × 75 µm fused silica capillary.
   c. Separation buffer: 50 m*M* sodium tetraborate/25 m*M* sodium phosphate buffer, pH 10.0.

Table 1
A Typical Capillary Electrophoresis—LIF Detection Method of Operation

Step	Time	Condition	Inlet vial	Outlet vial	Temperature
Prerinse 1	3 min	Water	33	10	33°C
Prerinse 2	3 min	Separation buffer	34	10	33°C
Injection	5 s	Pressure	20	10	33°C
Separation	30 min	13 kV	11	1	33°C
Postrinse 1	3 min	0.1*N* NaOH	32	10	33°C
Postrinse 2	3 min	Water	33	10	33°C

d. Detection: laser-induced fluorescence (Ex 488 nm, Em 520 nm).
e. Voltage: 13 kV (170–220 μA).
f. Temperature: 33°C.
g. Injection: 5.0 s (approx 30 nL).

Table 1 describes a typical program to run the CE method.

3.4. Electropherograms

1. A blank solution for FSE reagent is prepared exactly as for the derivatized sample, but water is used instead of the perfusate (*see* Note 6). (Electropherogram not shown.)
2. Each standard solution is also derivatized as described in Section 3.2. and loaded into the corresponding autosampler vial. Figure 1 shows three electropherograms (left panel) and the calibration curves (right panel) of the neuropeptides LHRH, NPY, and β-END (*see* Note 7). LHRH and NPY show individual peaks, β-END produces two peaks.
3. Figure 2 shows the electropherograms of endogenous samples. Samples were obtained during the early follicular phase before and after median eminence perfusion of an inhibitor of endopeptidase 24.15 using a PPC probe. An increased neuropeptide content was observed after perfusion of the inhibitor (*see* Notes 8 and 9).

4. Notes

1. All fluids are recommended to be filtered, and the pH and ionic strength of all solutions should be maintained in the appropriate range applicable to the study.
2. Confirm that the LIF ion source (argon ion laser Ex 488 nm) is connected to the detector unit through the fiberoptic cable prior to turning on the laser ion source.
3. When using an uncoated bare-fused silica capillary, it is generally good practice to perform a 5-min manual rinse using a cleaning (Pretreatment/Regenarant) solution (Beckman), or 0.1*N* sodium hydroxide, followed by a 5-min rinse with deionized water, and by a 5-min rinse with the run buffer.

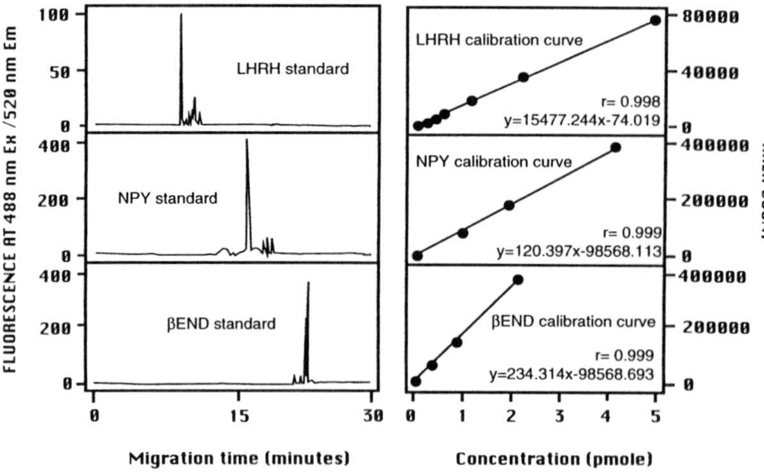

Fig. 1. Electropherograms and calibration curves of neuropeptide standards. Luteinizing-hormone releasing hormone (9–5000 fmol), neuropeptide Y (1000–4000 fmol), and β-endorphin (500–2000 fmol) were derivatized with a chromogenic substance (FSE). An aliquot of approx 30 nL of each derivatized neuropeptide was injected into the capillary and analyzed by CE-LIF. Left panel shows the typical electropherographic pattern of each neuropeptide. Right panel shows the corresponding calibration curves of each neuropeptide.

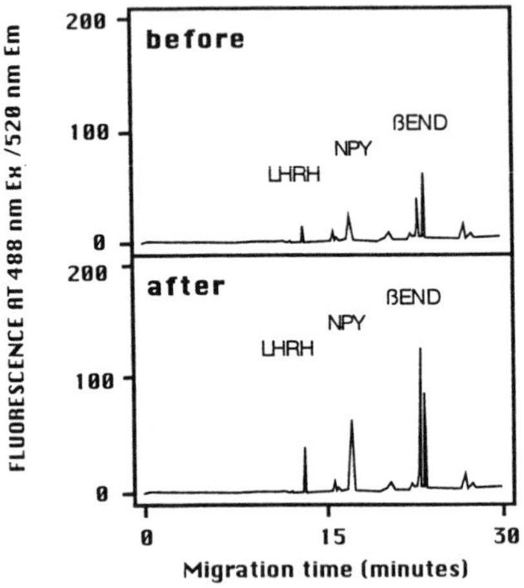

Fig. 2. Monitoring of the effect of an inhibitor of endopeptidase 24.15 by CE-LIF. Determination of analytes from the hypothalamic median eminence was carried out by using PPC and CE-LIF techniques. Simultaneous in vivo release samples were obtained from an ewe during the early follicular phase before and after median eminence perfusion with an inhibitor of EP-24.15 through the PPC probe. Endopeptidase 24.15 is the main enzymatic activity degrading LHRH and other neuropeptides. An increased neuropeptide content was observed in samples obtained after perfusion of the inhibitor, compared with those obtained immediately before.

4. Use the appropriate cartridge for fluorescence detection that will fit into the fluorescence detector. The cartridge should be set in gently in order to avoid breakage of the capillary during installation.
5. Since commercially available vials may contain residual surfactant, it is recommended to thoroughly wash the vials prior to use. Glass vials (used as buffer container) must also be well dried to prevent residual moisture that may influence changes in the concentration of the analytes. These vials placed in the autosampler should be filled with no more than 4.5 mL of liquid. The maximum recommended filling capacity of the vial is when the fluid meniscus reached the base of the threads. For proper operation, the fluid meniscus must not be lower than 1 cm below the threads.

   Do not fill vials higher than the bottom thread of the vial. When the vial rises and the capillary is inserted into the vial, liquid displacement occurs. If the vial is overfilled, liquid will be forced out of the slots in the vial cap into gas passages or onto the autosampler trays or the spill tray beneath. This liquid can lead to arcing when voltage is applied.

   Cover each vial with a silicone cap. The cap is designed with an O-ring seal on the top in order to seal the system and to avoid high voltage leakage. Each cap has a die-cut across the opening in the top, creating four triangular flaps. The caps must be seated as far down onto the vials as possible without causing the flaps to open. Caps must also be kept dry. This configuration allows the simultaneous allocation of the electrode and capillary in the contact with the solution in the vial.
6. The electropherogram of FSE control does not show the presence of a peak up to 22 min; however, several baseline peaks are detected after 22 min.
7. The electropherograms of commercially available standards show more than one peak, indicating the presence of contaminants.
8. It has been suggested that enzymatic degradation of the LHRH neuropeptide by endopeptidase 24.15 might play a role in the genesis of the preovulatory surge *(20,21)*. Thus, the possible role that inhibition of ME endopeptidase-24.15 might have on preovulatory in vivo ME-neuropeptide release has been studied in an attempt to generate endogenous samples to be tested in a CE-LIF based assay. PPC perfusate samples *(22–24)* were obtained both under basal conditions and during ME perfusion of an inhibitor of EP-24.15. This inhibitor prevents the activity of the endopeptidase that degrades LHRH and other neuropeptides *(20,21)*. The amount of neuropeptide was determined by CE-LIF assay (Figs. 1 and 2) and by specific radioimmunoassay (Fig. 3) in aliquots of each PPC sample. A good correlation was observed between CE-LIF and RIA results (Figs. 1 and 3). In vivo ME perfusion of the EP-24.15 inhibitor to early follicular ewes increased LHRH release to the extent that it advanced the onset of the preovulatory LH surge by 4–6 h. This is the first evidence that inhibition of ME endopeptidase 24.15 increases in vivo ME-LHRH release and causes a premature onset of the preovulatory LH surge. Furthermore, these data indicate that CE-LIF-based assay can be used to simultaneously detect ME endogenous release of multiple neuropeptides under defined physiologically relevant conditions.

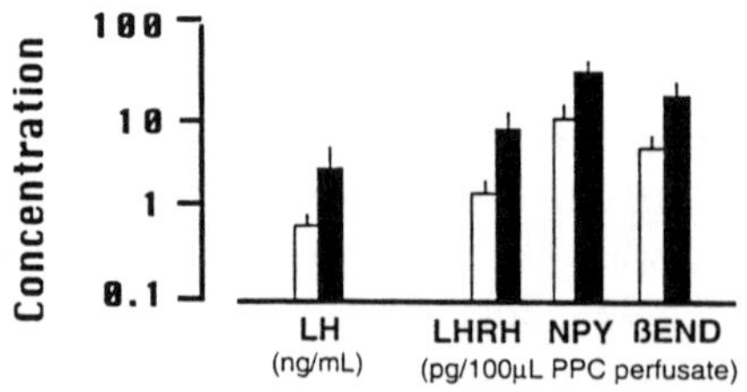

Fig. 3. Determination of peptides by radioimmunoassay. The levels of LH from plasma, and LHRH, NPY, and βEND from PPC perfusate were determined by RIA. Samples were obtained at 10-min intervals before (open bars) and after (filled bars) perfusion of the endopeptidase inhibitor.

9. Whereas the techniques of CE-LIF are still being refined, it is clear that these methods can overcome the current limitations of CE concentration sensitivity. Another useful technology currently in progress in our laboratory is the online preconcentration of analytes using the analyte concentrator. This technique utilizes an adsorptive phase within the CE capillary capable to concentrate samples that enable the detection of <200 fg/mL of an analyte *(25)*.

References

1. Grossman, P. D. and Colburn, J. C. (1992) *Capillary Electrophoresis: Theory and Practice.* Academic, San Diego.
2. Li, S. F. Y. (1992) *Capillary Electrophoresis: Principles, Practice and Applications.* Elsevier, Amsterdam.
3. Vindevogel, J. and Sandra, P. (1992) *Introduction to Micellar Electrokinetic Chromatography.* Hüthig Buch Verlag GmbH, Heidelberg.
4. Wiktorowicz, J. E. (1992) *Capillary Electrophoresis.* Academic, New York.
5. Foret, F., Krivánková, L., and Bocek, P. (1993) *Capillary Zone Electrophoresis.* VCH, Weinheim.
6. Guzman, N. A. (1993) *Capillary Electrophoresis Technology.* Dekker, New York.
7. Jandik, P. and Bonn, G. (1993) *Capillary Electrophoresis of Small Molecules and Ions.* VCH, Cambridge, UK.
8. Weinberger, R. (1993) *Practical Capillary Electrophoresis.* Academic, San Diego.
9. Camilleri, P. (1993) *Capillary Electrophoresis: Theory and Practice.* CRC, Boca Raton, FL.
10. Kuhn, R. and Hoffstetter-Kuhn, X. (1993) *Capillary Electrophoresis: Principles and Practice.* Springer-Verlag, Berlin.
11. Landers, J. P. (1994) *Handbook of Capillary Electrophoresis.* CRC, Boca Raton, FL.
12. Altria, K. D. (1995) *Capillary Electrophoresis Guidebook.* Chapman & Hall, London.
13. Baker, D. R. (1995) *Capillary Electrophoresis.* Wiley, New York.

14. Righetti, P. G. (1996) *Capillary Electrophoresis in Analytical Biotechnology.* CRC, Boca Raton, FL.
15. Hernandez, L., Escalona, J., Joshi, N., and Guzman, N. A. (1991) Laser-induced fluorescence and fluorescence microscopy for capillary electrophoresis. *J. Chromatogr.* **559,** 183–196.
16. Hernandez, L., Joshi, N., Verdeguer, P., and Guzman, N. A. (1993) Laser-induced fluorescence detection for capillary electrophoresis: a powerful analytical tool for the separation and detection of trace amounts of analytes, in *Capillary Electrophoresis Technology* (Guzman, N. A., ed.), Dekker, New York, pp. 605–614.
17. Pentoney, S. L., Jr. and Sweedler, J. V. (1994) Optical detection techniques for capillary electrophoresis, in *Handbook of Capillary Electrophoresis* (Landers, J. P., ed.), CRC, Boca Raton, FL, pp. 147–186.
18. Yao, Y. and Li, S. F. Y. (1996) Determination of erythrocyte porphyrins by epi-illumination fluorescence microscope with capillary electrophoresis. *J. Liq. Chrom. Rel. Technol.* **19,** 1–15.
19. Park, S. S., Guzman, N. A., Rabii, J., and Advis, J. P. (1994) Capillary electrophoresis (CE) coupled to fluorescence detection for the determination of *in vivo* neuropeptide release from the ewe median eminence (ME). 24th Annual Meeting Society for Neurosciences, November 13–18, Miami Beach, FL, Abstract 46.7.
20. Advis, J. P., Kuljis, R. O., and Dey, G. S. (1985) Distribution of LHRH and LHRH-degrading activity in the hypothalamus of the ewe. *Endocrinology* **116,** 2410–2418.
21. Molineaux, C. J., Ladsun, A., Michaud, C., and Orlowski, M. (1988) Endopeptidase 24.15 is the primary enzyme that degrades luteinizing hormone releasing hormone both *in vitro* and *in vivo*. *J. Neurochem.* **51,** 624–633.
22. Conover, C. D., Kuljis, R. O., and Advis, J. P. (1993) Use of a multiple guide cannula assembly and removable push-pull cannula probes to assess *in vitro* LHRH release from the posterior-lateral median eminence of the ewe. *Neuroendocrinology* **57,** 1119–1132.
23. Prasad, B. M., Conover, C. D., Sarkar, D. K., Rabii, J., and Advis, J. P. (1993) Feed restriction in prepubertal lambs: effect on *in vivo* release of LHRH, NPY and βEND from the median eminence and on puberty onset. *Neuroendocrinology* **57,** 1171–1181.
24. Conover, C. D., Kuljis, R. O., Sarkar, D. K., Rabii, J., and Advis, J. P. (1993) β-endorphin regulation of LHRH release in median eminence: immunocytochemical and physiological evidence in ewes. *Neuroendocrinology* **57,** 1182–1195.
25. Tomlinson, A. J., Guzman, N. A., and Naylor, S. (1995) Enhancement of concentration limits of detection in CE and CE-MS: a review of on-line sample extraction, cleanup, analyte preconcentration, and microreactor technology. *J. Cap. Elec.* **2,** 247–266.

12

Characterization of Neuropeptide Processing by Fast Atom Bombardment Mass Spectrometry

Jerzy Silberring

1. Introduction

Fast atom bombardment (FAB) was introduced as a new ionization technique *(1)* in 1981 by M. Barber and his coworkers. This was a breakthrough in the analysis of unstable and involatile compounds such as peptides, which were difficult to study by other ionization methods. FAB employs a particle beam consisting of the neutral atoms of xenon or argon. The gas atoms are first ionized and accelerated in the FAB gun and then neutralized on the counter-electrode. The beam, directed toward the sample that is deposited on the probe tip, ionizes and sputters charged particles that are then accelerated in the ion source and analyzed. The sample needs to be added to a viscous, chemically inert, and relatively involatile matrix, which allows a long-term analysis of a target compound. The most commonly used matrix remains glycerol. However, other substances have also been applied for the same purpose (*see* Section 3.5.). In contrast to electrospray-ionization mass spectrometry (ESI MS, *see* Chapter 13), the FAB technique produces molecular ions, corresponding to the mass of intact, charged molecules and designated as $[M+H]^+$ or $[M-H]^-$ (positive- or negative-ion mode, respectively). Interpretation of the data obtained is simpler than in ESI because transformation of the spectra is not necessary, and often fragment ions appear along the spectrum, providing sequence information sufficient to reveal the primary structure of the peptide. Compared to other ionization techniques such as ESI MS or matrix-assisted laser desorption ionization time-of-flight MS (MALDI TOF) (*see* Chapter 14), the major disadvantages of the FAB technique are: limited measuring range at a full accelerating voltage (*see* Note 1); and sensitivity (*see* Notes 1 and 2). Moreover, strong signals belonging to the matrix clusters may obscure analysis. Problems concerning

From: *Methods in Molecular Biology, Neuropeptide Protocols*
Edited by: G. B. Irvine and C. H. Williams Humana Press Inc., Totowa, NJ

sensitivity and possible connection to liquid chromatography/capillary electrophoresis systems have successfully been overcome by application of continuous-flow FAB (CF FAB, dynamic FAB) where the sample solution is loaded into the instrument via a fused silica capillary *(2)*.

Introducing the sample thus requires continuous flow of the carrier liquid, which, in most cases, consists of a 5% solution of glycerol in water. Because FAB operates at high vacuum, the amount of liquid entering the ion source is limited to about 5 μL/min and, therefore, either flow-split or capillary LC needs to be used in conjunction with this technique. Despite intensive development of techniques such as ESI MS and MALDI TOF MS, fast atom bombardment mass spectrometry and continuous-flow FAB MS still remain useful and reliable methods for biochemical research (*see* Note 2). Complex mixtures may be directly studied, but the final verification of sequences should be performed by MS/MS or other techniques.

During the past two decades, proteolytic conversion and degradation of neuropeptides have drawn increasing attention owing to the potential role of these processes as targets for development of new drugs. Proteolytic processing is believed to play a crucial role in the release of neuropeptides, which are biosynthesized in the form of larger inactive precursors, as well as in the termination of their biological actions. The enzymes involved in these processes often possess unusual specificity aimed at a limited number of cleavage sites. Recent applications of biochemistry and molecular biology have greatly advanced our understanding of the nature and mode of action of these proteases. As an example, conversion of dynorphins to enkephalins by neuropeptide peptidases is described in this chapter.

The general pathway of prodynorphin metabolism by known proteinases is presented in Fig. 1. Research on dynorphin convertases generating enkephalins, originated from work on the rapid metabolism of dynorphins in cerebrospinal fluid (CSF) *(3)*. A search for the possible tissue source of this enzyme revealed an entire family of new proteinases, cleaving dynorphins to shorter, bioactive fragments. The conversion of dynorphin-derived peptides to enkephalins is associated with changes in receptor specificity. For example, dynorphins A and B have pronounced affinity for kappa opioid receptors, while Leu-enkephalin-Arg^6 and Leu-enkephalin show preference for the delta receptor. Moreover, the Leu-enkephalin-Arg^6 sequence, the shortest distinctly deriving from the prodynorphin precursor, can therefore serve as a marker of the metabolic changes within this particular neuropeptide system. It is likely that the concentration of a particular peptide fragment strongly depends on the proteolytic activities acting on longer sequences and on the formed products. These proteinases are, in turn, controlled by endogenous inhibitors that are an integral part of metabolic pathways. Both proteinases and

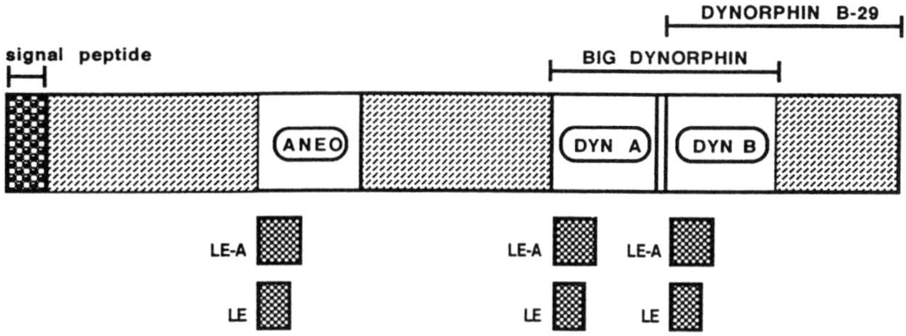

Fig. 1. Pre-prodynorphin and its major fragments with opioid activity. DYN A, dynorphin A; DYN B, dynorphin B; ANEO, α-neoendorphin; LE-A, Leu-enkephalin-Arg^{6}; and LE, Leu-enkephalin.

endogenous inhibitors have been considered and used as potential markers of various clinical states *(4)*.

This chapter shows how these enzymes might be studied with help of combined chromatographic/mass spectrometric techniques and provides experimental details.

2. Materials

2.1. Apparatus

1. A low pressure liquid chromatography system with gradient former.
2. Glass-teflon homogenizer.
3. Econo-Pac High Q cartridge, 5 mL (Bio-Rad, Hercules, CA).
4. Infusion pump (e.g., Harvard Apparatus, South Nattick, MA) or an HPLC syringe pump (e.g., Model 140 A, Applied Biosystems, Foster City, CA).
5. Gas-tight syringe, 2.5-mL with Luer lock (e.g., SGE, Austin, TX).
6. The Finnigan (Bremen, Germany) MAT 95Q mass spectrometer can be applied for this study with a FAB gun and an optional continuous-flow FAB probe. For static FAB, resolution is adjusted to 1400 (10% valley) and the total scan duration is set to 5 s. For the CF-FAB experiments, the syringe in the infusion pump or the HPLC pump is connected to the CF-FAB interface via a 25-cm long Peek capillary (0.25 mm id). Resolution is set at 500–1000 (10% valley) in order to increase sensitivity, and the total scan duration is set at 2 s. Spectra acquisition is performed in a profile mode (magnetic scan). The scan range depends on the substances to be analyzed (*see* Note 1).

2.2. Chemicals

1. Solvent A: 20 m*M* Tris-HCl, pH 7.4.
2. Solvent B: 1.0*M* NaCl in Tris-HCl, pH 7.4.

3. Dynorphin B (e.g., Bachem, Bubendorf, Switzerland or Peninsula, St. Helens, UK).
4. 40 m*M* phenylmethylsulfonyl fluoride (PMSF, serine proteinase inhibitor) in isopropanol or MeOH; Sigma (*see* Note 3).
5. 4 m*M* EDTA (metalloproteinase inhibitor) in water; Fluka, Buchs, Switzerland.
6. 10 m*M* *p*-hydroxymercuribenzoic acid (PHMB, thiol-dependent proteinase inhibitor) in 0.05*M* NaOH; Sigma (*see* Note 4).
7. 100 μ*M* pepstatin (aspartic proteinase inhibitor) in MeOH; Boehringer Mannheim, Mannheim, Germany (*see* Note 5).
8. Glycerol anhydrous; Fluka.
9. 1-Thioglycerol; Fluka.
10. Acetic anhydride; Sigma.

3. Methods

3.1. Enzyme Extraction

Neural tissues and body fluids perform a variety of proteolytic activities that cleave neuropeptides *(5)*. It is imperative to ascertain which enzymes are responsible for metabolizing the peptide of interest before performing other experiments. The rough characterization of soluble neuropeptide peptidases includes rapid purification/concentration steps and evaluation of the optimal inhibitory cocktail that blocks the unwanted proteolysis. Such a procedure allows the measurement of the enzyme activity directly in the mixture. The protocol is presented below.

All procedures should be performed at 4°C.

1. Homogenize tissue (0.1–5 mg tissue/1 mL of 20 m*M* Tris-HCl, pH 7.4) in a glass-teflon homogenizer.
2. Centrifuge (12–15,000*g* for 10 min).
3. Separate supernatant on a short anion-exchange cartridge (Econo-Pac High Q, 5 mL). The major advantages of the Econo-Pac cartridges are their easy usage, relatively good resolution, and low price. Being robust, they are suitable for the first step separation. This cartridge can conveniently be directly coupled to the HPLC system using adaptors; however, it should be placed in an ice beaker and the fraction collector should also be cooled. Use the following gradient of NaCl: 0–10 min: 100% A; 10–40 min: 0–50% B; 40–60 min: 50–100% B; and 60–70 min: 100–0% B. Flow rate is maintained at 1 mL/min. Collect 1-mL fractions without delay because basic proteins (having a higher isoelectric point, pI) will not be retained under such conditions and will elute in the void volume (3–4 mL). It is recommended to divide fractions into smaller portions (e.g., 0.1–0.25 mL) before freezing (preferably at –80°C), to avoid repetitive thawing. A similar procedure applies for the body fluids; however, the ionic strength should be considered before separation of the sample on the ion-exchange column. For example, plasma should be diluted at least 5 times before application on the column (*see* Note 6).

3.2. Enzyme Assay

1. Prepare stock solution of the peptide substrate (1 mg/mL) in water.
2. Dilute 10 times the small aliquots (5–10 µL) of every second fraction (first 40 tubes) with two different buffers: 20 m*M* Tris-HCl, pH 7.4 and 50 m*M* ammonium acetate, pH 5.5 and proceed as follows:
   a. Add 10 µL of fraction (diluted 1:10) to an Eppendorf tube (0.5 mL).
   b. Add 10 µL of appropriate buffer.
   c. Add 2 µg peptide.
   d. Withdraw 1 µL at 0 min (zero-time sample).
   e. Incubate the mixture for 30 min.
   f. Withdraw 1–1.5-µL aliquots from each tube.
   g. Mix with 1 µL of acetonitrile/0.1%TFA and add 2 µL of glycerol.
   h. Analyze directly by FAB MS.

If there are no cleavage products, continue incubation for an additional 30 min or longer, unless approx 50% of the substrate will be converted (*see* Notes 6 and 7). The amount of the substrate in the incubation mixture (step 2C, above) may be decreased to approx 1.0 µg/assay for CF-FAB measurements. Similarly, chromogenic or other short synthetic substrates may be applied.

A schematic (simulated) chromatogram is presented in Fig. 2 in which three distinct proteolytic activities were detected both at neutral and acidic pH.

Quantitation (CF-FAB) or semiquantitation (static FAB) should be performed using peak area integration of particular ions, characteristic for each peptide fragment and the appropriate standard.

3.3. Characterization by Class-Specific Inhibitors

Select peak fractions of every proteolytic activity (fractions 6, 19, 29, Fig. 2) for further tests. Each enzyme should then be characterized using each of the class-specific synthetic inhibitors in turn. The necessary reagents (given at final concentrations in the incubation mixture) are: 1 m*M* PMSF, 1 m*M* EDTA, 0.25 m*M* PHMB, and 1.0 µ*M* pepstatin (*see* Notes 3–8). The assay is similar to that described previously:

1. Add 10 µL fraction to an Eppendorf tube (0.5 mL).
2. Add 10 µL of appropriate buffer.
3. Add 10 µL water or 10 µL of one of the inhibitors (*see* Note 3).
4. Preincubate for 30 min at 37°C.
5. Add 2 µg of peptide and withdraw a 1–1.5-µL aliquot (0 min).
6. Incubate for 30 min at 37°C.
7. Withdraw the 1–1.5-µL aliquot, mix with 1 µL acetonitrile/0.1% TFA.
8. Add 2 µL of glycerol and analyze by FAB MS.

The proteolytic pattern shown in Fig. 2 can be expected to change after addition of at least one of the inhibitors. The reagents listed here cover the four

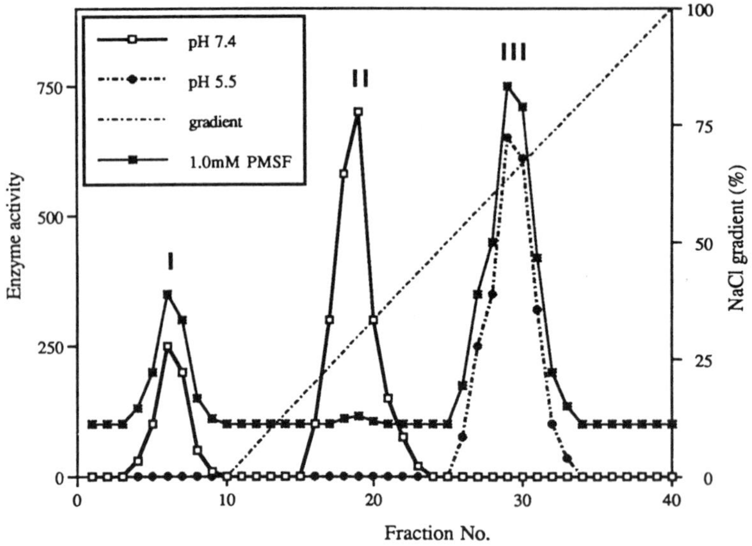

Fig. 2. Separation of the tissue extract on the anion-exchange column. Peak fractions No. 6, 19, and 29 were further tested with the major class-specific inhibitors. Column: Econo-Pak Q (5 mL) connected to the HPLC system; column load: extract corresponding to 0.5 g tissue; flow-rate, 1 mL/min; fractions: 1 mL.

major classes of proteolytic enzymes, but detailed studies on the active site require more comprehensive experiments and the necessary details may be found elsewhere *(6, 7)*.

3.4. Direct Measurements of Enzymatic Activity in Body Fluids

Detection in a body fluid will be given as an example, but the procedure can also be applied to tissue homogenates. The body fluid should be collected on ice, divided into smaller aliquots and kept frozen at –80°C. The protocol is similar to that described in the preceding text. It should be noted that direct measurement of the enzyme activity in certain body fluids (e.g., CSF), compared to the concentrated material obtained after ion-exchange chromatography, may require longer incubation time with the substrate. In certain cases (e.g., plasma) it is necessary to dilute samples before assay, otherwise the mass spectral signal may be significantly affected by other components or the enzymatic activity will be too high (*see* Notes 6 and 9). Collect the blood (10 mL) in the Vacutainer tube (100 × 16 mm, Becton Dickinson, Meylan Cedex, France), containing EDTA (0.12 mL, 0.34*M*). Separate plasma by centrifugation, dilute 5 times with water, and add the following:

1. Add 5 μL plasma (diluted 1:5) to the Eppendorf tube (0.5 mL).
2. Add 5 μL water.
3. Add 2 μL (2 μg) DYN A.
4. Collect 1–1.5-μL aliquot (zero-time).
5. Incubate at 37°C and withdraw 1-μL aliquots after 30, 45, and 90 min.
6. Mix the sample with an equal amount of glycerol and measure directly in the mass spectrometer. Alternatively, dilute 10 times with the mobile phase (usually 5–10% glycerol) and inject via a continuous-flow FAB interface.

3.5. Selection of the FAB Matrix

The most commonly used matrix for FAB experiments is glycerol *(8,9)*. Static FAB also works well with 1-thioglycerol or a mixture of both and these substances are recommended as a first choice during optimization of the assay. As a rule, we use pure 1-thioglycerol for the analysis of molecules above 3000 Dalton on the magnetic sector instrument. The presence in the analytes of disulfide bridges, which will be reduced in the presence of 1-thioglycerol, should be considered. An improved sensitivity and better signal-to-noise ratio are achieved by dilution of the sample in 3 μL of 50% acetonitrile/0.04% TFA and by addition of equal amounts of glycerol or 1-thioglycerol. A list of glycerol and 1-thioglycerol clusters (aggregates) is given in Note 10. Static FAB is less suitable for quantitation owing to the difficulties in reproducing identical droplets on the probe tip when liquid matrices are used. However, under well-defined conditions, the measurement of neuropeptides can be successfully achieved for the endogenous enkephalins *(10)*.

3.6. Estimation of Kinetic Parameters

Complete characterization of the enzymatic activity requires determination of certain physicochemical parameters, such as initial velocity, Michaelis-Menten constant, and a turnover number. These values can also be calculated with the help of MS. The theory of enzyme kinetics is beyond the scope of this paper and is available elsewhere *(6)*. Here, we will focus on the measurement of the initial velocity, which gives a rough estimation of the processing rate. A typical progress curve is presented in Fig. 3. Formation of the product is linear, as a function of time, only at the initial portion of the plot. Deviation from linearity along the time scale may have several explanations, but is usually caused by the depletion of substrate in the incubation mixture. Initial velocity is calculated as a tangent to the asymptote, drawn at the origin of the graph. The parameter is expressed in terms of activity units/min or units/s and is often recalculated per microgram of protein or microliter of the body fluid. Estimation of other kinetic parameters, such as the Michaelis constant (K_m), is not recommended unless the enzyme is purified to apparent homogeneity or other

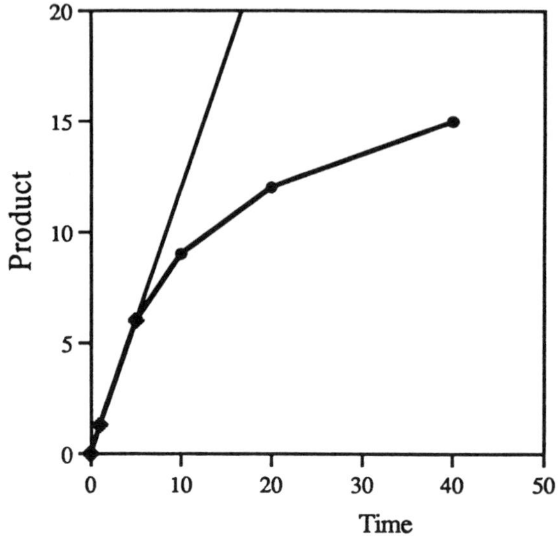

Fig. 3. Progress curve: product formation (arbitrary units) in function of time. The initial velocity is calculated as a tangent of the asymptote drawn at the origin.

proteinases, present in the incubation mixture, are completely inhibited. The physiological significance of the initial velocity is that it corresponds to the conversion rate of a given peptide in a body fluid or tissue. This, in turn, provides new insight into potential diagnostic methods such as measurement of endogenous peptides in body fluids, where the entire peptide pool might be rapidly degradated as soon as it enters the blood or CSF *(11)*.

3.7. Practical Suggestions for the Preparation of the CF-FAB Set-Up

Preparation of the stainless steel probe tip is performed as follows:

1. Pull the tip over a very fine sand paper and flush with MeOH. Ensure that the entire surface has been treated. The gold-plated target requires no treatment other than washing.
2. Cover the target with few drops of concentrated HCl and leave for 5 min. This procedure is called etching and is time-dependent!
3. Wash off the acid with MeOH and evaporate with dry air.
4. Fused-silica capillary transferring the mobile phase should terminate not more than 0.3–0.5 mm from the probe tip.
5. A thin film of the mobile phase evenly distributed over the entire surface should be observed.

6. Absorbing wicks should be used to withdraw excess matrix and protect against droplet formation on the FAB target.
7. The vacuum within the MS is high enough to drive the fluid into the ion source, even without the use of a pump. It is necessary to use a fused-silica capillary of internal diameter not larger than 50 μm to limit this effect.
8. The maximum concentration of acetonitrile is limited because this solvent may form an emulsion with glycerol, present in the matrix.

Additional technical suggestions are described in Notes 11 and 12 and also in ref. *12*.

3.8. Acetylation of Peptides

This reaction was originally applied for peptide sequencing by FAB MS *(13)*. In certain cases, it may be important to study cleavage products in the presence of the peptide inhibitor. For example, dynorphins or enkephalins have four identical N-terminal amino acids (Tyr-Gly-Gly-Phe-Leu, leucine-enkephalin or Tyr-Gly-Gly-Phe-Met, methionine-enkephalin, respectively). If the potential inhibitor contains the same sequence, it will be difficult to differentiate where the cleavage product comes from. It is, however, possible to derivatize the N-terminus of one peptide by, e.g., acetylation. If, for example, Leu-enkephalin is formed during enzymatic cleavage of both peptides, then the one derived from the acetylated peptide will have higher molecular mass (42 mass units apart). Figure 4 shows an example where N-acetylated α-neo-endorphin was tested as a potential inhibitor of the conversion of another opioid peptide, dynorphin B, by a peptidase derived from the U-1690 cell line. The only fragments released by the enzyme and detected along the spectrum were Leu-enkephalin and Leu-enkephalin-Arg^6 at m/z of 556.1 and 712.2 and not the Ac-Leu-enkephalin (m/z 598) or Ac-Leu-enkephalin-Arg^6 at m/z 754. This experiment clearly suggests that the N-terminal hexapeptide is released from DYN B.

Reaction is performed as follows:

1. Dilute acetic anhydride 1:3 with MeOH.
2. Prepare peptide solution in water (1 mg/mL).
3. To an Eppendorf tube (0.5 mL), add 20 μg peptide and 1 μL of diluted acetic anhydride.
4. Mix well and evaporate in the vacuum centrifuge.
5. Check the obtained product quality by MS (*see* Note 13).

The above procedure may also be expedient in differentiating between N- or C-terminal cleavages of the peptides flanked by the same residue at both ends, e.g., neuropeptide Y (tyrosines) or bradykinin, fringed by arginines.

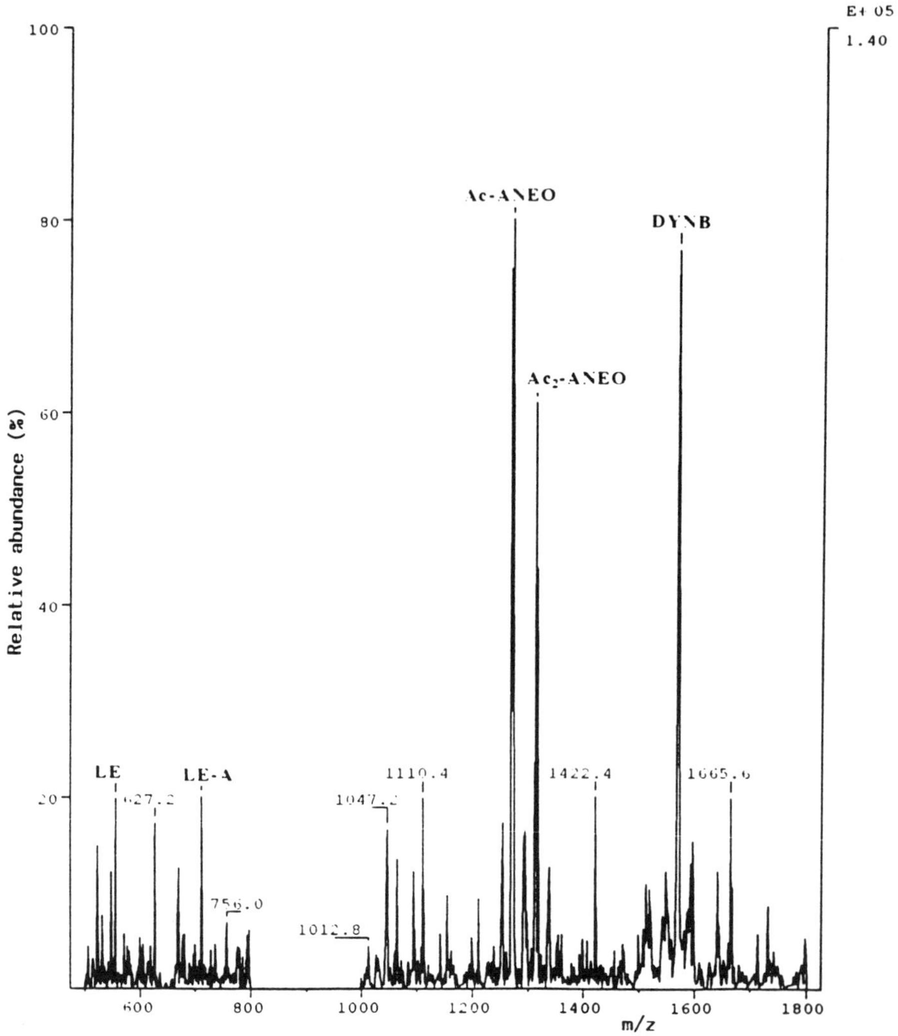

Fig. 4. Identification of the conversion fragments formed from the two similar peptides present in the incubation mixture. N-acetyl-α-neoendorphin was tested as a potential inhibitor and dynorphin B served as a substrate. Both peptides have identical sequence at the N-terminus (first 6 amino acids). Partial spectrum shows that Leu-enkephalin (LE) and Leu-enkephalin-Arg^6 (LE-A) are formed exclusively from the underivatized DYN B (m/z of 556.1 and 712.2, respectively) and not from the acetylated ANEO (calculated m/z of 598 and 754, respectively). Abundant ions at m/z 1272.8, 1314.4, and 1570.4 were identified as belonging to Ac-α-neoendorphin (Ac-ANEO), Ac_2-α-neoendorphin (Ac_2-ANEO), and dynorphin B (DYN B), respectively.

3.9. Liquid Chromatography–Mass Spectrometry

The combined LC-MS technique *(14,15)* is sometimes advantageous over direct FAB MS, particularly when small amounts of samples need to be analyzed. The peak intensity in FAB MS depends on the presence of other components in the mixture (so-called signal suppression), which will be discussed in the following paragraphs. Liquid chromatography linked to MS may often provide a solution to this problem by allowing preseparation of the constituents. The LC-MS also concentrates dilute samples before they enter the MS. It should be noted that high concentrations of acetonitrile (>30–40%), present in the mobile phase, may affect stability of the CF-FAB MS signal. It is therefore recommended to use reversed-phase columns of lower hydrophobicity (e.g., C_4 or C_8) rather than C_{18}, or to apply ion-exchange chromatography and volatile buffers (*see* Note 14). Application of capillary HPLC columns in combination with mass spectral detection has become very popular because the flow rates are compatible with CF-FAB (3–5 µL/min). A precolumn flow-splitting is required if the separation in the gradient mode is to be performed. Another possibility is the use of an analytical column (4.6 mm id) and postcolumn flow-splitting at the expense of sensitivity. The details on splitter calculations are described in Chapter 13.

Most reversed-phase columns tolerate approx 5–10% glycerol in the mobile phase. Our experience shows that it is beneficial to add glycerol to the postcolumn flow via a zero-dead volume tee (e.g., Alltech, Deerfield, IL) rather than to add it to the eluents *(16)*. We prefer this set-up rather than a coaxial flow as it is much simpler to prepare *(17)*. A brief description of such construction has been given elsewhere *(18)*. Ion-exchange columns and elution with volatile buffers, such as ammonium acetate or ammonium bicarbonate, work well and do not affect the liquid film properties on the probe tip to the same extent as acetonitrile *(18)*. Both ionic strength or pH gradients may be applied.

3.10. Structural Information

Low-resolution MS analysis of peptides is not sufficient for detailed examination of a particular sequence. Selected MS/MS techniques are discussed in Chapter 13. Other methods are, however, available for this purpose. A method has been presented where the C-terminal peptide sequence was obtained by limited hydrolysis with HCl *(19)*. In a similar approach, pentafluoropropionic acid, as a hydrolyzing agent, was found optimal in producing abundant sequence fragments *(20)*. Such simple techniques may complement Edman degradation and are helpful to assign MS/MS data. Another useful application is the use of exopeptidases, cleaving sequentially all amino acid residues from either the N- or C-terminus *(21)*. For example, carboxypeptidase Y was successfully applied to assign the C-terminal part of naturally occurring LVV-hemorphin *(22)*. All of the above techniques require relatively pure peptide preparations, thus separation of the mixture must be accomplished before any of these reactions can be performed. When enkephalin/dynorphin fragments are to be studied, application of C-terminal sequencing is recommended, because these peptides contain identical amino acids at the N-terminus (enkephalin sequence). Only sequential grade enzymes, preferentially bound to a solid support, and volatile buffers (usually 0.1*M* ammonium bicarbonate) should be used for these purposes.

4. Notes

1. The drawbacks of FAB and CF-FAB are: limitations in the mass range, strong dependence on the hydrophobicity of substances, sensitivity to salts, and easy contamination of the ion source. The nominal mass range varies, depending on the instrument. Magnetic sector instruments provide a larger measuring range owing to the possibility of decreasing the acceleration voltage in the ion source. Quadrupole instruments are restrained to the relatively low mass range of commercial mass analyzers (approx 3000 Dalton). The majority of the instruments have an option (bake) that burns out most of the glycerol and other contaminants present in the source, and helps in maintaining the equipment for about 2 wk (CF-FAB), unless a more extensive cleaning is necessary. For CF-FAB experiments, the temperature of the ion source is adjusted experimentally and should be around 45–60°C. This is particularly important when a cold trap (liquid nitrogen) is used to improve the vacuum, because too low temperature may cause freezing of the mobile phase at the CF-FAB target.
2. The FAB MS is known to be susceptible to the presence of hydrophobic substances. This effect is owing to the surface chemistry (glycerol and most of the other matrices are hydrophilic) and can be reduced by using continuous-flow FAB or by addition of small amounts of acetonitrile to the probe tip. Extraction cartridges and reversed-phase columns based on a silica matrix always release some stationary phase in soluble or even insoluble forms (so-called column bleeding). This column packing or soluble silica may strongly affect the quality of the MS analysis. Contamination by the stationary phase is difficult to remove even during further purification procedures, including HPLC. This may result in a high noise in the MS signal, disguising the expected information. Thus it is recommended to test suitability of the selected strategy using a defined amount of standard substances at comparable concentrations. The solution to such a problem may be the application of polymer-based packings or continuous beds *(23,24)*. Buffers at higher ionic strength should be avoided whenever possible and replaced by volatile solutions (ammonium acetate or ammonium bicarbonate).
3. PMSF (as well as other protease inhibitors) is highly toxic; use necessary precautions during preparation. This reagent should be prepared as a stock solution (40 m*M*) in isopropanol or MeOH and kept at 4°C. Before each experiment, an ali-

quot should be diluted with water to the working concentration of 4 m*M*. Addition of a control sample containing the same amount of solvent but no inhibitor is required, as the enzymatic activity may be affected by the presence of organic additives.

4. PHMB used at higher concentrations (>1 m*M* in the assay) may react nonspecifically with other residues, essential for the enzyme activity. This reagent should be prepared as a stock solution (10 m*M*) in 0.05*M* NaOH and further diluted with water to 1 m*M*.
5. Pepstatin should be prepared as a stock solution (100 μ*M*) in MeOH and further diluted with water to 4 μ*M*.
6. High amounts of the tissue material present in the assay mixture may contribute to the markedly increased background. To minimize this problem, the preparation should be further diluted and the incubation time reoptimized.
7. Certain CNS structures contain high amounts of aminopeptidases. Their proteolytic activity may mask other cleavages within the peptide sequence. Therefore, it might be necessary to preincubate the tissue preparation with, e.g., amastatin.
8. Inhibitors present in the incubation mixture may give an abundant signal in the mass spectrum to suppress intensities of the peptide fragments. This effect can be omitted by setting the appropriate scanning mass range.
9. The nominal sensitivity of static FAB and CF-FAB is much better than that estimated from the protocols given here. Subpicomolar detection level can be achieved when standard or extracted peptides are to be analyzed. In complex mixtures such as tissue extracts it is difficult to detect fragments if the absolute amount of substrate is below 1 μg/30 μL without any preseparation procedure (e.g., LC-MS). Addition of greater amounts of substrates during the preliminary steps is recommended, as contaminating components present in the extracts may otherwise affect the spectrum quality. The commonly used solvent dimethylsulfoxide (DMSO) may influence the spectrum baseline. It is also advisable to test pure substrates before applying them for enzyme tests; for example the butyloxycarbonyl- protecting group (Boc-) may decompose in the MS according to the so-called McLafferty rearrangement *(25)*.
10. Partial list of selected glycerol and 1-thioglycerol clusters (positive ions).

No.	Glycerol	1-Thioglycerol
1	93	108
2	185	216
3	277	324
4	369	432
5	461	540
6	553	648
7	645	756
8	737	864
9	829	972
10	921	1080

11. We observe no significant difference between stainless steel and copper tips for static FAB. A stainless steel tip is preferred as it is easier to clean. A gold-plated target for the CF-FAB is advantageous because of the greater wettability and the lower necessary concentration of glycerol in the matrix (<5%), which, in general, increases the sensitivity of the measurements.
12. Sensitivity of the continuous-flow FAB technique decreases with increasing molecular mass, and above 2.5 kDa the advantages of CF-FAB over static FAB are no longer beneficial, unless the substances need to be quantitated.
13. Acetic anhydride, when used at higher concentrations, may derivatize basic residues (Lys at a first instance) within the peptide sequence. In such a case, it is recommended to prepare further dilution of the reagent and repeat the procedure to achieve monoacetylated peptide. This is a harmful reagent; use necessary precautions.
14. During work with the capillary columns it is important to set the scan parameters properly. Components are eluted relatively quickly, and the total scan time should not exceed 2 s.

Acknowledgments

The author gratefully acknowledges Lars Terenius for stimulating discussions and Delphi Post for preparation of the manuscript. This work was partly supported by grants from The Swedish Alcohol Research Fund and the National Institute on Drug Abuse, Rockville, MD.

References

1. Barber, M., Bordoli, R. S., Sedgwick, R. D., and Tyler, A. N. (1981) Fast atom bombardment of solids (F.A.B.): a new ion source for mass spectrometry. *J. Chem. Soc., Chem. Commun.* 325–327.
2. Caprioli, R. M. (1990) Continuous-flow fast atom bombardment mass spectrometry. *Anal. Chem.* **62,** 477A–485A.
3. Terenius, L. and Nyberg, F. (1988) Neuropeptide-processing, -converting, and -inactivating enzymes in human cerebrospinal fluid. *Int. Rev. Neurobiol.* **30,** 101–121.
4. Mignatti, P. and Rifkin, D. B. (1993) Biology and biochemistry of proteinases in tumor invasion. *Physiol. Rev.* **73,** 161–195.
5. Turner, A. J. and Barnes, K. (1994) Neuropeptidases: candidate enzymes and techniques for study. *Biochem. Soc. Trans.* **22,** 122–127.
6. Tipton, K. F. (1992) Principles of enzyme assay and kinetic studies, in *Enzyme Assays; A Practical Approach. The Practical Approach Series.* (Eisenthal, R. and Danson, M. J., eds.), IRL, Oxford, UK, pp. 1–58.
7. Barrett, A. J. (1994) Proteolytic enzymes: serine and cysteine peptidases. *Methods Enzymol.* **244,** 1–15.
8. De Pauw, E., Agnello, A., and Derwa, F. (1991) Liquid matrices for liquid secondary ion mass spectrometry-fast atom bombardment: an update. *Mass Spectrom. Rev.* **10,** 283–301.

9. Gower, J. L. (1985) Matrix compounds for fast atom bombardment mass spectrometry. *Biomed. Mass Spectrom.* **12,** 191–196.
10. Lovelace, J. L., Kusmierz, J. J., and Desiderio, D. M. (1991) Analysis of methionine enkephalin in human pituitary by multi-dimensional reversed-phase high-performance liquid chromatography, radioreceptor assay, radioimmunoassay, fast atom bombardment mass spectrometry, and mass spectrometry-mass spectrometry. *J. Chromatogr.* **562,** 573–584.
11. Chou, J. Z., Kreek, M. J., and Chait, B. T. (1994) Matrix-assisted laser desorption mass spectrometry of biotransformation products of dynorphin A in vitro. *Am. Soc. Mass Spectrom.* **5,** 10–16.
12. Kokkonen, P., Schröder, E., Niessen, W. M. A., Tjaden, U. R., and Van Der Greef, J. (1990) Important parameters in liquid chromatography-continuous-flow fast atom bombardment mass spectrometry. *J. Chromatogr.* **511,** 35–47.
13. Morris, H. R., Panico, M., Karplus, A., Lloyd, P. E., and Riniker, B. (1982) Elucidation by FAB MS of the structure of a new cardioactive peptide from Aplysia. *Nature* **300,** 643–645.
14. Caprioli, R. M., Moore, W. T., DaGue, B., and Martin, M. (1988) Microbore high-performance liquid chromatography-mass spectrometry for the analysis of proteolytic digests by continuous-flow fast atom bombardment mass spectrometry. *J. Chromatogr.* **443,** 355–362.
15. Caprioli, R. M. (1988) Analysis of biochemical reactions with molecular specificity using fast atom bombardment mass spectrometry. *Biochemistry* **27,** 513–521.
16. Pleasance, S., Thibault, P., Moseley, M. A., Deterding, L. J., Tomer, K. B., and Jorgenson, J. W. (1989) Continuous flow fast atom bombardment with packed microcolumns: a comparison of precolumn versus coaxial matrix delivery. *J. Am. Soc. Mass Spectrom.* **1,** 312–319.
17. Moseley, M. A., Deterding, L. J., and Tomer, K. B. (1991) Nanoscale packed-capillary liquid chromatography coupled with mass spectrometry using coaxial continuous-flow fast atom bombardment interface. *Anal. Chem.* **63,** 1467–1473.
18. Li, Y.-M., Brostedt, P., Hjertén, S., Nyberg, F., and Silberring, J. (1995) Capillary liquid chromatography-fast atom bombardment mass spectrometry using a high-resolving cation exchanger, based on a continuous chromatographic matrix. Application to studies on neuropeptide peptidases. *J. Chromatogr. B* **664,** 426–430.
19. Knierman, M. D., Coligan, J. E., and Parker K. C. (1994) Peptide fingerprints after partial acid hydrolysis-analysis by matrix-assisted laser desorption ionization mass spectrometry. *Rapid. Commun. Mass Spectrom.* **8,** 1007–1010.
20. Tsugita, A., Takamoto, K., Kamo, M., and Iwadate, H. (1992) C-terminal sequencing of protein. A novel partial acid hydrolysis and analysis by mass spectrometry. *Eur. J. Biochem.* **206,** 691–696.
21. Woods, A. S., Huang, A. Y., Cotter, R. J., Pasternack, G. R., Pardoll, D. W., and Jaffe, E. M. (1995) Simplified high-sensitivity sequencing of a major histocompatibility complex class I-associated immunoreactive peptide using MALDI MS. *Anal. Biochem.* **226,** 15–25.

22. Glämsta, E.-L., Nyberg, F., and Silberring, J. (1992) Application of fast-atom bombardment mass spectrometry for sequencing of a hemoglobin fragment, naturally occurring in human cerebrospinal fluid. *Rapid Commun. Mass Spectrom.* **6,** 777–780.
23. Kassel, D. B., Shushan, B., Sakuma, T., and Salzman, J.-P. (1994) Evaluation of packed capillary perfusion column HPLC/MS/MS for the rapid mapping and sequencing of enzymatic digests. *Anal. Chem.* **66,** 236–243.
24. Silberring, J., Li, Y.-M., and Hjertén, S. (1994) Strategies in studies on neuropeptide processing using mass spectrometry. *Biochem. Soc. Trans.* **22,** 136–140.
25. McLafferty, F. W. and Turecek, F. (1993) *Interpretation of Mass Spectra.* University Science Books, Mill Valley, CA, pp. 51–83.

13

Analysis of Neuropeptides by Size-Exclusion HPLC Linked to Electrospray Ionization Mass Spectrometry

Jerzy Silberring

1. Introduction

Atmospheric pressure ionization (API) interfaces have become powerful and popular tools for sample ionization (so-called soft ionization) and are primarily used in the analysis of polar and thermolabile compounds (e.g., peptides, proteins). Detailed descriptions of the most important features of this technique have been published elsewhere *(1,2)*. The API technique is based on the ionization of the sample in solution passing a thin capillary, and applying high potential (2–5 kV) at the capillary tip. This process causes formation of the small, charged droplets, containing both solvent and the sample. After evaporation of the solvent (desolvation), the dry ions in the gas phase are directed to the ion source, accelerated, and analyzed. Every mass spectrometer analyzes mass-to-charge ratio (m/z) rather than the molecular mass of the compound. In other words, addition of one proton to the molecule of a nominal molecular weight of 2000 Dalton, will result in detection of a singly-charged ion at m/z of 2001/1 = 2001. Attraction of, e.g., 4 protons to this molecule will cause detection of the quadruply-charged ion at m/z of 2004/4 = 501. This observation leads to the conclusion that, using API interfaces, it would be possible to detect molecules like proteins or longer polypeptides having molecular masses that are far beyond the scanning range of instruments, i.e., >3–5 kDa. The maximal number of attracted charges depends on the amount of polar amino acids within the analyzed molecule. In fact, molecules are represented by several multiply charged species, constituting mass spectrum (Fig. 1A), which, in turn, needs to be transformed to the molecular mass (so-called deconvolution process) as shown in Fig. 1B.

From: *Methods in Molecular Biology, Neuropeptide Protocols*
Edited by: G. B. Irvine and C. H. Williams Humana Press Inc., Totowa, NJ

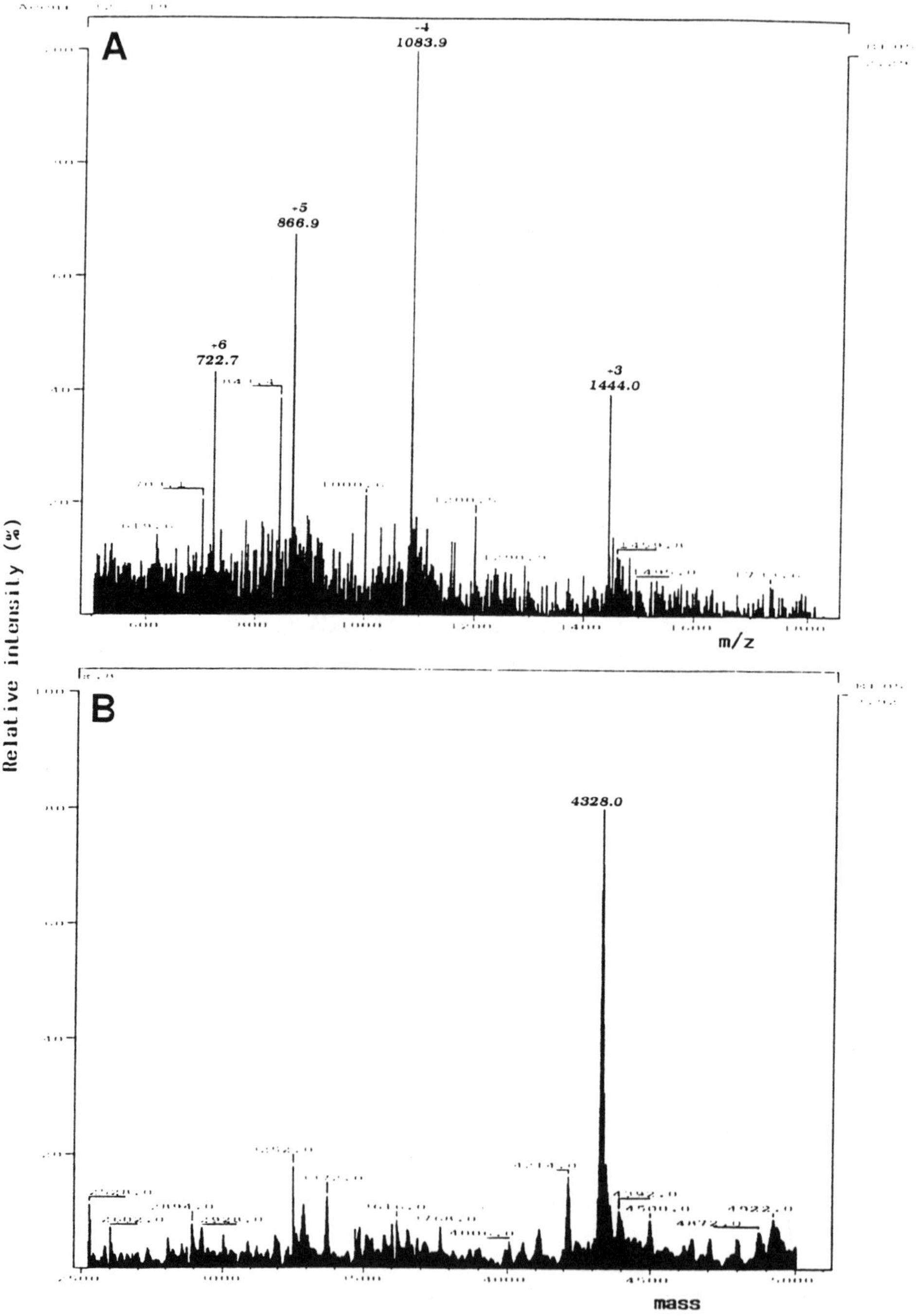
A
+4
1083.9
+5
866.9
+6
722.7
+3
1444.0
m/z
Relative intensity (%)
B
4328.0
mass

API interfaces may utilize sheath gas and/or nebulizing gas (usually nitrogen) to maintain better stability of the signal and to speed the ion evaporation process. Depending on the manufacturer and the technique used, the API interfaces may accept flow rates from 1 μL up to several hundred microliters (or more) per minute, thus being compatible with liquid chromatography or capillary electrophoresis systems. The name of a particular technique may vary owing to the various ways of producing ions in the API source: electrospray ionization (ESI), atmospheric pressure chemical ionization (APCI), ion-spray (pneumatically-assisted ESI).

Electrospray ionization mass spectrometry (ESI MS) offers several advantages over other mass spectrometric techniques. Owing to full compatibility with liquid chromatography and capillary electrophoresis techniques, it can serve as a sensitive and structure-specific online detector for the analysis of complex peptide mixtures. The most commonly applied separations utilize reversed-phase columns (RPC) of various sizes that, in certain cases, are not adequate. Recently, we introduced size-exclusion liquid chromatography (SEC) linked to ESI MS for peptide separation *(3)*. The columns are capable of separating peptides within the range of 0.5–7 kDa. Although SEC cannot provide as high resolution as RPC, its application as a prepurification step in liquid chromatography-mass spectrometry (LC-MS) offers some unique advantages: isocratic elution; online elimination of proteins, salts, and other low-mol-wt substances present in, e.g., body fluids; verification of the molecular mass, obtained after spectra deconvolution; and a possibility to separate molecules of extremely high hydrophobicity. The SEC combined with MS also reduces identification problems owing to crossreactions and microheterogeneity sometimes encountered with antibody-based techniques.

2. Materials

2.1. Instrumentation

The SMART system (Pharmacia, Uppsala, Sweden) is used, but any HPLC/FPLC instrument capable of providing accurate and pulse-free flow may also be applied. The SMART instrument has been designed for micro-

---

Fig. 1. *(previous page)* ESI mass spectrum of the synthetic amyloid β protein fragment (1–40) containing multiply charged ions (so-called envelope) between +3 and +6 **(A)**. The number of positive charges attracted to the molecule is indicated above each ion. Note that the scanning range was between m/z of 400 and 2000, whereas the monoisotopic molecular mass of this fragment is 4328.1 Dalton. **(B)** Deconvoluted spectrum with calculated molecular mass (and not m/z ratio) of the oligopeptide. Note that the horizontal axes are different: m/z for the raw spectrum (A) and mass for the deconvoluted spectrum (B). Sample load: 0.1 μg (2 μL) via injector (5 μL loop). Mobile phase: 30% acetonitrile/0.1% formic acid. Flow rate: 30 μL/min.

purification of biomolecules, in particular peptides. A novel size-exclusion column Superdex Peptide PC 3.2/30 (3.2 × 300 mm, Pharmacia) is connected to the system. This column is designed for use with SMART system, but can be adapted to any HPLC/FPLC system with help of column holder (Pharmacia). Superdex Peptide is also available in a larger column (10 × 300 mm).

The Finnigan (Bremen, Germany) MAT 95Q instrument can be used for the experiments *(3)* with an electrospray source (ESI-2). A 40-cm long Peek capillary (0.17 mm id) connects the HPLC instrument to the mass spectrometer inlet. When the entire flow from the HPLC column is directed into the MS, removal of the conductivity meter (gradient monitor) to minimize peak broadening is recommended (*see* Note 1). The voltage at the sprayer tip is held at 2.5 kV and nitrogen is used as a sheath gas. Temperature of the heated capillary is set to 250°C. Resolution is set at 1400 (10% valley) and the calibration is tested with polypropylene glycol 400. Spectra are taken in a positive-ion mode (magnetic scan of m/z 200–1200) and the scanning time is set to 5 s/decade. The flow rate through the entire LC-MS system is maintained at 100 µL/min. The majority of modern ESI interfaces accept flow rates of 3–200 µL/min or even higher. Should the splitter be necessary for, e.g., simultaneous fraction collection during preparative runs, the reader may refer to the formula given in Section 3.1.

2.2. Chemicals, Buffers, and Chromatography Products

1. Peptides can be purchased from Bachem Feinchemikalien (Bubendorf, Switzerland). Other reagents are available from commercial sources. Peptides are dissolved in the mobile phase or in water.
2. The mobile phase consists of 20% acetonitrile in water, supplemented with 0.1% trifluoroacetic acid (TFA) (*see* Notes 2 and 3).
3. The necessary connectors, capillaries, tubings, and so on, may be purchased from, e.g., Alltech (Deerfield, IL) or from Upchurch Scientific (Oak Harbor, WA).

2.3. Sample Preparation

Basically, SEC-ESI MS requires no special preparation of samples. It is, however, highly recommended to keep the samples in Eppendorf tubes (0.5-mL capacity). The conical shape of the bottom simplifies sample handling, particularly when low amounts are to be used. Glass or polycarbonate tubes should be avoided for peptide storage owing to high adsorption by tube walls. Particles should be removed by centrifugation, e.g., in an Eppendorf microfuge (around 12,000*g*) for 5 min or by filtering the sample through 0.45-µm filter unit (Ultrafree MC, Millipore, Bedford, MA). The maximum volume of the sample should not exceed 50 µL when injected on the Superdex Peptide PC 3.2/30 column. Preparations in larger volumes should be concentrated using vacuum centrifuge.

3. Methods

3.1. Flow Splitting

$$\frac{\text{MS-flow}}{\text{SPLIT-flow}} = \frac{L_{SPLIT} \times (ID_{MS})^4}{L_{MS} \times (ID_{SPLIT})^4}$$

MS-flow, flow rate to MS;
SPLIT-flow, flow-rate to fraction collector or waste;
L_{SPLIT}, length of capillary to fraction collector or waste (in mm);
L_{MS}, length of capillary to mass spectrometer (in mm);
ID_{MS}, inner diameter of capillary to MS (in mm); and
ID_{SPLIT}, inner diameter of capillary to fraction collector or waste (in mm).

In cases in which the chromatographic peaks are tightly spaced, it is recommended to remove the conductivity meter (gradient monitor) to minimize postcolumn peak broadening. The reader may refer to further details, published by Pharmacia (*see* Note 1) with detailed description of interfacing the SMART System to the MS *(4)*.

3.2. Separation of Neuropeptides

The UV trace at 214 nm and a corresponding mass chromatogram are presented in Fig. 2, in which a peptide test mixture was separated on the Superdex Peptide PC 3.2/30 column. The stationary phase is obtained by the covalent bonding of dextran to cross-linked agarose beads. It has been found that the retention time of various peptides is dependent on the acetonitrile concentration in the eluent (Winter, personal communication), a property that can significantly improve resolution, when properly used. It is therefore advisable, by trial-and-error, to optimize the acetonitrile content. When size estimation is of importance, 20–30% acetonitrile seems optimal for this purpose.

Figure 3 illustrates that size-exclusion column can be successfully applied for the purification/separation of highly hydrophobic peptides such as amyloid beta-peptide (Aβ) fragments. These peptides are difficult to elute from reversed-phase columns *(5)*. Figure 3 shows purification of crude Aβ(1–42) on the Superdex Peptide column, performed online with UV and mass spectral detection. The peak, eluted at around 10–11 min, belongs to the synthesized peptide. Other UV visible peaks present along the chromatogram were not identified. Owing to extreme insolubility in common solvents, some of the Aβ fragments should be dissolved in a small volume of 70% formic acid and then diluted with an appropriate mobile phase.

3.4. Structural Information

Modern mass spectrometry of peptides provides accurate mass measurement, but it is also used for structural information, i.e., amino acid sequence of

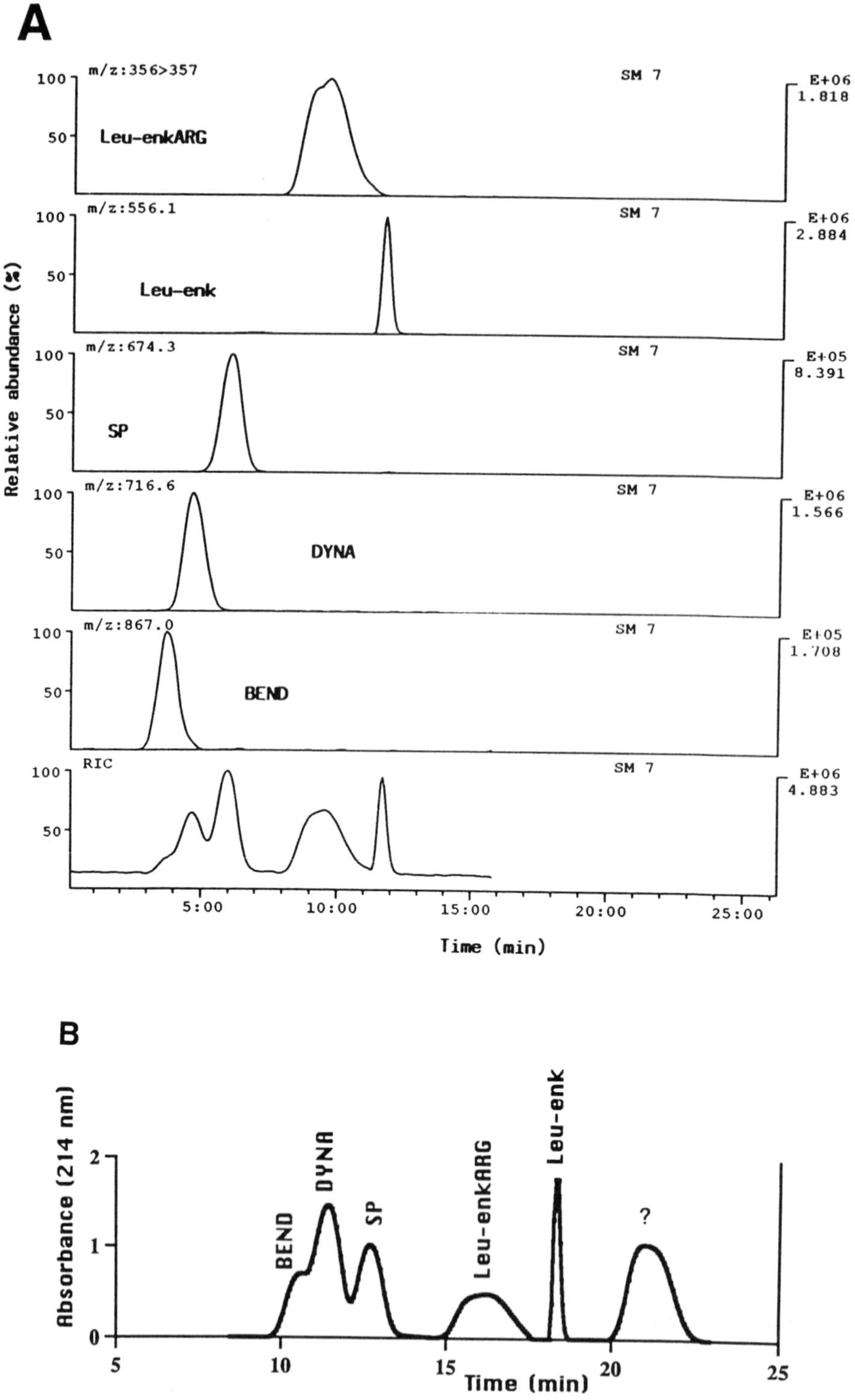

A
m/z:356>357
Leu-enkARG
SM 7
E+06
1.818
m/z:556.1
Leu-enk
SM 7
E+06
2.884
m/z:674.3
SP
SM 7
E+05
8.391
m/z:716.6
DYNA
SM 7
E+06
1.566
m/z:867.0
BEND
SM 7
E+05
1.708
RIC
SM 7
E+06
4.883
Relative abundance (%)
100
50
5:00
10:00
15:00
20:00
25:00
Time (min)
B
Absorbance (214 nm)
BEND
DYNA
SP
Leu-enkARG
Leu-enk
?
0
1
2
5
10
15
20
25
Time (min)

the substances. Detection of peptide fragments at low resolution may often render ambiguous results, e.g., when cleavage products are studied. Even a relatively short peptide such as dynorphin A (17 amino acids) may give rise to two peptides having similar molecular weights: fragment *(1–6)* at m/z 712.4 or fragment *(5–9)* at m/z 713.5. Differentiation between these two peptides cannot be made by a simple inspection of the spectra, but needs confirmation by other methods like MS/MS (collision-induced dissociation, CID). The choice of CID depends on the instrument used, but often facilities to perform sequencing are not available. Recently, a novel technique has been developed: a so-called source CID that involves fragmentation in the electrospray ion source and does not require tandem mass spectrometric instrumentation *(6,7)*. The principle of this technique is based on the fact that API sources are operating at the atmospheric pressure, thus there is enough gas pressure inside the source to function as a collision chamber. Ions introduced to the API region are additionally accelerated by simply adjusting the potentials applied to the capillary exit and skimmer. This feature may be utilized by any ESI interface and works well for substances up to about 1 kDa. The major drawback of this technique is that fragmentation is obtained from all components present in the mixture, which may obscure further interpretation, making preseparation essential. An example of this technique is shown in Fig. 4, in which Leu-enkephalin present in the peptide mixture was simultaneously separated by SEC and sequenced. Assignment of the fragment ions and database searches can be performed with the aid of software listed in Note 4.

3.5. Interpretation of the Obtained Mass Spectra

The presence of multiply charged ions in complex mixtures of peptides may sometimes obscure interpretation of the raw data. In these instances, deconvolution software often considers various components as belonging to the same envelope, thus calculating a false molecular mass. Therefore, preseparation of the sample components according to their masses allows the investigator to verify the mass calculations done by the computer program. An example has been recently described *(3)* in which the singly- and doubly-charged ions of

---

Fig. 2. *(previous page)* ESI mass chromatogram **(A)** and UV trace **(B)** of the peptide mixture, separated by size exclusion chromatography. The selected ion current profiles represent the particular peptides (labeled with names) at various charge states. BEND, β-endorphin; DYNA, dynorphin A; SP, substance P; Leu-enkARG, Leu-enkephalin-Arg^6; Leu-enk, Leu-enkephalin. The component labeled by a question mark is probably a refractive index artifact. Sample load: 5 μg of each peptide (2 μg of Leu-enkephalin). Sample volume: 50 μL. Column: Superdex Peptide PC (3.2/30). Eluent: 20% acetonitrile supplemented with 0.1% TFA. Flow: 100 μL/min.

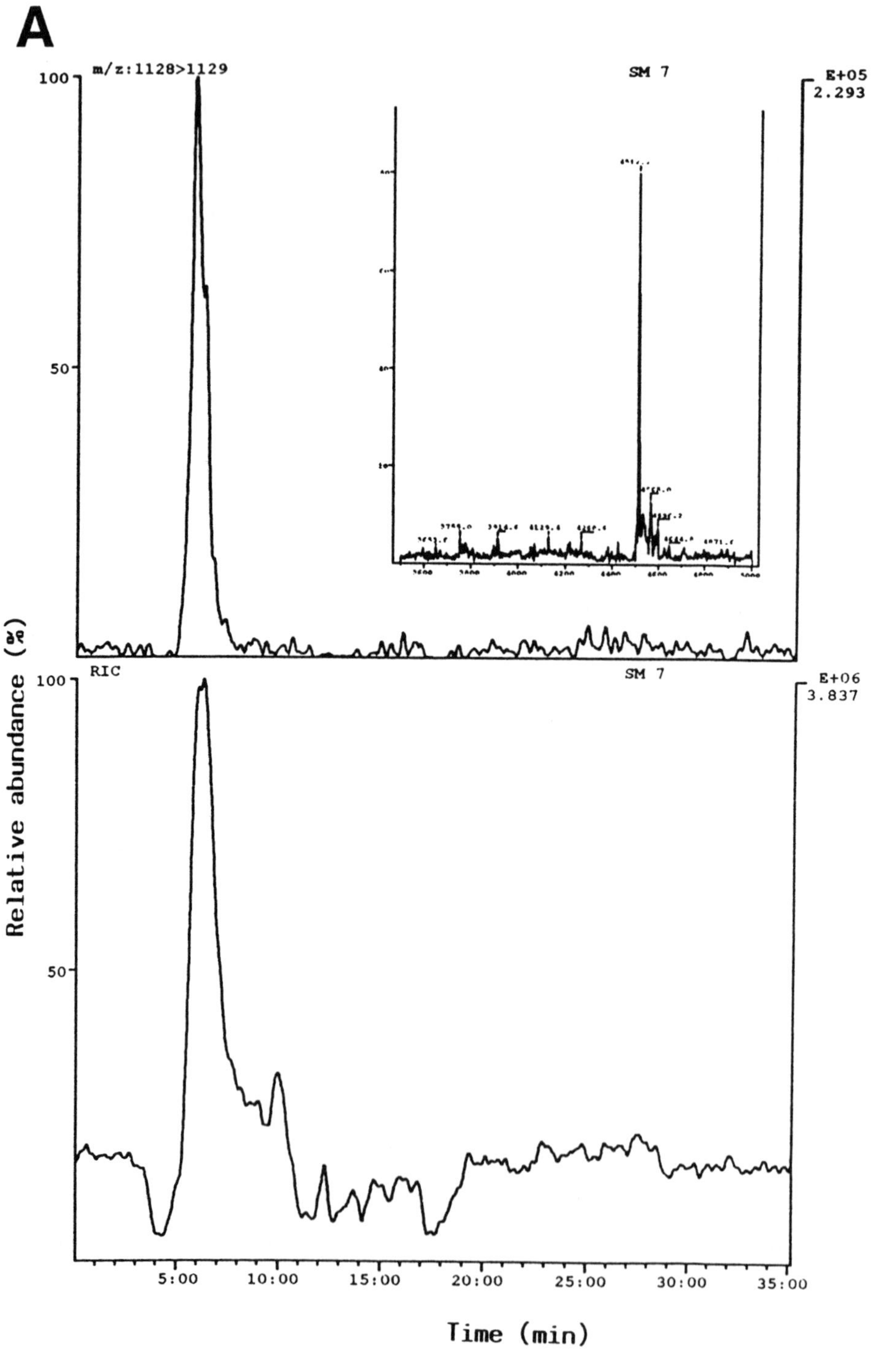
A
m/z:1128>1129
SM 7
E+05
2.293
100
50
RIC
SM 7
E+06
3.837
100
50
Relative abundance (%)
5:00
10:00
15:00
20:00
25:00
30:00
35:00
Time (min)

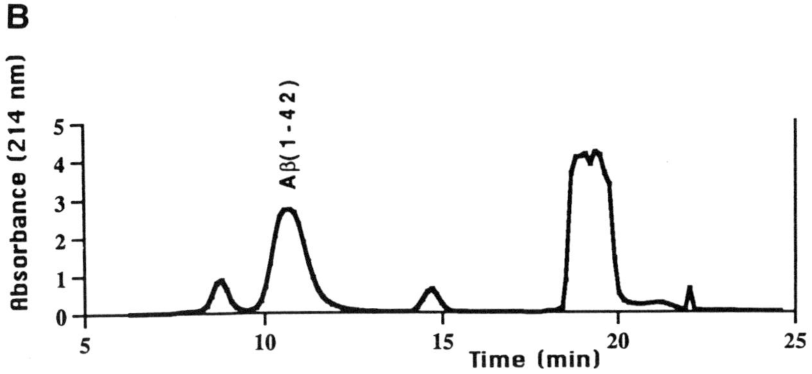

Fig. 3. Separation of crude Aβ(1–42) by size-exclusion chromatography linked to the ESI MS. Partial mass chromatogram **(A)** and the UV trace **(B)**. Separation conditions and solvent composition were identical to that described in Fig. 1. Sample load: 15 μg (30 μL) of crude peptide. 100 μg of the preparation were dissolved in 10 μL of 70% formic acid and then diluted to 200 μL with the mobile phase.

Leu-enkephalin-Arg^6 were counted together with the singly-charged ion of the protease inhibitor amastatin, and the final result showed an incorrectly deconvoluted mass. Figure 5 presents another spectrum of a mixture of dynorphin A (triply-charged ion at m/z 716.9) and amastatin (singly-charged ion at m/z 475.3). Automatic transformation (deconvolution) of the ions belonging to different components leads to false mass value estimation. For unknown substances, it is therefore highly recommended to verify the obtained results by another technique allowing proper mass assignment (e.g., SEC, sequencing, amino acid analysis).

3.6. Final Remarks

SEC (as a complementary technique to reversed-phase HPLC) linked to the mass spectrometer has potential applications for, e.g., detailed analysis of proteolytic processing pathways or for the preseparation of components present in body fluids. This is particularly important when an electrospray interface has been installed, because salts and other low-mol-wt components present in the body fluids strongly affect the quality of the analysis *(8)*. SEC is the only chromatographic technique in which salts migrate as the last component and can easily be directed to the waste by column switching. Alternatively, samples may be separated offline and, after fractionation and concentration, analyzed by the mass spectrometer. A similar approach utilizing SEC technique linked to the ESI MS and another type of HPLC column has been applied to studies on noncovalent dimers of leucine zipper peptides *(9)*.

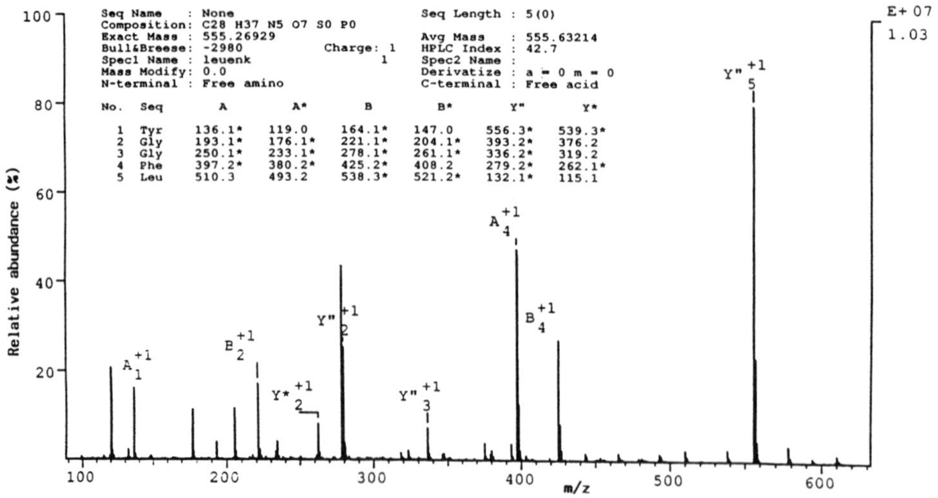

Fig. 4. Fragmentation of Leu-enkephalin by collision-induced dissociation, performed in the ESI source. The peptide was sequenced during elution from the Superdex Peptide PC column. *See* legend to Fig. 1 for separation details. Voltages applied to the heated capillary and tube lenses were increased by 70 V each. Other source parameters remained unchanged. Fragments were automatically assigned according to Roepstorff and Fohlman nomenclature *(12)* by the Pepmatch software, supplied by Finnigan MAT. All fragment ions detected by the program are listed in the table and labeled with an asterisk.

4. Notes

1. Removal of the conductivity meter (gradient monitor), to avoid peak broadening, may affect split ratio because of the changed back pressure. The reader should refer to the application note (18-1104-38) released by Pharmacia *(4)*.
2. The Superdex Peptide columns were eluted with solvents, supplemented with 0.1% TFA. Selection of this acid was a method of choice, though it decreases the mass spectral signal approx 5 times compared to formic acid, a commonly used additive in MS. Replacement of TFA with other acids such as formic acid in the eluent may significantly affect retention times or hamper peptide elution from the column.
3. Addition of TFA to the mobile phase increases the spray current. This in turn may affect the preprogrammed spray high voltage or even switch it off (automatic safety switch). It is therefore recommended to discuss the separation technique with the MS operator in order to keep the spray high voltage at a lower level (approx 2.5 kV).
4. Assignment of the sequences, fragmentation patterns, and other calculations can be obtained with the help of the softwares MacProMas *(10)* or GPMA *(11)*. Search for the necessary sequences in databases (gene or protein data bank) is freely available via Internet: http://www.public.iastate.edu/~pedro/ research_tools.html, under common name: Pedro's biomolecular tools. Other useful

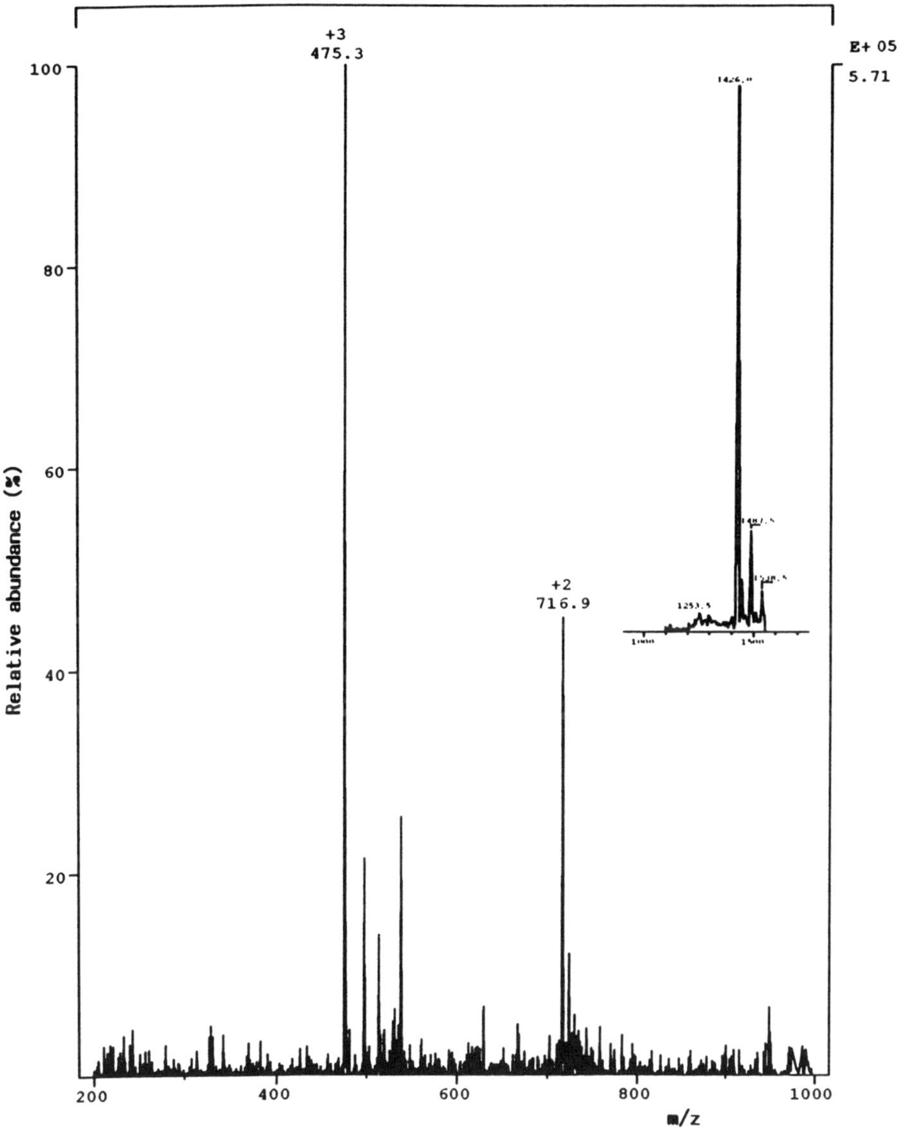

Fig. 5. Partial spectrum of the mixture of two components and the false deconvolution to incorrect mass at 1426.0 (insert). Amastatin at m/z 475.3 (singly-charged) and dynorphin A at m/z 716.9 (triply-charged). The software considers both components as belonging to the same envelope and assigns inaccurate charges to the ions.

addresses containing software suitable for mass spectrometry purposes are: http://mac-mann6.embl-heidelberg.de/ and http://rafael.ucsf.edu/, which provide further connections to other Internet user groups and resources.

Acknowledgments

The author is indebted to Lars Terenius for helpful discussion during preparation of this manuscript and to Pharmacia Biotech AB for technical aid. This work was supported by the National Institute on Drug Abuse, Rockville, MD and the Swedish Alcohol Research Fund.

References

1. Mann, M. and Wilm, M. (1995) Electrospray mass spectrometry for protein characterization. *TIBS* **20,** 219–224.
2. Smith, R. D., Light-Wahl, K. J., Winger, B. E., and Goodlett, D. (1995) Electrospray ionization, in *Biological Mass Spectrometry. Present and Future.* (Matsuo, T., Caprioli, R. M., Gross, M. L., and Seyama, Y., eds.), Wiley, Chichester, UK, pp. 41–74.
3. Nylander, I., Tan-No, K., Winter, A., and Silberring, J. (1995) Processing of prodynorphin-derived peptides in striatal extracts. Identification by electrospray ionization mass spectrometry linked to size-exclusion chromatography. *Life Sci.* **57,** 123–129.
4. Technical Note No. 18-1104-38. (1994) Electrospray LC-MS using SMART System. Pharmacia Biotech AB, Uppsala, Sweden.
5. Näslund, J., Schierhorn, A., Hellman, U., Lannfelt, L., Roses, A., Tjernberg, L. O., Silberring, J., Gandy, S., Winblad, B., Greengard, P., Nordstedt, Ch., and Terenius, L. (1994) Relative abundance of Alzheimer Aβ amyloid peptide variants in Alzheimer disease and normal aging. *Proc. Natl. Acad. Sci. USA* **91,** 8378–8382.
6. Meng, C. K., McEwen, C. N., and Larsen, B. S. (1990) Peptide sequencing with electrospray ionization on a magnetic sector mass spectrometer. *Rapid Commun. Mass Spectrom.* **4,** 151–155.
7. Loo, J. A., Udseth, H. R., and Smith, R. D. (1988) Collisional effects on the charge distribution of ions from large molecules, formed by electrospray-ionization mass spectrometry. *Rapid Commun. Mass Spectrom.* **2,** 207–210.
8. Beavis, R. C. and Chait, B. T. (1990) Rapid, sensitive analysis of protein mixtures by mass spectrometry. *Proc. Natl. Acad. Sci. USA* **87,** 6873–6877.
9. Li, Y.-T., Hsieh, Y.-L., Henion, J. D., Senko, M. W., McLafferty, F. W., and Ganem, B. (1993) Mass spectrometric studies on noncovalent dimers of leucine zipper peptides. *J. Am. Chem. Soc.* **115,** 8409–8413.
10. Lee, T. D. and Vemuri, S. (1990) MacProMass: a computer program to correlate mass spectral data to peptide and protein structures. *Biomed. Mass Spectrom.* **19,** 639–645.
11. Højrup, P. (1990) General protein mass analysis (GPMA), a convenient program in studies of proteins by mass analysis, in *Ion Formation from Organic Solvents* (Hedin, A., Sundqvist, B. U. R., and Benninghoven, A., eds.), Wiley, Chichester, UK, pp. 61–66.
12. Roepstorff, P. and Fohlman, J. (1984) Proposal for a common nomenclature for sequence ions in mass spectra of peptides. *Biomed. Mass Spectrom.* **11,** 601.

14

Identification of Peptides by Matrix-Assisted Laser Desorption Ionization Time of Flight Mass Spectrometry (MALDI-TOF-MS) and Direct Analysis of the Laterobuccal Nerve from the Pond Snail *Lymnaea stagnalis*

Glenn Critchley and Belinda Worster

1. Introduction

This method describes how, by mixing peptides with a UV-absorbing matrix, their masses can be determined using matrix-assisted laser desorption ionization time of flight mass spectrometry (MALDI-TOF-MS).

The MALDI-TOF-MS makes use of a very simple principle to determine the mass of ions generated in the MALDI process. This is based on the time *(t)* taken for an ion of mass *(m)* and known kinetic energy ($eV = 1/2mv^2$) to travel a distance *(l)* in a field-free region. This time *(t)* is proportional to the square root of the mass *(m)* of the ion.

$$t = l\sqrt{m/2eV} \qquad (1)$$

First described by Karas and Hillenkamp in 1988 *(1)*, MALDI-TOF-MS provided the biochemist with a rapid and sensitive technique for mass determination of biomolecules over a very large mol-wt range (up to 500 kDa). Established techniques, such as fast atom bombardment (FAB) mass spectrometry often failed to give intact molecular ions and were limited in mass range to <25 kDa.

The MALDI-TOF-MS technique is also simple in terms of preparing the sample, provided that a sample preparation procedure has been established for the class of sample being analyzed. What is not fully understood are the energetic processes taking place that give rise to the MALDI effect *(2)*. Nevertheless, there is a ready acceptance of new sample preparation procedures to be adopted if the results are an improvement on existing methods.

From: *Methods in Molecular Biology, Neuropeptide Protocols*
Edited by: G. B. Irvine and C. H. Williams Humana Press Inc., Totowa, NJ

In this method, the analysis of peptides (*see* Notes 1 and 2) is described, in particular the direct MALDI-TOF analysis of the laterobuccal nerve from the pond snail *Lymnaea stagnalis (3)*. This nerve innervates the buccal mass and contains neuropeptides that are involved in the control of feeding behavior in the snail.

2. Materials

2.1. Linear Instrument

Figure 1 shows a schematic representation of the of the Micromass TofSpec MALDI-TOF-MS. The N2 UV laser produces a 4-ns pulse width at a 337-nm wavelength; each pulse has an energy of approx 180 μJ. Because the energy produced by the laser is fixed and the energy required for the MALDI process varies, it is necessary to adjust the amount of energy arriving at the sample surface. This is effected by the use of a variable, neutral density filter in conjunction with a continuously variable iris. The matrix, a chromophore, which is normally acidic, absorbs the energy of the laser, causing some of the sample to be ionized, usually by protonation in the case of positive ion peptides. The desorbed ions are accelerated in the source (typically operated at 25,000 V), and are then also allowed to drift over a fixed, field-free distance. Part of the laser beam is diverted in a beam splitter (a half-silvered mirror) and is detected with a photodiode. This is used as the starting time for the clock used to measure the flight time of the ions. The time taken for the ions to arrive at the detector is measured by sampling the detector output at a frequency of 500 MHz. The detector is a hybrid detector that comprises an electron multiplier preceded by a single microchannel plate. As in all mass spectrometers, it is essential to maintain the source and mass analyzer under high vacuum to minimize ion/molecule interactions.

2.2. Reflectron Instrument

Figure 2 is a schematic diagram of the Micromass TofSpec fitted with a reflectron lens assembly. The reflectron is located in front of the linear detector and reflects the ions back along the instrument axis to an annular microchannel plate detector, thus increasing the flight path in the mass analyzer.

Figure 3 illustrates the effect on the ions in the mass analyzer when the reflectron is switched off, and the instrument is said to be in the linear mode of operation and all species from the ionization event arrive at the linear detector. It will be seen from Eq. (1) that ions that have small differences in energy will have small differences in flight time, and hence the mass peak will be broadened to some extent. In addition to the precursor ions, there are fragment ions or neutral species resulting from metastable dissociation of the precursor ion while traveling in the mass analyzer. It is this mixture of species arriving at the detector that contributes to the peak broadening associated with the linear mode.

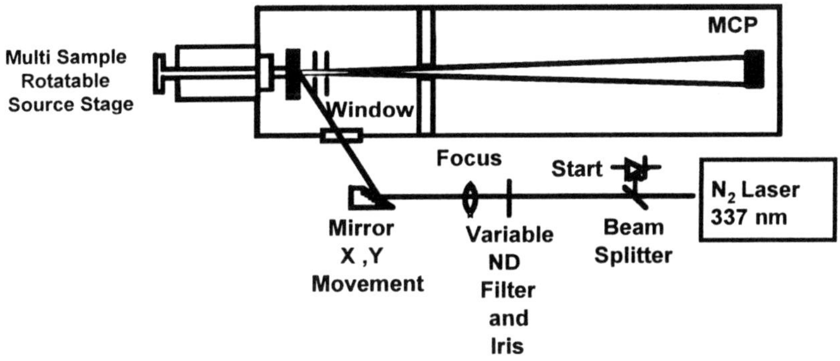

Fig. 1. Schematic diagram of the linear time of flight mass spectrometer.

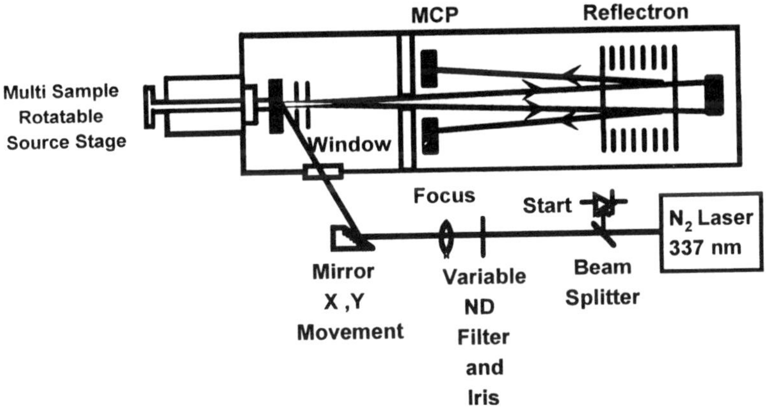

Fig. 2. Schematic diagram of the reflectron time of flight mass spectrometer.

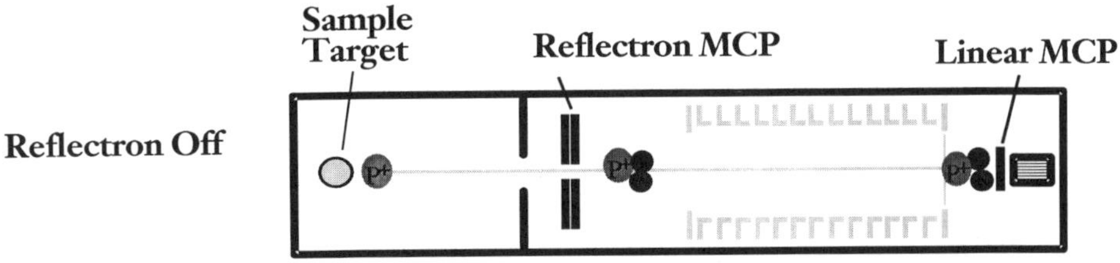

Fig. 3. The behavior of species in the linear mode of operation in MALDI-TOF MS.

When the reflectron is held at a potential slightly higher than the source acceleration voltage, the reflectron acts as an ion mirror, and by design it can be arranged that all ions of the same mass arrive back at the annular detector at the same time, regardless of any small differences in energy they have. The reflectron is said to be energy-focused for the precursor ions. Figure 4 shows that the precursor ions are reflected by the reflectron to the annular reflectron detector. The neutral species are unaffected by the reflecting field and the metastable fragment ions, though reflected, now arrive at the detector at a completely different time. Thus, the other main cause of precursor peak broadening is removed and better resolution is obtained.

This energy-focusing effect of the reflectron is illustrated in Fig. 5. Consider two ions of the same mass, but with an energy difference that arises during the ionization process such that ion E2 has a greater energy than ion E1. Because E2 has greater energy, this ion penetrates the reflecting field to a greater extent than E1, allowing E1 to catch up with E2. Thus, both ions arrive at the reflectron detector simultaneously. This results in a reduction in peak broadening and therefore higher resolution (*see* Note 2).

2.2.1. Postsource Decay (MS-MS)

In the normal operation of the reflectron, the precursor ion is energy-focused. By reducing the reflectron voltage and maintaining a constant source voltage, the precursor ions have too much energy to be reflected back to the reflectron detector (*see* Fig. 6). However, fragment ions that have energies in proportion to their masses in which $E_f = Ep.m_f/m_p$, are reflected back to the reflectron detector. For example, if the reflectron voltage is reduced to half the voltage of that when the precursor ion is energy focused, a fragment ion of half the mass of the precursor ion becomes energy-focused by the reflectron and detected by the reflectron detector. Thus, by stepping the reflectron voltage in small increments, the full range of fragment ions can be observed. This fragment ion spectrum is referred to as postsource decay (PSD) and can be used to assist sequence determination of peptides (*see* Note 4).

2.2.2. Precursor Ion Selection

If the analyte is a mixture of peptides, for example when looking at a protein digest, it is necessary to select the ion of interest using an ion gate *(4)* for PSD studies. When the gate is switched off, the ions are allowed to pass into the mass analyzer. Thus, an ion of a certain mass can be transmitted into the mass analyzer by turning off the ion gate at the time of arrival at the ion gate of the required ion (*see* Note 3). This selection method is capable of transmitting only ions of a 10-Dalton mass range at mass 1000 Dalton, or a 20-Dalton mass range at 2000 Dalton.

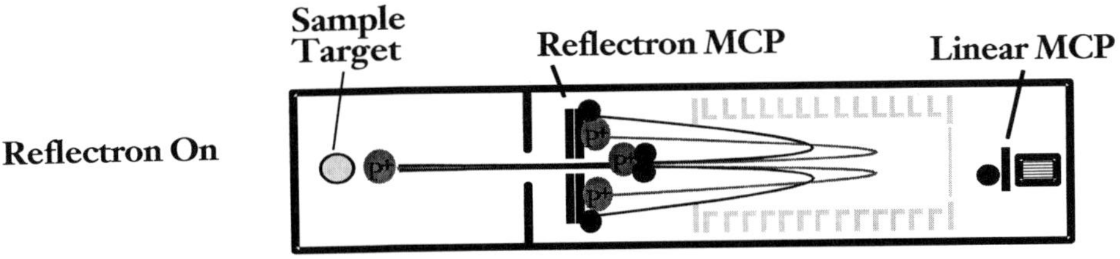

Fig. 4. The behavior of species in the reflectron mode of operation in MALDI-TOF MS.

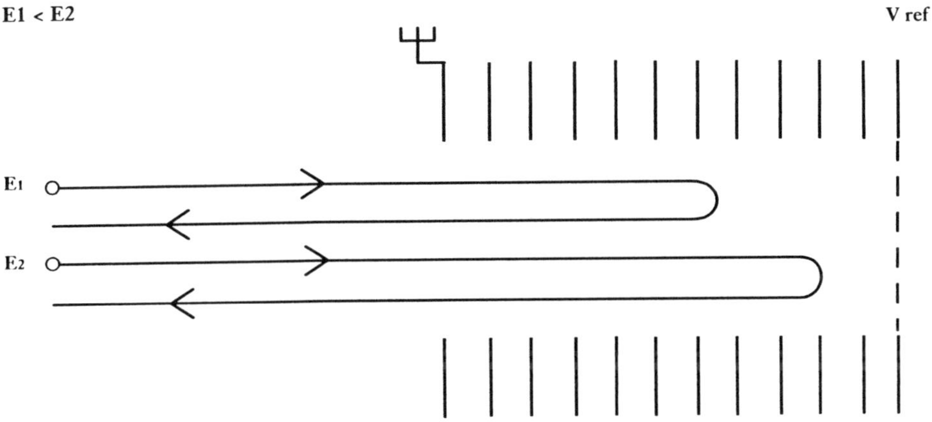

Fig. 5. Energy-focusing of ions of the same mass in the reflectron mode of operation.

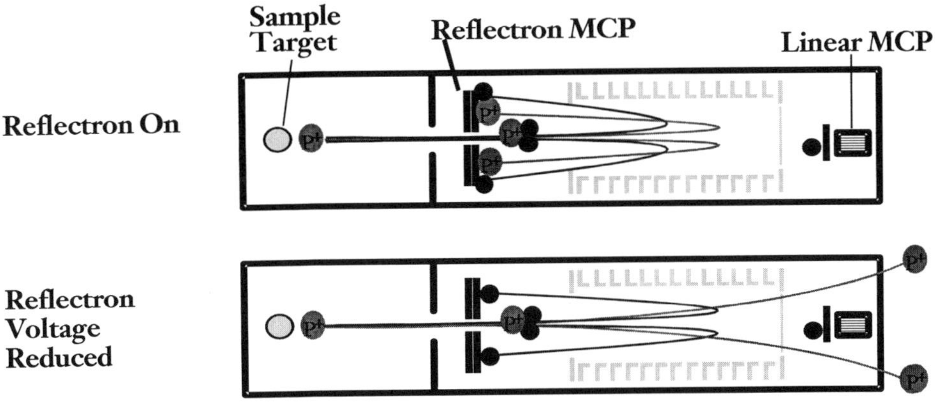

Fig. 6. Reduction of the reflectron voltage to focus metastable species at the reflectron detector.

Table 1
Levels of Contaminants Permitted When Acquiring Spectra for Proteins in Sinapinic Acid Matrix

Contaminant	Concentration (max)
Phosphate buffer	20 m*M*
Tris buffer	50 m*M*
Detergents	0.1%
SDS	0.01%
Alkali metal salts	1*M*
Glycerol	2%
Ammonium bicarbonate	30 m*M*
Guanidine	1*M*
Sodium azide	1%

2.3. Sample Preparation

Typically a solution is mixed of sample in the pmol/μL range with a UV-absorbing matrix solution in the m*M* range. One microliter of the mixture is spotted on one of the target positions and allowed to dry. The technique will work with impurities present at concentrations up to those listed in Table 1 when running proteins in sinapinic acid matrix.

2.4. Chemicals

There are a wide range of matrices available in MALDI-TOF-MS *(5)*; the most commonly used for the analysis of peptides is alpha-cyano-4-hydroxy-cinnamic acid.

1. Calibrants: Angiotensin I, and adrenocorticotropic hormone (ACTH) (18–39 clip) (Sigma, St. Louis, MO), each prepared at 10 pmol/μL in 0.1% trifluoroacetic acid (Aldrich, Milwaukee, WI).
2. Matrix: Alpha-cyano-4-hydroxycinnamic acid (Aldrich) at 10 mg/mL in 7:3 (v/v) mixture of 0.1% trifluoroacetic acid/acetonitrile.
3. Sample: A 0.5-mm length of laterobuccal nerve dissected from the pond snail *Lymnaea stagnalis*.
4. PSD Calibrant: ACTH (18–39 clip) prepared at 10 pmol/μL in 0.1% trifluoroacetic acid (Aldrich).

3. Method

3.1. Sample Preparation

1. Take a multiposition sample target and load the calibrant solution by mixing 2 μL of the calibrant solution with 2 μL of the matrix solution, and depositing 1 μL of the mixture on one of the target positions. Allow to dry.

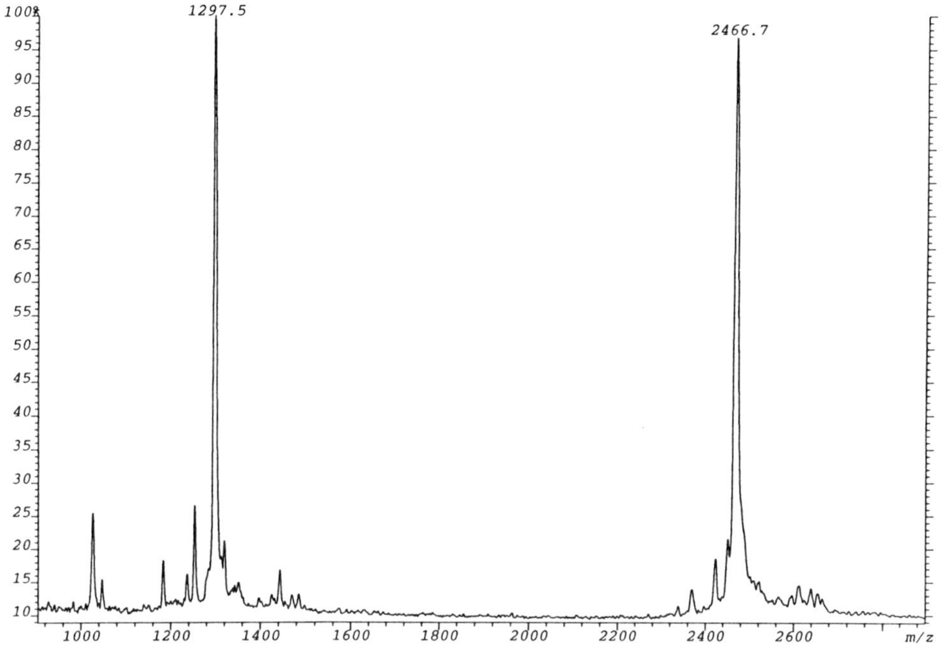

Fig. 7. Linear MALDI-TOF mass spectrum of Angiotensin I and ACTH (18–39 clip) using the alpha-cyano-4-hydroxycinnamic acid matrix.

2. Place the sample (dissected nerve) in 1 μL of matrix solution previously deposited on a new target position and crush the nerve. Allow to dry.

3.2. Mass Spectrometry

3.2.1. Linear Acquisition Mode

1. Insert dried sample stage in Micromass TofSpec or similar MALDI-TOF-MS instrument, and allow the vacuum to achieve the normal operating pressure, i.e., 1 × E-7 atm.
2. Select a source extraction voltage of 25,000 V and a detector voltage of 3500 V in the linear mode.
3. Select the calibration spot.
4. Fire the laser in repetition mode and reduce the laser energy to the point just before ionization ceases to take place, i.e., threshold ionization, using a combination of the neutral density filters and the iris. Average approx 50 shots into one spectrum.
5. Generate a two-point calibration using the known masses of angiotensin I (average [M+H] 1297.5 Dalton) and ACTH (18–39 clip) (average [M+H] 2466.7 Dalton) from the spectrum acquired in Section 3.2.1., step 4 (*see* Fig. 7).
6. Select the nerve sample spot.
7. Repeat step 4 in Section 3.2.1.

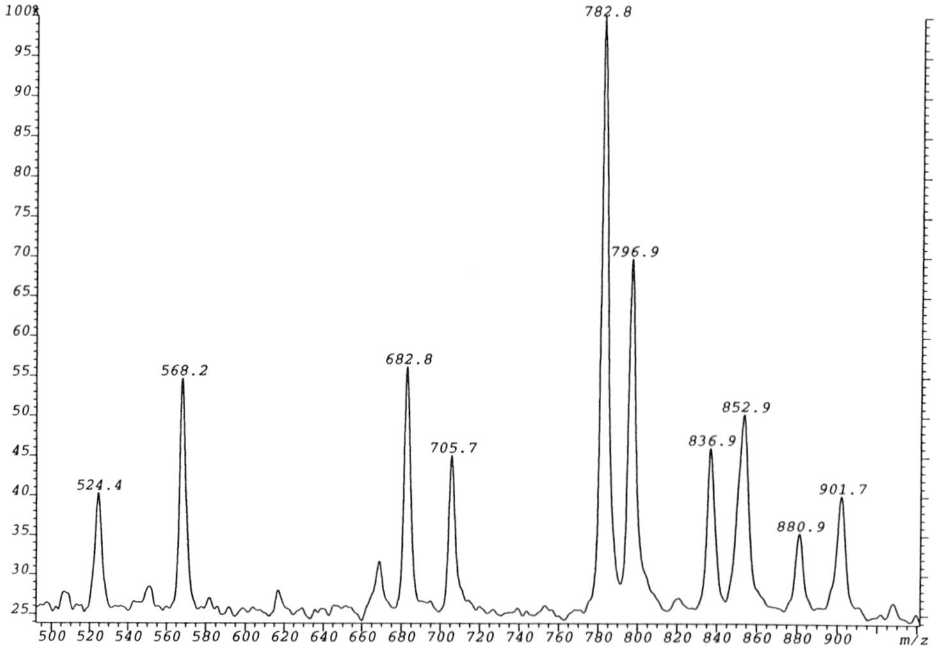

Fig. 8. Linear MALDI-TOF mass spectrum of the expected neuropeptides in the laterobuccal nerve from the pond snail *Lymnaea stagnalis* using the alpha-cyano-4-hydroxycinnamic acid matrix.

8. Display the spectrum (*see* Note 1) acquired in Section 3.2.1., step 7, using the calibration file generated in Section 3.2.1., step 5, as an external calibration.

3.2.2. Reflectron Acquisition Mode

Repeat steps 2–8 in Section 3.2.1. in reflectron mode except in step 5 use the monoisotopic masses for the calibrants (Angiotensin I monoisotopic [M+H] 1296.7 Dalton, ACTH [18–39 clip] [M+H] 2465.2 Dalton) (*see* Note 2).

3.2.3. Postsource Decay (MS-MS)

1. Set up an external calibration using the PSD calibrant so that the reflectron is calibrated in PSD mode according to the instrument manufacturer's recommendations.
2. Select the nerve sample spot.
3. Select ion at 783 Dalton (*see* Note 3) using the ion gate.
4. Acquire a PSD spectrum of the selected ion, ensuring that the greatest signal-to-noise ratio is achieved at each voltage step. Usually more laser energy is required to observe the smaller fragments.
5. Stitch the data together acquired at each voltage step (*see* Note 4) against the external calibration obtained in Section 3.2.3., step 1.

Table 2
Observed Masses Compared with Expected Masses for Neuropeptides Present in the Laterobuccal Nerve from the Pond Snail *Lymnaea stagnalis*

Observed [M+H] Dalton	Expected [M+H] Dalton	Neuropeptide sequence
682.8	682.8	TLFRFa
705.7	705.9	GTLLRFa
782.8	782.9	GGSLFRFa
796.9	796.9	NTLFRFa

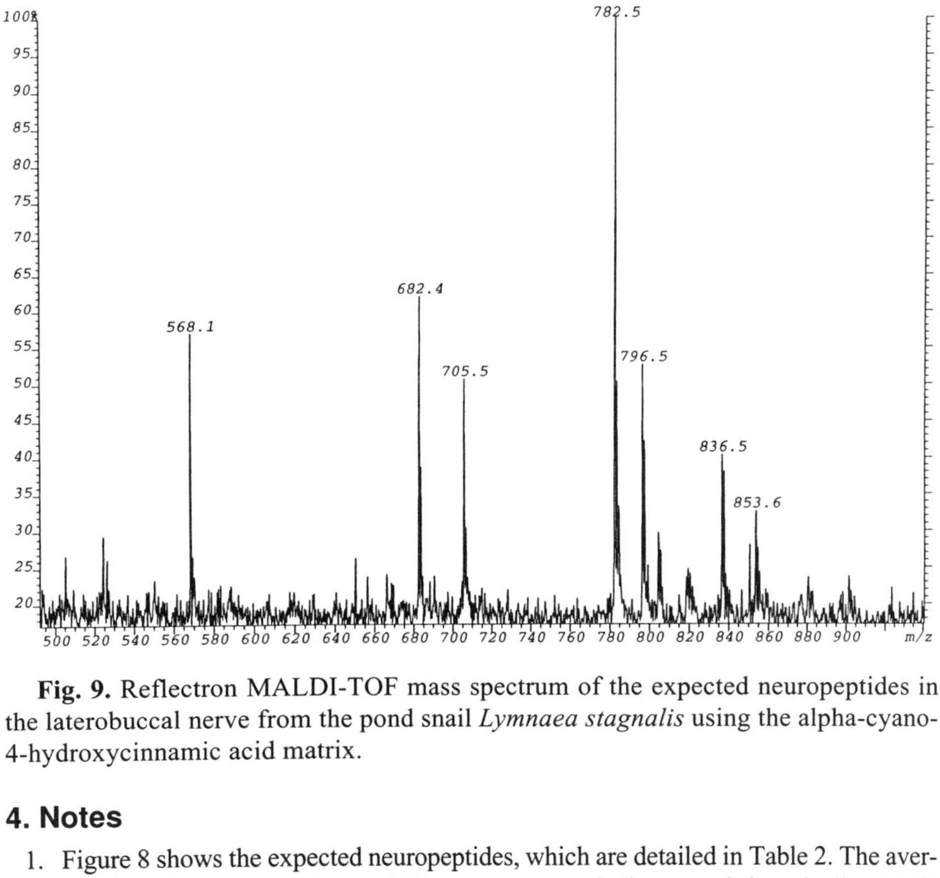

Fig. 9. Reflectron MALDI-TOF mass spectrum of the expected neuropeptides in the laterobuccal nerve from the pond snail *Lymnaea stagnalis* using the alpha-cyano-4-hydroxycinnamic acid matrix.

4. Notes

1. Figure 8 shows the expected neuropeptides, which are detailed in Table 2. The average [M+H] species are observed. The mass accuracy in linear mode is typically >0.2%.
2. Figure 9 shows the reflectron spectrum, again showing the neuropeptides expected. A closer inspection of the spectrum reveals the isotope patterns (*see* Fig. 10) illustrating the greater resolving power of the reflectron mode (bottom spectrum) over the linear mode (top spectrum). In addition, a greater mass accuracy is achieved in reflectron mode, typically >0.02%.

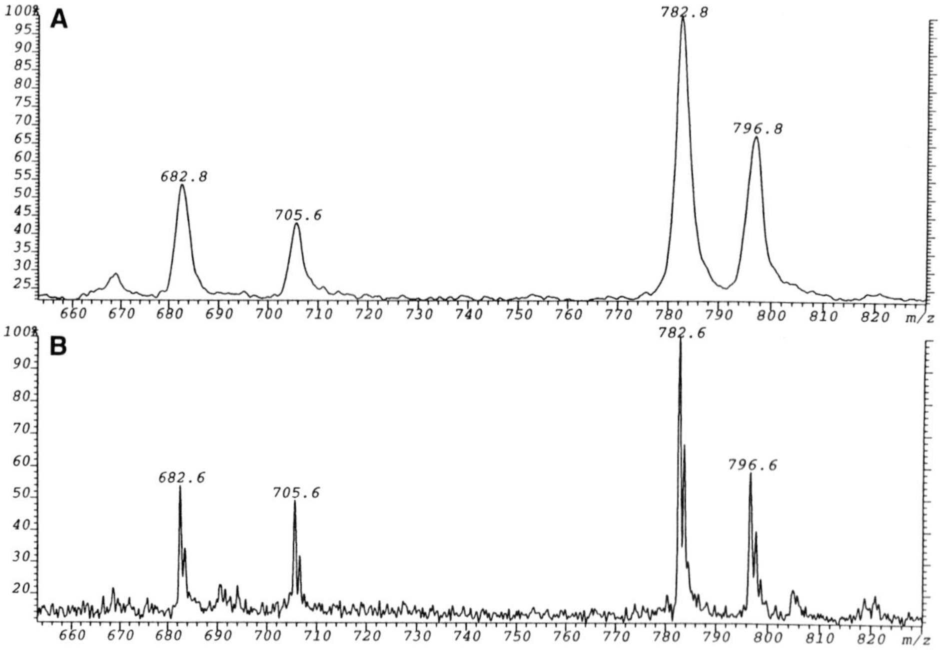

Fig. 10. Comparison between the linear and reflectron data showing enhanced resolution in the reflectron mode (lower spectrum).

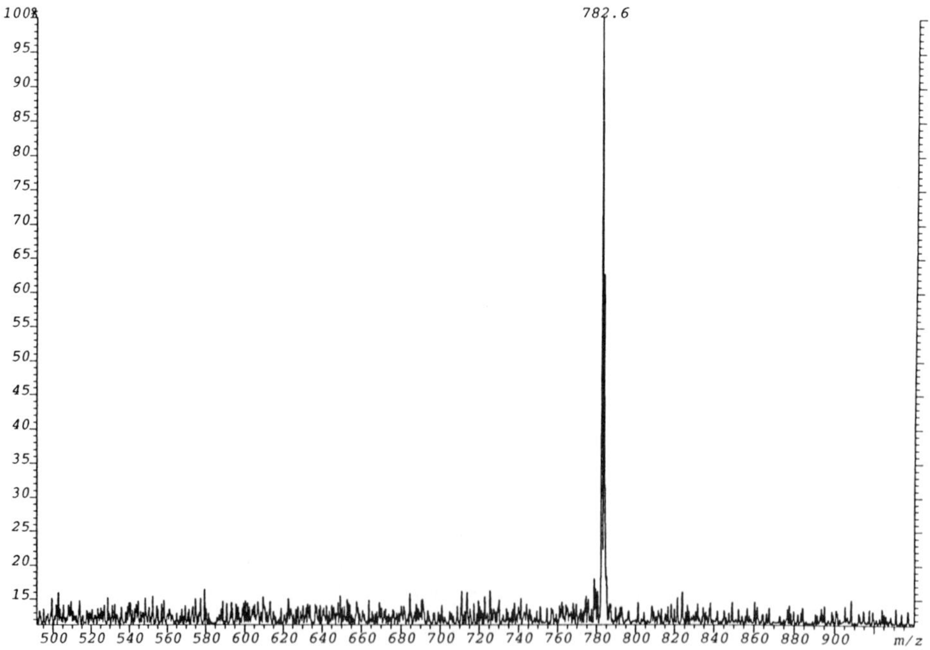

Fig. 11. Use of the ion gate for precursor ion selection of neuropeptide GGSLFRFa.

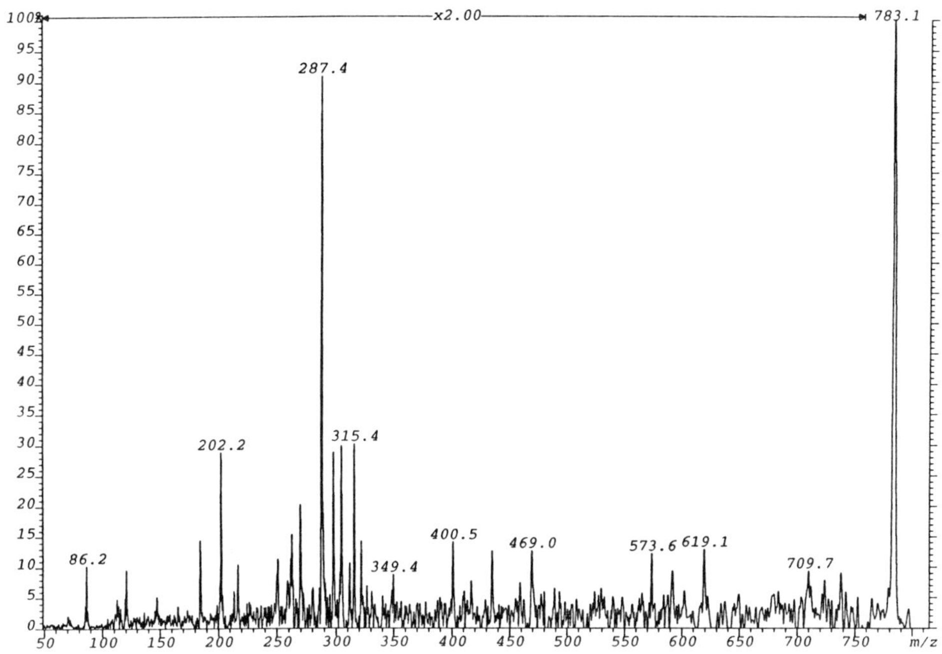

Fig. 12. PSD spectrum of neuropeptide GGSLFRFa.

Table 3
Observed Masses Compared with Expected Masses for Metastable Ions of the Precursor Ion at 783 Dalton from Neuropeptide Sequence GGSLFRFa

Observed [M+H] Dalton	Expected [M+H] Dalton	Metastable ion
86.2	86.1	i4
120.2	120.1	i5/i7
202.2	202.1	b3
287.4	287.2	a4
315.4	315.2	b4
434.5	434.2	a5
573.6	573.3	a6-NH_3

3. Figure 11 shows how the ion gate can be used to select a precursor ion from a mixture of ions (Fig. 9). Each ion can be selected in turn for PSD studies. For this example, the largest ion was selected at 783 Dalton.
4. A PSD spectrum of the 783 Dalton ion is shown in Fig. 12. The metastable ions observed are consistent with the expected fragments for the neuropeptide GGSLFRFa (Table 3).

Acknowledgments

Many thanks to all colleagues at Micromass UK, Ltd., Floats Road for their expert technical assistance.

References

1. Karas, M. and Hillenkamp, F. (1988) Laser desorption ionisation of proteins with molecular masses exceeding 10,000 Daltons. *Anal. Chem.* **60(20),** 2299–2301.
2. Beavis, R. C. and Chait, B. T. (1989) Factors affecting the ultraviolet laser desorption of proteins. *Rap. Comm. Mass Spectrom.* **3(7),** 233–237.
3. Li, K. W., Hoek, R. M., Smith, F., Jimenez, C. R., Van der Schors, R. C., Chen, S., Van der Greef, J., Parish, D. C., Benjamin, P. R., and Geraerts, W. P. M. (1994) Direct peptide profiling by mass spectrometry of single identified neurons reveals complex neuropeptide-processing pattern. *J. Biol. Chem.* **269,** 30,288–30,292.
4. Bradbury, N. E. and Nielsen, R. A. (1936) Absolute values of the electron mobility in hydrogen. *Phys. Rev.* **49(5),** 388–393.
5. Beavis, R. C. and Chait, B. T. (1989) Cinnamic acid derivatives as matrices for ultraviolet laser desorption mass spectrometry of proteins. *Rap. Comm. Mass Spectrom.* **3(12),** 432–435.

15

Use of Circular Dichroism to Determine Secondary Structure of Neuropeptides

Laszlo Otvos, Jr.

1. Introduction

Circular dichroism (CD) spectroscopy is currently the method of choice to study the conformation of proteins in low resolution *(1)*. Chiral substances absorb right and left circularly polarized lights to different extents, demonstrating differences in absorbance (ΔA) and molar extinction coefficients ($\Delta \varepsilon$). A typical CD curve is the wavelength dependence of $\Delta \varepsilon$, or of the molar ($[\Theta]_M$) or mean residue ($[\Theta]_{MR}$) ellipticities *(2)*. Protein CD concentrates on the amide chromophore between 180 and 260 nm, where the $n\pi^*$ and $\pi\pi^*$ transitions occur. Although a single amide group is achiral, amide–amide interactions (chiral perturbation) give rise to a positive CD in proteins *(3)*. Proteins and polypeptides with a single type of secondary structure such as α helix, parallel or antiparallel β pleated sheets, regularly repeating β turns, or unordered structure exhibit characteristic CD curves *(4,5)* (*see* Note 1). Many algorithms for curve analysis have been developed *(6–9)*, but they offer pitfalls *(10)*, and some exceed the capabilities of commercial instrumentation *(11)*.

The CD behavior of short- and medium-sized peptides is similar to that of proteins to some extent, although considerable differences exist. Peptides are highly mobile systems. For example, type U spectra of peptides are likely to correspond to unordered conformation (peptide structure without conformational preferences), whereas the same spectra for proteins can reflect a left-handed polyproline II helix *(12)*. Conformations of peptides in aqueous media are not fully relevant to the secondary structures of regions of proteins in the partially folded states, since the microenvironments of the latter (including the

From: *Methods in Molecular Biology, Neuropeptide Protocols*
Edited by: G. B. Irvine and C. H. Williams Humana Press Inc., Totowa, NJ

hydrophobic surfaces of the proteins themselves) are less polar than water. The most common procedures for lowering the dielectric constant of water is mixing it with alcohols (usually with trifluoroethanol *[13]*), with other organic solvents such as acetonitrile, or with detergents *(14)*. The influence of additives or organic solvents on the basic CD spectra renders the computer analysis of the peptide CD curves largely irrelevant. Moreover, different types of β turns provide different CD spectra *(15)*, a phenomenon unknown for proteins. Given these variables, although a single CD curve is still indicative of the conformation of a peptide, the technique is more useful for studying the conformational stability or conformational transitions of an individual peptide in different solvent systems, or comparing the conformation of analog peptides after modifications of the amino acid composition, side-chain substitutions, and so on (*see* Notes 2 and 3). Figures 1–3 demonstrate selected applications of CD spectroscopy in the secondary structural analysis of neuropeptides. A set of characteristic, basic CD spectra of synthetic peptides for which the proposed conformations were verified by other spectroscopical methods is shown in Fig. 1. Figure 2 presents a type U → type C spectral transition, that reflects an unordered → type I (III) β turn conformational transition of a conformationally, highly mobile neurofilament fragment during trifluoroethanol titration. The varying conformation of the amyloid β-peptide, as it is modulated by the solvent composition (environment) or pH, is illustrated in Fig. 3. Other common variables include peptide concentration, sample temperature, and the presence of cations, anions, or other additives (*see* Note 4).

2. Materials

2.1. Equipment

The basic instrument is a computer-controlled spectropolarimeter connected to an online plotter. The computer must have a math coprocessor installed. Currently four manufacturers provide CD instruments: Jasco (Hachioji, Japan) (two models), Jobin Yvon (Longjumeau, France), AVIV (Lakewood, NJ), and Olis (Bogart, GA) (the last two manufacturers use refurbished Cary spectrometers). At the time of the purchase, this author found that the Jasco J-720 outperformed the others in terms of available wavelength range for peptide conformational analysis, signal-to-noise ratio, reliability, ease of use, and service. The CD spectra and protocols featured in this article were produced using the J-720.

To determine the exact peptide concentration, we use reversed-phase high performance liquid chromatography (RP-HPLC) (*see* Note 5). If this method is selected, a liquid chromatograph is also needed.

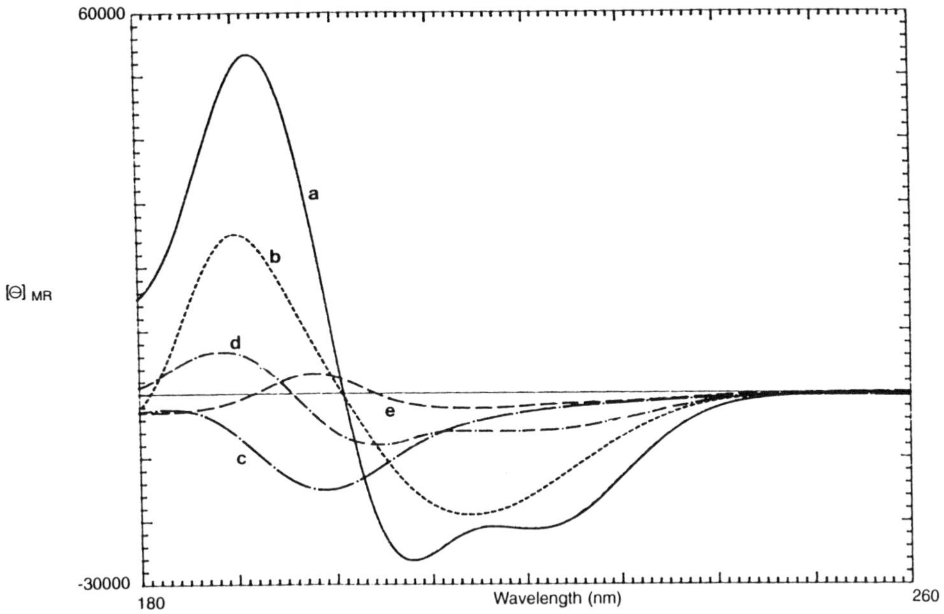

Fig. 1. Basic CD curves representing the most common secondary structures of synthetic peptides. Curve a (solid trace) is the human amyloid β-peptide in trifluoroethanol representing α helices. Curve b (dots) is the same peptide in 32% aqueous trifluoroethanol representing β pleated sheets *(14)*. Curve c (dots and dashes) is the multiphosphorylation repeat of the middle-sized human neurofilament protein (HNFM 1-17) in water representing unordered conformation. Curve d (dots and two dashes) is the same peptide in trifluoroethanol (type C CD curve) representing type I (III) β turns *(17)*. Curve e (dashes) is a peptide analog corresponding to the principal neutralizing determinant tip of the V3 loop of gp 120 of the human immunodeficiency virus in trifluoroethanol (type B CD curve), representing pure type II β turns *(18)*.

2.2. Chemicals

1. Double-distilled, or distilled and ion-exchanged water.
2. 2,2,2-Trifluoroethanol, 99.5+%, NMR grade (Aldrich, Milwaukee, WI).
3. Ethyl alcohol, reagent, spectrophotometric grade (Aldrich).
4. Acetonitrile, 99.5+%, spectrophotometric grade (Aldrich).
5. n-Octyl β-D-glucopyranoside >98% (Sigma, St. Louis, MO).
6. Additives: phosphate-buffered saline, pH 7.2 (150 m*M* sodium chloride and 150 m*M* sodium phosphate) (Sigma); calcium perchlorate, 99% (Pfaltz & Bauer, Germany); magnesium chloride, anhydrous 98% (Sigma).
7. Calibration standards: ammonium d-10-camphorsulfonate, Katayama Chemicals, Jasco standard; D-(-)-pantolactone, Jasco standard (*see* Note 6).

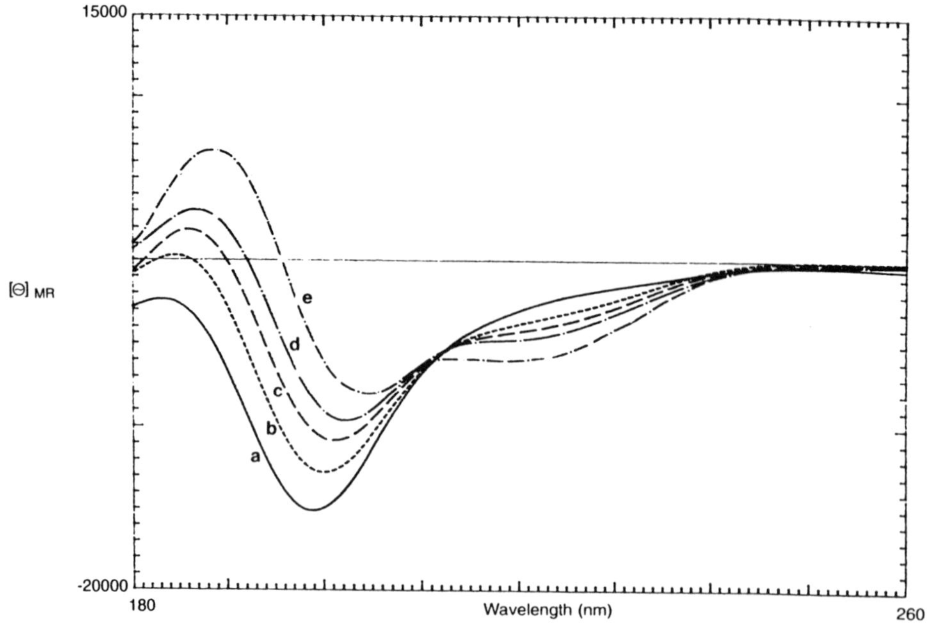

Fig. 2. The aqueous conformation of peptide HNFM 1-17 can be easily influenced by the addition of organic solvents. Trifluoroethanol titration reveals a linear unordered → type I (III) β turn conformational transition. Curve a (solid trace) is in water, curve b (dots) is in 25%, curve c (dashes) is in 50%, curve d (dots and dashes) is in 75%, and curve e (dots and two dashes) is in 100% trifluoroethanol. Reprinted from ref. *17* with permission from Elsevier Science B.V.

3. Methods

1. Turn on and warm up the instrument according to the manufacturers' instructions. Make sure that the computer, the CD instrument, and the plotter are in proper working order (*see* Note 7).
2. Dissolve 0.5-mg aliquots of the peptide in 1-mL amounts of water (*see* Note 8), phosphate-buffered saline (*see* Note 4), 2% β-octyl glucoside, trifluoroethanol, or acetonitrile. Always use freshly prepared solutions.
3. Set the parameters for data acquisition. We generally set parameters as follows: wavelength range: 178–260 nm (needed if curve analyzing algorithms are selected for the estimation of secondary structure); band width: 1 nm; sensitivity: 50 millidegree; response time: 2 s; scan speed: 20 nm/min; step resolution: 0.2 nm (keep it below 0.5 nm for the curve analysis programs). Usually four spectra are averaged out.
4. Record the background spectra. Use different background spectra for each solvent composition. For reasons of memory-saving and sample preparation, complete the studies in one solvent before moving to other. For each solvent, fill the

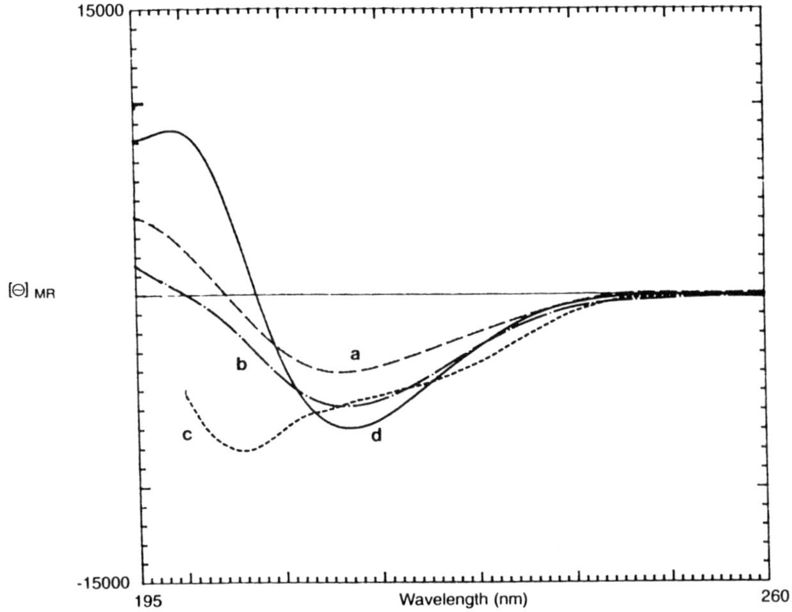

Fig. 3. The effect of different environmental conditions on the conformation of the human amyloid β-peptide. Addition of a detergent or an organic cosolvent (membrane-mimicking solvent systems) increases the band intensities, which may represent stabilization of the β pleated sheet structure. In contrast, increasing the pH loosens the extended structure and results in the formation of other secondary structures, in agreement with infrared data *(14)*. Curve a (dashes) is the peptide in phosphate-buffered saline, pH 7. Curve b (dots and dashes) is the peptide in 50% aqueous acetonitrile at pH 7; curve c (dots) is in 50% aqueous acetonitrile at pH 10. Curve d (solid trace) is the peptide in 2% aqueous β-octyl glucoside solution.

light-path of a 0.2-mm water-jacketed cell. Eighty μL of solution is usually enough. Make sure that there are no bubbles in the cuvet. Close the cell with the stoppers, and place it into the cell holder. Do not touch the surface to be exposed to the light. Place the holder into the sample compartment of the instrument. Mark the location of the cell holder in the carriage. Always use the same spot so that the distance of the sample from the light source remains identical.

5. Store the background spectrum in one of the memories.
6. Remove the cell holder and the cell. Pipet out the solution, and air dry the cuvet.
7. Fill the cuvet with 80 μL of peptide solution in the solvent of the previous background spectrum (*see* Note 4).
8. Place the cell in the holder and the holder in the sample compartment, and record the CD of the peptide under conditions identical to those of the solvent alone (*see* Note 9).

9. Store the peptide spectrum in another memory.
10. Subtract the background spectrum.
11. Store the resulting spectrum (expressed in $\Delta\varepsilon$) on a floppy disk.
12. Input the peptide concentration and molecular weight and transform the data into molar or mean residue ellipticity. Convert the molar CD ($\Delta\varepsilon$) to mean residue ellipticity ($[\Theta]_{MR}$) by calculating in the path length of the cell and the peptide concentration expressed in mg/mL divided by the peptide molecular weight/amide bond ratio. The last unit in the equation, the mean residue weight, of most unmodified (unphosphorylated, unglycosylated, fatty acid-free, and so on) peptides is estimated as 110 (*see* Note 10).
13. Store the converted data on a floppy disk.
14. If the peptide spectra are too noisy, use the instrument's smoothing algorithm. Retry the smoothing process until the overlaid smoothed/unsmoothed spectra exhibit good correlation. Store the smoothed spectrum on a floppy disk.
15. Plot the results.
16. Remove the cell holder and the cell, pipet the peptide solution out, and wash the cuvet three times with water, trifluoroethanol, and methanol. Air dry before the next solvent spectra are taken.

4. Notes

1. Type III β turns, which share their torsion angles with single units of 3_{10} helices, closely resemble type I β turns. Consequently, type I and type III β turns or 3_{10} helices cannot be distinguished by CD spectroscopy, all three exhibit type C CD curves *(16)*. Although pure type II β turns can be identified based on the type B CD character of their CD (*see* Fig. 1), β turn mixtures with considerable type I character again exhibit type C CD curves *(15)*.
2. If a conformational transition involves only two basic secondary structures, the CD curves cross each other in a single intersection, called the isodichroic point (as in Fig. 2). The presence of more than one intersection indicates the coexistence of more than two basic secondary structures (*see* additional peptide spectra in ref. *17*).
3. Generally, a shift of the location of the bands to longer wavelengths (red shift) indicates stabilized conformations, whereas a shift to shorter wavelengths (blue shift) indicates destabilized conformations compared with the basic or reference spectra. A good example is the loosening of an α helix. The exciton splitting of the $\pi\pi^*$ transition in an ideal α helix produces a negative band at 208 nm. Destabilization of the α helix leads to blueshift of the band to 203–204 nm, which is characteristic for type C CD spectra and strong reverse-turns. Further blue shift to 201–202 nm indicates a distorted turn structure (type D CD curve), and ultimately a blue shift to 200 nm or below indicates a completely unordered conformation. The intensity changes of the bands are not so clear-cut. The amplitude of the 215-nm band of the β sheets, for example, depends on the number and relative position of the sheets, as well as on the length and width of the sheets *(2)*. Nevertheless, the band intensities of the same peptide in different solvents or

analog peptides in the same solvent may indicate an increase or decrease of the extended structure. A decrease in the intensity of the otherwise very strong 200-nm negative band of unordered peptides almost always indicates the appearance of some kind of regular structure. The α helix content for proteins (and usually for peptides) can be estimated based on the amplitude of the 208-nm band by the formula ($[\Theta]_{M/MR208}$-4000) divided by 29,000 *(4)*. By the same token, the intensity of the positive ππ* band often reflects the chain length rather than the α helix percentage *(2)*.

4. When the effect of inorganic salt additives on the conformation of neuropeptides is studied, the peptide is usually dissolved in the salt solution (e.g., phosphate-buffered saline, 150 m*M*), the aqueous spectra are taken, and then the solution is diluted with ethanol or trifluoroethanol as long as the salt remains in the alcohol solution. Calcium perchlorate is fully soluble in trifluorethanol. To study effect of multivalent cations on peptide conformation, the cations (Mg, Ca, Al) are usually added in a 1–100*M* excess calculated on the net negative charges of the peptides.
5. The determination of the exact peptide concentration is extremely important in evaluating peptide conformations either by comparison with known peptide spectra or by computational methods. The peptide concentration is determined each time by quantitative RP-HPLC. For this purpose, the peak integration units on the particular column are calibrated at 214 nm with known amounts of standard peptides that have mean residue mass divided by amide bond ratio of approx 110. In our experience, this method provides more accurate and reproducible results than amino acid analysis or simple wavelength-dependent UV absorbance measurements when the constituents are chromatographically not separated.
6. The CD instrument needs to be calibrated regularly. It is usually done by using ammonium d-10-camphorsulfonate ($[\Theta]_{M290.5} = +7910$) or, in the far UV region with D-(-)-pantolactone ($[\Theta]_{M219} = -16140$).
7. The CD instrument needs very little maintenance. The high-energy lamp of the J-720 needs to be cooled by both dry nitrogen, and running water. It is a good idea to use distilled or recirculated and cooled water because salts from hard tap water may clog the tubing in the lamp house. The lamp needs to be replaced after 500 h. One should be careful; the xenon lamp contains gas at high pressure. The mirrors should be adjusted after lamp replacement. We usually repolish or replace the mirrors every year. Other, less frequent maintenance instructions are found in the instrument manuals.
8. Some peptides are not soluble in distilled water. Wet these lyophilized samples with a drop of trifluoroethanol. Most peptides will remain soluble after dilution with water.
9. Some peptides are prone to aggregate, especially in the very narrow (with large surface area) cell. If the peptide spectrum is altered after the solution stands in the cuvet for some time, aggregation is very likely the reason. Wash the cuvet with diluted trifluoroacetic acid and rinse well with water and methanol to remove all deposited peptides.

10. If the use of computer algorithms of the peptide spectra is chosen for the analysis of secondary structure, perform it offline with the Softspec software package of Softwood Company. Softspec is a file conversion program, and comes with all the currently used secondary structural analysis algorithms (Prov, Varselec, Selcon, Yang, CCA).

Acknowledgment

Thanks are owed to Gyorgyi I. Szendrei and Hildegund C. J. Ertl for expert reading of the manuscript.

References

1. Johnson, W. C. (1988) Secondary structure of proteins through circular dichroism spectroscopy. *Ann. Rev. Biophys. Biophys. Chem.* **17,** 145–166.
2. Woody, R. W. (1985) Circular dichroism of peptides, in *The Peptides, vol. 7, Conformation in Biology and Drug Design* (Hruby, V. J., ed.), Academic, San Diego, pp. 15–114.
3. Johnson, W. C. (1990) Protein secondary structure and circular dichroism: a practical guide. *Proteins* **7,** 205–214.
4. Greenfield, N. and Fasman, G. D. (1969) Computed circular dichroism spectra for the evaluation of protein conformation. *Biochemistry* **8,** 4108–4116.
5. Pribic, R., van Stokkum, H. M., Chapman, D., Haris, P. I., and Bloemendal, M. (1993) Protein secondary structure from Fourier Transform infrared and/or circular dichroism spectra. *Anal. Biochem.* **214,** 366–378.
6. Provencher, S. W. and Glockner, J. (1981) Estimation of globular protein secondary structure from circular dichroism. *Biochemistry* **20,** 33–37.
7. Yang, J. T., Wu, C.-S. C., and Martinez, H. M. (1986) Calculation of protein conformation from circular dichroism. *Methods Enzymol.* **130,** 208–269.
8. Perczel, A., Hollosi, M., Tusnady, G., and Fasman, G. D. (1991) Convex constraint analysis: a natural deconvolution of circular dichroism curves of proteins. *Protein Eng.* **4,** 669–679.
9. Sreerama, N. and Woody, R. W. (1993) A self-consistent method for the analysis of protein secondary structure from circular dichroism. *Anal. Biochem.* **209,** 32–44.
10. Venyaminov, S. Y., Baikalov, I. A., Wu, C.-S. C., and Yang, J. Y. (1991) Some problems of CD analyses of protein conformation. *Anal. Biochem.* **198,** 250–255.
11. Toumadje, A., Alcorn, S. W., and Johnson, W. C. (1992) Extending CD spectra of proteins to 168 nm improves the analysis for secondary structures. *Anal. Biochem.* **200,** 321–331.
12. Woody, R. W. (1992) Circular dichroism and conformation of unordered polypeptides. *Adv. Biophys. Chem.* **2,** 37–79.
13. Lehrman, S. R., Tuls, J. L., and Lund, M. (1990) Peptide α-helicity in trifluoroethanol: correlations with predicted α-helicity and the secondary structure of the corresponding regions of bovine growth hormone. *Biochemistry* **29,** 5590–5596.

14. Otvos, L., Jr., Szendrei, G. I., Lee, V. M.-Y., and Mantsch, H. H. (1993) Human and rodent Alzheimer β-amyloid peptides acquire distinct conformations in membrane-mimicking solvents. *Eur. J. Biochem.* **211,** 249–257.
15. Perczel, A., Hollosi, M., Sandor, P., and Fasman, G. D. (1993) The evaluation of type I and type II β-turn mixtures. Circular dichroism, NMR, and molecular dynamics studies. *Int. J. Pept. Protein Res.* **41,** 223–236.
16. Smith, J. A. and Pease, L. G. (1980) Reverse turns in peptides and proteins. *CRC Crit. Rev. Biochem.* **8,** 315–401.
17. Lang, E., Szendrei, G. I., Lee, V. M.-Y., and Otvos, L., Jr. (1994) Spectroscopic evidence that monoclonal antibodies recognize the dominant conformation of medium-sized synthetic peptides. *J. Immunol. Meth.* **170,** 103–115.
18. Laczko, I., Hollosi, M., Urge, L., Ugen, K. E., Weiner, D. B., Mantsch, H. H., Thurin, J., and Otvos, L., Jr. (1992) Synthesis and conformational studies of N-glycosylated analogues of HIV-1 principal neutralizing determinant. *Biochemistry* **31,** 4282–4288.

16

^1H Nuclear Magnetic Resonance (NMR) in the Elucidation of Peptide Structure

David J. S. Guthrie

1. Introduction

^1H nuclear magnetic resonance (NMR) has proven to be a uniquely powerful tool for studying the structure of peptides in solution. I will concentrate on the type of structural information that is obtainable from NMR, the types of spectra needed to get this information, and how to interpret these spectra. I will assume that the spectra will be recorded by an experienced operator. What can be achieved in a particular case will depend not only on the sample, but also on the time available on the spectrometer.

NMR spectroscopy studies transitions between energy levels that arise from the interaction of a magnetic field with nuclei (e.g., ^1H, ^2H, ^{13}C, ^{14}N, ^{15}N, ^{17}O, ^{19}F, ^{31}P) that possess spin, and therefore a magnetic dipole. Other nuclei, e.g., ^{12}C and ^{16}O, have no spin, no magnetic dipole, and no NMR spectra. The most commonly studied nucleus is ^1H, and I will deal solely with ^1H spectra.

There are advantages in using as strong a magnet as possible, e.g., increased spectral dispersion, sensitivity, and ease of interpretation. Spectrometers used to study neuropeptides have magnetic field strengths (or rather, flux densities) of 9.4–17.6 T (tesla). The spacing of the energy levels (spin states) produced by a magnetic field is directly proportional to the field but is very small, even with strong fields (*see* Note 1). The associated transition frequencies lie in the radiofrequency region of the spectrum and are measured in megahertz (MHz). It is usual to describe NMR spectrometers, not by their field strength, but by the resonance frequencies of protons in their magnetic fields, i.e., not 9.4 T but 400 MHz.

From: *Methods in Molecular Biology, Neuropeptide Protocols*
Edited by: G. B. Irvine and C. H. Williams Humana Press Inc., Totowa, NJ

1.1. The Fourier Transform

Modern NMR spectrometers operate by exciting all the protons in a sample simultaneously using a short radiofrequency pulse. The emissions, as the protons relax back to the ground state, are captured and stored by computer. The resulting interferogram, called the free induction decay (FID), is converted to the normal spectrum by a mathematical operation, the Fourier transform *(1)*. The data are said to be transformed from the time domain to the frequency domain. This mode of operation has several advantages. As the data are computerized, they can be massaged mathematically to enhance either resolution or sensitivity. Several FIDs can be acquired from a sample (data acquisition takes only seconds) and their sum transformed to a spectrum. Random noise cancels out as the FIDs are added, improving sensitivity. Further, instead of a single pulse, multipulse sequences can be used to obtain a variety of spectra (*see* Section 1.7.).

1.2. Chemical Shift

A proton in a magnetic field experiences a local field, which is the sum of the applied field and fields induced in the surrounding electrons by the applied field (these generally oppose the applied field and are said to shield the proton). Because of this electronic shielding, each chemically distinct proton experiences a different local field and so has a different resonance frequency.

The chemical shift, or δ value of a proton is the difference between its resonance frequency and that of a standard. For easy comparison of spectra recorded at different field strengths, δ values are usually quoted in parts per million (ppm), with the standard arbitrarily set to zero. Thus, a proton with a δ value of 1 ppm would be 250 Hz away from the standard in a spectrum recorded at 250 MHz , but 600 Hz from it in one recorded at 600 MHz. The usual standard for ^{1}H NMR is tetramethylsilane (TMS). Most protons give signals in the range of 0–10 ppm from TMS. Although the relationship between chemical shift and structure is not a simple one, some generalizations are possible. For example, amide protons fall in or near the range 7.0–8.0 ppm, α protons in or near the range of 4.0–5.0 ppm and side-chain protons between 0.0 and 4.0 ppm *(2)*. Amide protons are said to be deshielded relative to α and side-chain protons. Aromatic rings exert a strong influence on chemical shifts, strongly shielding protons held above or below the plane of the ring, but deshielding protons in the plane of the ring. As an example, the ^{1}H spectrum of a heptapeptide is shown in Fig. 1.

1.3. Spin–Spin Coupling

In Fig. 1, most signals display a splitting or fine structure owing to the protons sensing slightly different magnetic fields from neighboring protons in their

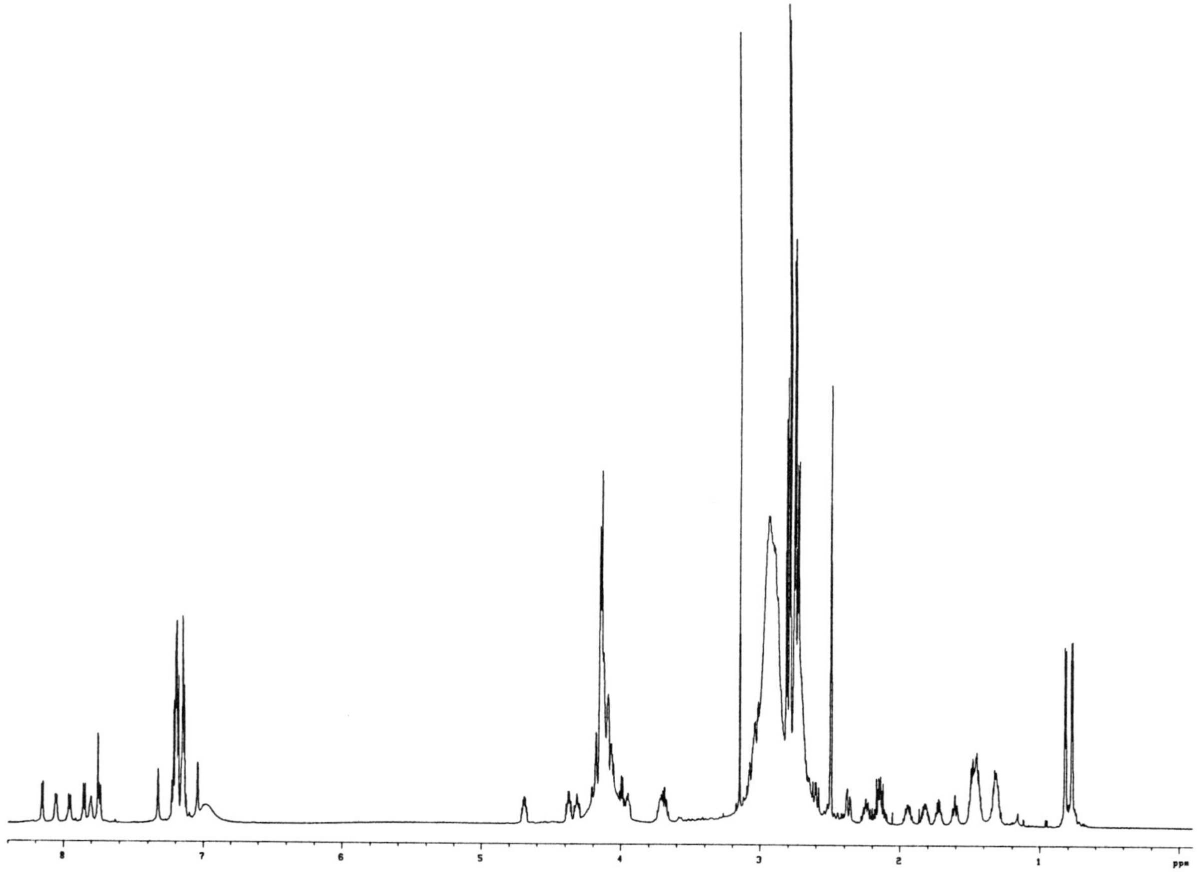

Fig. 1. One-dimensional ^{1}H NMR spectrum of the invertebrate neuropeptide pyroGlu.Asp.Pro.Phe.Leu.Arg.Phe.NH_2. The spectrum was recorded from a solution in $[^{2}H_{6}]$DMSO/H_2O (2:1), at 600 MHz. Thanks are owed to I. H. Sadler and his staff at the Edinburgh University Ultra High Field NMR Service.

ground and excited states. This is called spin–spin, scalar, or J coupling. It is characterized by a coupling constant J, which in simple cases is the observed splitting (more complex couplings occur and can be analyzed *[3]*). J is independent of field strength but depends on the local geometry.

Scalar coupling occurs through bonds and the magnitude of J decreases as the number of bonds increases. Typical values for proton couplings over two or three bonds (e.g., H-C-H or H-N-C-H), are 0–20 Hz and 0–10 Hz, respectively. Longer, smaller couplings exist, but cannot be detected with the resolution commonly used for peptide spectra. It follows that scalar coupling is observed only between protons within a single residue and not between residues, since the shortest path between residues, from an α proton (H^{α}) to an amide proton (H^N), is over four bonds. Coupling constants are written as, e.g., $^3J_{HN\alpha}$. The superscript indicates the number of bonds over which coupling occurs, and the subscript indicates the protons that are coupled. For vicinal (^3J) coupling, empirical correlations (called Karplus equations) of J with the torsion angle about the central bond have been established, as shown for $^3J_{HN\alpha}$ in Fig. 2.

1.4. Relaxation

In most forms of spectroscopy, pathways and rates of relaxation are of little interest. However, two relaxation pathways play crucial roles in NMR spectroscopy. The return of excited protons to the ground state by loss of energy to their surroundings (the lattice), called spin-lattice (or longitudinal) relaxation, maintains the small excess population of the ground state and is characterized by a relaxation time T_1 (a relaxation time is the reciprocal of the relaxation rate). The sharing of energy with other protons, called spin–spin (or transverse) relaxation, is characterized by T_2 and this largely controls the linewidths of signals (half height linewidth $\propto 1/T_2$). Paramagnetic species, e.g., oxygen and some transition metal ions, cause line broadening by drastically shortening relaxation times. T_1 and T_2 are both strongly dependent on the rate of molecular tumbling.

1.4.1. The Nuclear Overhauser Effect (NOE)

A major pathway for sharing energy between protons (cross relaxation) is the coupling of the protons' magnetic dipoles. Perturbing the signal from one proton (e.g., by saturating it) in a group of cross-relaxed (dipolar coupled) protons often alters the intensities of signals from other protons in the group. This intensity change, usually expressed as a percentage, is the NOE. Perturbing a proton's signal alters that proton's spin state populations, and these in turn, by dipolar coupling, perturb the populations of the spin states of neighboring protons and thus the intensities of transitions between these states.

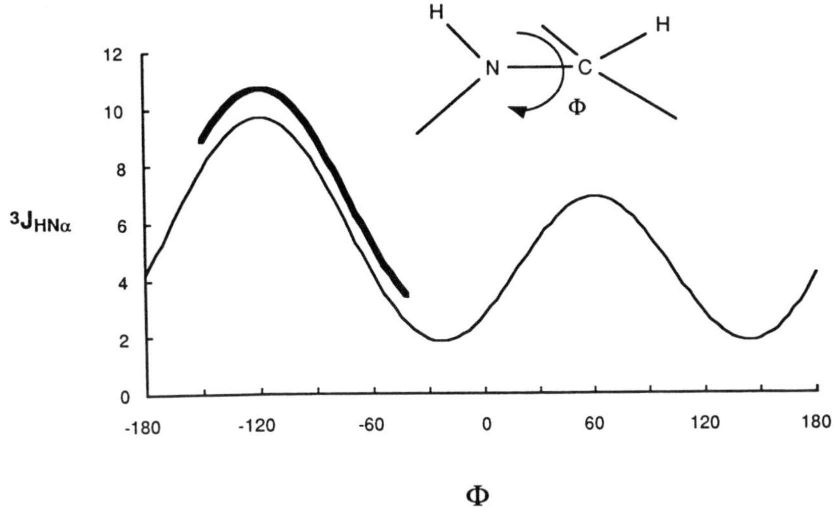

Fig. 2. Variation of the coupling constant $^3J_{HN\alpha}$ with the torsion angle Φ, according to the equation $J = 6.4\cos^2\theta - 1.4\cos\theta + 1.9$ ($\theta = |\Phi - 60|$) *(4)*. The heavy line indicates the values of Φ that occur most frequently for amino acids other than glycine.

The magnitude of the NOE observed depends on the internuclear distance (r) and the rate of molecular tumbling. Dipolar coupling is effective only over very short distances, varying with the inverse sixth power of the internuclear distance ($\propto r^{-6}$). An NOE is observed therefore only between protons that are very close (≤4–5 Å). Such protons need not be neighbors within a molecule; unlike scalar coupling, the NOE is propagated directly through space (Fig. 3). The effect of the rate of molecular tumbling on dipolar coupling is quite complicated. The interproton NOE has opposite signs for rapidly tumbling (small) molecules and slowly tumbling (large) molecules. For molecules of intermediate size, a point occurs (corresponding to a mass of approx 1000) when the NOE is zero, no matter how close the protons. Because of such complexities, failure to observe an NOE between protons is not proof that they are far apart.

An NOE builds up over a short but finite time as spin state populations change and the rate of build-up of the NOE is a better guide to the interproton distance than is the final level reached. Pulse sequences designed to measure NOE include a mixing time, τ_m, during which the NOE develops, and the initial slope of a plot of NOE against τ_m depends only on distance. Long values of τ_m give unreliable NOEs, as by further cross-relaxation, the perturbed spin populations of the protons showing the NOE gradually perturb the spin populations and hence signal intensities of additional protons, an effect called spin diffusion. Dipolar coupling is more effective in larger, more slowly tumbling

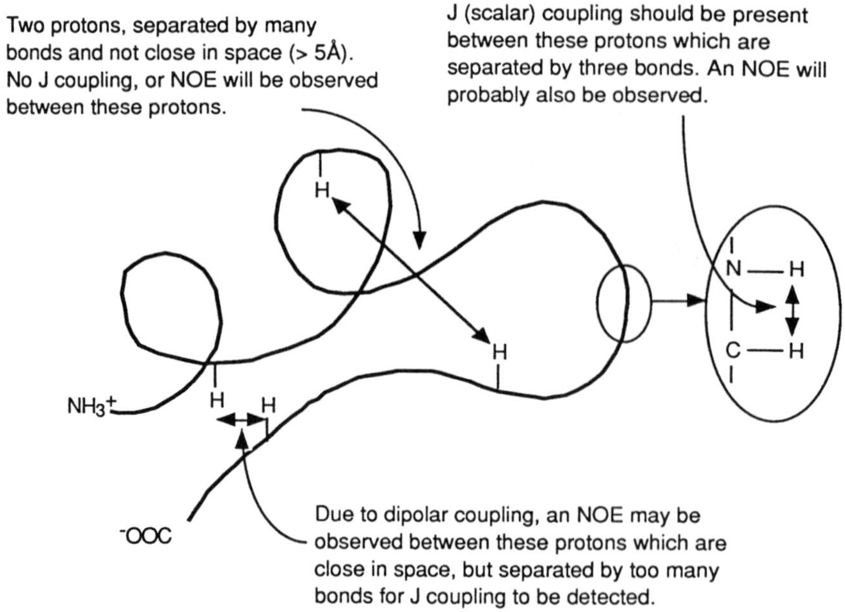

Fig. 3. Diagram illustrating the dependency of dipolar coupling (NOE) and scalar (spin–spin) coupling on peptide structure.

molecules. Two consequences of this are that short peptides require a longer τ_m than larger peptides to produce usable levels of NOE and that spin diffusion is more of a problem in large molecules. Dissolved oxygen provides an alternative relaxation pathway, reducing the sensitivity of NOE, and is best removed, though this step is sometimes ignored.

1.5. Chemical Exchange

NMR may be said to have a similar timescale to many chemical reactions. Because of the lifetime of the excited state and the relatively small dispersion of resonance lines, an NMR spectrum is affected if a molecule under study undergoes any exchange reaction. If the exchange is slow compared with the NMR timescale, separate signals will be seen from all the species involved, whereas, with fast exchange, only a weighted average signal will be seen. At intermediate rates, separate but broadened signals will be present.

Exchange reactions relevant to peptides are equilibria between conformers (*see* Note 2) and exchange of labile protons. Amide protons form an important class of labile proton and their exchange with solvent protons can give useful structural information (*see* Note 3).

1.6. Temperature Dependence of H^N Chemical Shifts

Solvent molecules are small and their motions are affected by temperature to a greater extent than those of most peptides. Not unexpectedly, δ values of peptide H^Ns H-bonded to solvent show a marked variation with temperature. Those of H^Ns that are buried or involved in internal H–bonding show much less variation. In dimethyl sulfoxide (DMSO) and in water, exposed H^Ns have temperature coefficients ($\Delta\delta/\Delta T$) of 6–10 parts per billion/°C (ppb/°C), whereas shielded H^Ns have reduced coefficients, <2–3 ppb/°C. Note that reduced values can also result from H-bonding to side-chains as well as backbone groups.

1.7. Two-Dimensional Spectra

If a multipulse sequence is used to produce a set of FIDs, the interval between two pulses in the sequence being varied between each FID, the result is a data set that displays two time dependencies. The time-dependent data in each FID are said to be a function of time t_2, but each FID also differs from each of the other FIDs because of the altered pulse timing and this variation is said to be a function of t_1. By applying a double Fourier transform, this two-dimensional data set can be transformed in both time dimensions (t_1 and t_2) to give a spectrum with two frequency dimensions (f_1 and f_2).

Two-dimensional spectra are usually presented as intensity contour plots (Fig. 4). Two types of peak occur, diagonal peaks with the same chemical shift in both dimensions (which represent a projection of the normal one-dimensional spectrum), and crosspeaks, each of which connects two diagonal peaks. The significance of the connections depends on the pulse sequence. The major benefit of two-dimensional spectra is the removal of the overcrowding of signals common in one-dimensional spectra, even at high fields, for all but the smallest peptides. In general, crosspeaks will be well resolved for small to medium peptides. Spectra should be acquired in the phase-sensitive mode as this gives better resolution (*see* Note 4).

1.7.1. COrrelation SpectroscopY (COSY)

A COSY spectrum is a two-dimensional spectrum in which the crosspeaks connect protons that are scalar (spin–spin) coupled. Characteristic patterns of crosspeaks are expected from the different residues; e.g., *see* Fig. 4 *(1,3,13,14)*.

In the basic COSY spectrum, the peaks are rather broad and the resolution, particularly of crosspeaks close to the diagonal, is poor. A modified COSY, the double quantum-filtered, phase-sensitive COSY (DQF-COSY) is often preferred. The double quantum filter means that signals from protons without scalar coupling disappear, so that, e.g., methyl peaks and most solvent peaks will

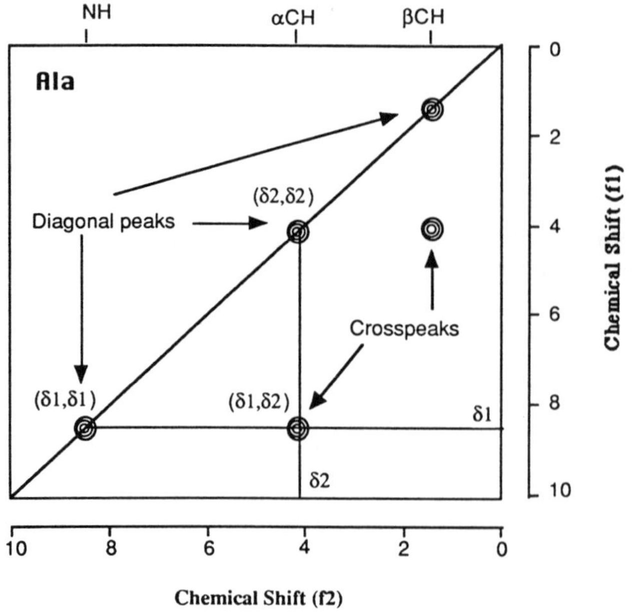

Fig. 4. Schematic representation of a two-dimensional NMR spectrum as an intensity contour plot. Each crosspeak connects two diagonal peaks. The pattern shown is that expected from alanine residues in a COSY spectrum.

be eliminated (*see* Section 3.2.2.). The phase-sensitive mode means crosspeaks generally show fine structure from scalar coupling. Two further variants on COSY, HOmonuclear HArtmann-Hahn (HOHAHA) and TOtal Correlation SpectroscopY (TOCSY), are used. These give equivalent results, with crosspeaks between all the protons of a spin system (*see* Fig. 5), and greatly simplify identification (*see* Note 5).

1.7.2. Nuclear Overhauser Effect SpectroscopY (NOESY)

A NOESY spectrum is a two-dimensional spectrum in which the crosspeaks connect cross-relaxed (dipolar coupled) protons, i.e., those that are close in space.

A potential problem in NOESY spectra is that the exchange of magnetization involved in the relaxation processes leading to the NOE cannot be distinguished from that owing to chemical exchange. However, in practice, chemical exchange will not be a likely explanation for many of the crosspeaks that appear. Another problem for small peptides, is that no NOEs at all may be observed. The use of low temperatures or viscous solvents may help, as these cause slower molecular tumbling and hopefully lead to detectable NOEs. An

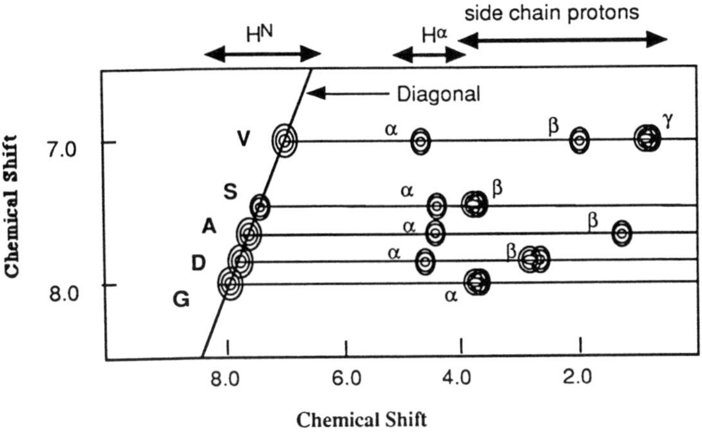

Fig. 5. Schematic representation of a TOCSY (HOHAHA) spectrum showing crosspeaks between the H^N protons and the H^α and side-chain protons that would be expected from five amino acid residues in a peptide.

alternative is another pulse sequence, ROtating frame Overhauser Effect SpectroscopY (ROESY). With ROESY, there is no sign reversal between slow and fast tumbling molecules, so an NOE can always be observed. Further, the ROESY pulse sequence is less susceptible to the effects of spin diffusion than is the NOESY sequence, but there can be problems with interpretation *(5)* (*see* Note 6).

1.8. Summary

Several NMR parameters can give information about peptide conformation. In order to use such information, it is necessary to assign, as completely as possible, all the resonance signals to protons in the molecule. This task is made easier by the use of two-dimensional spectra.

Of the various introductory texts on NMR in print, I recommend those by Rattle *(1)* and Abraham et al. *(3),* because in addition to basics, they include an introduction to peptide and protein NMR. More advanced, general treatments of solution NMR, including two-dimensional spectra and NOE, are available *(5–8).*

2. Materials

Various NMR spectrometers are available, with different strengths of magnet and, as mentioned in Section 1, the strongest magnetic field is normally to be preferred. NMR spectra of peptides are usually acquired from a sample in solution in a glass tube that is inserted into a part of the spectrometer called the probe. Several different probes may be available for a spectrometer to allow the study of different nuclei and to accommodate different volumes of solvent (*see* Note 7). In ^1H NMR, it is usual to use solvents in which most if not all ^1H have been replaced by ^2H to avoid problems with strong signals arising from the solvent. With modern spectrometers, the level of deuteration should be at least 99.8–99.9%.

3. Methods

3.1. Sample Preparation

To keep data acquisition times to a minimum (a two-dimensional spectrum may need >12 h), it is usual to work at high concentration. The minimum concentration would normally be reckoned as 1 m*M*, but spectrometer operators will appreciate 5–10 m*M* (*see* Note 8).

The sample solution must contain no solid particles that would distort the local magnetic field, causing line broadening. The solution can be centrifuged or filtered, e.g., by drawing it through a plug of clean cotton wool in the tip of a Pasteur pipet, before transfer to the NMR tube. Removal of dissolved oxygen, either to improve NOE measurements or to prevent oxidation of susceptible residues, can be achieved in one of two ways. Either purge the solution with helium, using a nonmetallic needle to avoid contamination by metal ions, or use a series of freeze–thaw cycles, when the sample is successively frozen, evacuated, and thawed under an inert gas (at least 5 times). With aqueous solutions, this may be performed in a vial to avoid cracking the NMR tube.

3.2. Choice of Solvent

Water is a major component of all living systems and would seem to be the obvious choice. However, other considerations can be important.

3.2.1. Solubility

The peptide may be insufficiently soluble in water to allow the necessary concentration for NMR, and other solvents must be tried. DMSO is a good solvent for many peptides and is used widely for this reason.

3.2.2. Suppression of Solvent Signal

Even at the high peptide concentrations used in NMR spectroscopy, solvent protons outnumber sample protons and steps must be taken to avoid a large solvent signal that would dominate the spectrum. Most commonly, deuterated solvents are used. Although ^2H has spin, it is not excited by the pulses used to study ^1H. However, a ^2H signal is produced by auxiliary circuits and used as a lock to ensure the field-frequency stability of the spectrometer. Commercial deuterated solvents contain traces of protons that have characteristic δ values and can be used as secondary chemical shift standards. To obtain a spectrum from aqueous solution, it is not possible simply to use ^2H_2O as the ^2H atoms would exchange with H^N so that the latter would disappear. Nor, since ^2H atoms are needed for a lock signal, can H_2O be used alone. It is normal to use a mixture of H_2O and ^2H_2O (about 9:1). This leaves a strong solvent signal that can be reduced by use of a solvent-suppression pulse sequence *(9)*.

3.2.3. Choice of pH

Most peptides are exposed to acid during synthesis or purification and, when dissolved in water, give a pH of about 3.0–3.5. Carboxylic acid groups in, e.g., Asp and Glu residues, are then protonated and this can affect conformation. A good way to explore the effect of ionization is to acquire spectra over a pH range. Sudden changes in the chemical shift values of protons, particularly H^N, point to a change in conformation. If carboxyl groups are required to be ionized, the pH can be adjusted to 5.5–6.0 (remember that H^N signals vanish above about pH 6.5) before NMR studies are started. For NMR in nonaqueous solvents, the pH can be adjusted in an aqueous solution that is then lyophilized and the residue dissolved in the chosen solvent.

3.2.4. Mimicking a Biological Environment

The main interest with neuropeptides is the conformation when bound to a receptor, not that in solution, which may be different. The interior of a binding site is less polar than aqueous solution, and various suggestions have been made as how best to mimic this. Polar solvents such as DMSO or CH_3CN have been proposed as models, but this is perhaps an attempt to rationalize the use of solvents really chosen for reasons of solubility. More plausibly, it has been argued that water near to a protein surface is more viscous than bulk water and that this increased viscosity reduces conformational mobility and filters an otherwise flexible peptide into a single, dominant conformation. Mixtures of water and DMSO (or ethylene glycol) have a higher viscosity than either solvent alone and have been suggested as appropriate solvents for conformational studies, on account of both polarity and viscosity *(10–12)*. Some neuropeptides interact with membranes, possibly as a prebinding step prior to receptor binding, and methanol, trifluoroethanol, hexafluoropropan-2-ol, and SDS micelles have all been used to simulate a membrane environment. Further discussion on studying the conformation of a peptide either at its receptor or in a membrane-like environment will be found in Chapter 17.

3.3. Spectral Assignment

The sequential assignment strategy as developed by Wüthrich *(13)* consists of two steps: identifying the systems of spin (scalar) coupled resonances, and assigning each spin system to a particular residue in the peptide. As explained earlier, spin systems are confined to a single residue but some amino acids possess two spin systems, e.g., Phe or Trp, as there is no detectable coupling between the H^{β} and the protons of the aromatic rings.

3.3.1. Identification of Spin Systems

1. Acquire DQF-COSY and TOCSY spectra for the sample. Like NOESY, the TOCSY pulse sequence incorporates a mixing time and a value of 70–90 ms would seem appropriate for most neuropeptides, to show long-range connections.
2. Check that the spectra are of good enough quality for conformational studies. If significantly fewer than the expected number of H^N–H^{α} crosspeaks (one per residue except for proline and the first residue) are observed, consult with the spectrometer operator about alternative strategies (*see* Note 9).
3. Begin identifying spin systems. Traditionally, this has been achieved using a COSY or DQF-COSY spectrum. Beginning with each H^N in turn, trace via the crosspeaks, the connection to the H^{α}, then to the H^{β}, and so on. In practice, it may be psychologically more satisfying to begin with easily identifiable resonances, e.g., methyl groups or the H^{β} of Val, Asp, or Ser. The identification of spin systems is achieved more easily using a TOCSY spectrum, when crosspeaks should link each H^N with all the other protons of its spin system (Fig. 5). The individual residues giving rise to some of the spin systems will now be known.

   As peaks are being connected, their chemical shifts should be recorded. These may be available from lists provided with the spectra. If not, they must be obtained from a plotted spectrum, by careful use of a ruler or preferably by using a digitizing tablet if one is available.
4. Attempt to complete assigning the spin systems to amino acid types. Some amino acids have unique spin systems, i.e., Gly, Ala, Val, Ile, Leu, and Lys *(1,3,13,14)*, and the type can be uniquely identified. Others require more data, e.g., Asp, Asn, Cys, His, Phe, Ser, Tyr, and Trp have identical spin systems (H^N, H^{α}, and two H^{β}). In some cases, the chemical shift allows identification to be made (only Ser has H^{β} near δ 3.6), but, in most cases, a correlation with the sequence will be necessary (*see* Section 3.3.2.). This is also true when several copies of an amino acid residue are present.

3.3.2. Sequence Specific Assignment

1. Acquire a NOESY spectrum with an appropriate mixing time (80 ms for a long peptide, 200 ms for a short one). If possible, several NOESY spectra should be acquired with mixing times of 50–400 ms, to help distinguish genuine NOEs from the effects of spin diffusion.
2. Assess the quality of the spectrum. If few H^N - H^N or H^N - H^{α} peaks are present, use a longer mixing time, or a ROESY spectrum.
3. In a NOESY spectrum with as short mixing time as possible, examine the region containing crosspeaks between H^N and H^{α}. Identify intraresidue crosspeaks for these protons (these are also present in the COSY spectrum) and then look for

crosspeaks linking the H^{α} of one residue to the H^N of the next. In this way one can walk along the peptide sequence, stepping from intraresidue to interresidue to intraresidue crosspeaks (Fig. 6). It is not necessary to start at the N- or C-terminus; the walk can be built up in sections. Proline residues will break the sequence. If a connection is missing for this or any other reason, attempt to re-establish it using H^N - H^N or H^N/H^{α}-side-chain crosspeaks.

4. In the event that any ambiguity remains owing either to overlapping or missing crosspeaks, one can try spectra acquired at longer mixing times, or at a slightly different temperature.

3.4. Gathering Data for Conformational Studies

Having assigned as fully as possible all the observed resonances, the next step is to look for data that relates directly, or indirectly, to conformation.

3.4.1. Quantitative or Semiquantitative NOEs

The object is to note all the significant crosspeaks in NOESY spectra and to make some estimate of their relative intensities. Take each residue in turn and identify all the NOEs between its protons and those of all the other residues.

3.4.1.1. Classification of NOEs

NOEs are classified as short-, medium-, and long-range. Long-range NOEs, between protons on residues 6 or more positions apart in the sequence, are very useful in defining conformation but are rarely found in the spectra of small- to medium-sized peptides. Short-range NOEs are common but have a low information content. Those occurring within residues have almost no value, since protons within a residue are almost always close enough to give NOEs. Sequential NOEs, between protons in neighboring residues, are individually also of little use, but patterns of sequential NOEs (and their intensities) over a peptide may be informative (*see* Section 3.5.). Medium-range NOEs, between protons on residues 2–5 positions apart, yield useful information about secondary structure (*see* Section 3.5.). The nomenclature used to describe NOE connectivities between protons is illustrated in Fig. 7.

3.4.1.2. Quantitation of NOEs

For large peptides or small proteins, NOE intensities can be measured by integration of crosspeak volume, using the NMR processing software. Remember that the rate of build-up of the NOE is the best measure of interproton distance. Pairs of protons separated by a known, fixed distance (e.g., H^{α}s in Gly or H^{β}s in Phe) can be used to calibrate the NOE for that molecule. Then all the NOEs can be converted to distances that can be used in

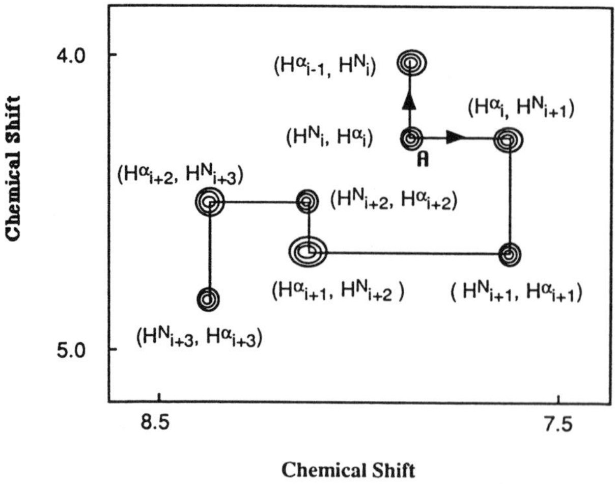

Fig. 6. Schematic representation of part of the H^N - H^α region of the NOESY spectrum of a peptide, illustrating the walk between inter- and intraresidue crosspeaks. Starting with the peak labeled "A," identified as the intraresidue $d_{N\alpha}$ of residue i (from the presence in COSY spectra of a crosspeak at this position) it is possible to identify NOESY crosspeaks between H^Ni and H^αi-1, and between H^αi and H^Ni+1.

computer modeling. For smaller, more flexible peptides (<30 residues), such effort is not justified and it is usual to estimate the NOE crosspeak intensity from its height by counting contours. The NOEs are classified as weak, medium, or strong and can be used to define distance constraints for computer modeling *(13,14)* or simply to identify the various conformations present and estimate their populations.

3.4.2. Measurement of Coupling Constants

In order to measure accurately coupling constants, e.g., $^3J_{HN\alpha}$, of the order of 4–8 Hz, the digital resolution of the spectrum must at least 0.2 Hz/point. This is easily attained in one-dimensional spectra if 32K data points are collected, but if peaks overlap, then two-dimensional spectra must be used. In principle, DQF-COSY spectra have the resolution to allow measurement of coupling constants, but for purposes of assignment are often acquired with a resolution of only a few Hz/point, which is insufficient. Therefore, a dedicated DQF-COSY spectrum must be run, with at least 4K data points in t_2. Measurement of $^3J_{\alpha\beta}$ is more difficult, and more specialized forms of COSY, e.g., Exclusive COSY (E.COSY) or Primitive Exclusive COSY (PE.COSY), may be useful on occasion *(14)*. It may also be advantageous to exchange all amide protons by ^{2}H, to simplify the coupling pattern at H^α.

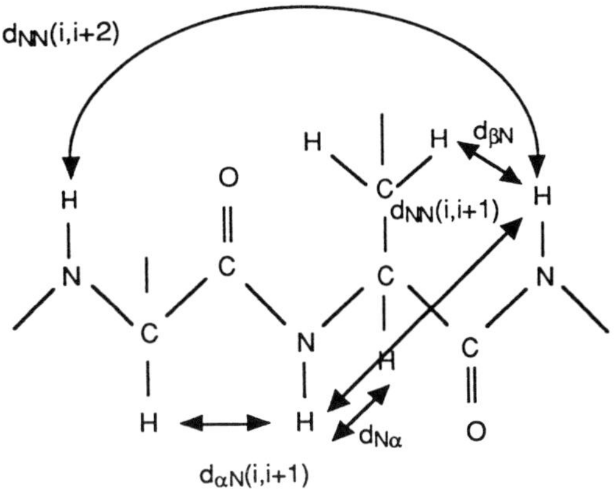

Fig. 7. Section of a peptide showing some NOE connectivities that may be observed, and the nomenclature used to describe these connectivities. If a particular distance, e.g., between two H^N is short, then a strong NOE may be observed between those protons, and a d_{NN} is said to be present.

3.4.3. Amide Proton Exposure to Solvent

If large peptides or small proteins are dissolved in ^2H_2O, those H^N that are exposed to solvent will disappear from the spectrum. Buried or H-bonded H^N will also disappear with time, as molecular breathing motions gradually allow solvent into the core of the molecule. Recording spectra over a period of time ranging from a few minutes to a day (or days) after dissolution will allow estimation of the degree of exposure of protons to solvent. For peptides of less than 30 residues, it is unlikely that any core exists, and all H^N may have exchanged before a spectrum can be obtained. In this case, a better approach is to measure temperature coefficients for the chemical shifts of H^N. Spectra can be recorded at 5–10° intervals over the range 20–55°C. The chemical shifts of exposed H^N should show a large variation with temperature. In addition, the resulting data should give a good linear fit; lack of this is strong evidence for a change in conformation over the temperature range examined.

3.5. Conformational Data Analysis

The complete *de novo* calculation of a three-dimensional structure, using NMR data, has been carried out in a few cases (Section 3.5.2.). A more modest approach, trying to match the patterns observed in the NOE, coupling constant, and solvent exposure data with those expected from known structures, using listings such as Table 1, is more widely applicable. Any conclusions must be consistent with all the data, including that from other techniques such as circular dichroism (CD) (*see* Chapter 15) and Fourier transform infra-red (FTIR)

Table 1
NMR Parameters Useful for Secondary Structure Identification: Interproton Distances[a], Coupling Constants, and Amide Proton Accessibility to Solvent

Parameter	β-Antiparallel sheet	α Helix	3_{10} Helix	Type I β turn[b]	Type II β turn[b]	γ turn
$d_{N\alpha}(i,i)$	2.8	2.6	2.6	2.7, 2.9	2.8, 2.3	2.3
$d_{NN}(i,i+1)$	4.3	2.8	2.6	2.7, 2.5	4.5, 2.5	3.9
$d_{NN}(i,i+2)$		4.2	4.1	3.8	4.3	
$d_{\alpha N}(i,i+1)$	2.2	3.5	3.4	3.4, 3.2	2.2, 3.3	3.6
$d_{\alpha N}(i,i+2)$		4.4	3.8	3.6	3.3	
$d_{\alpha N}(i,i+3)$		3.4	3.3			
$d_{\alpha N}(i,i+4)$		4.2	>4.5			
$d_{\alpha\beta}(i,i+3)$		2.5–5.0	3.1–5.0			
$^3J_{HN\alpha}$ (Φ)	≥9.0 Hz (–139°)	3.9 Hz (–57°)	4.2 Hz (–60°)	4 Hz (–60°) 9 Hz (–90°)	4 Hz (–60°) 5 Hz (90°)	6.7 (–80°)
NH exchange/ Δδ/ΔT	slow/low for alternate NHs	slow/low	slow/low	slow/low for residue 4	slow/low for residue 4	slow/low for residue 3

[a]All distances quoted in Å.

[b]Where two distances are quoted, the first refers to the distance between residues 2 and 3 of a turn, the second to that between residues 3 and 4. The distances $d_{\alpha N}(i,i+2)$ and $d_{NN}(i,i+2)$ are between residues 2 and 4 and the two J values refer to residues 2 and 3, respectively.

spectroscopy. Detailed accounts of peptide and protein NMR and analysis are available *(13–16)*.

3.5.1. Analysis of NMR Data

The first step is to look for patterns in the $d_{\alpha N}(i,i+1)$ and $d_{NN}(i,i+1)$ crosspeaks in NOESY spectra obtained with as short a mixing time as possible. Further points concern *cis/trans* isomers and side-chain conformations.

3.5.1.1. A Run of Strong Sequential $d_{\alpha N}(i,i+1)$ Crosspeaks

This is indicative of an extended structure that may be β sheet if the following criteria are met:

1. There are several such sequences.
2. $^3J_{HN\alpha} \geq 9.0$ Hz for the residues in each sequence.
3. $d_{NN}(i,i+1)$ crosspeaks are weak or unobserved over the sequences.
4. Low amide exchange rates and/or temperature coefficients are observed for alternate residues in the strands on the edge of the sheet.

In addition, there should also be $d_{\alpha N}$, $d_{\alpha\alpha}$, and d_{NN} crosspeaks between residues on neighboring strands of the β sheet.

3.5.1.2. A Run of Strong Sequential $d_{NN}(i,i+1)$ Crosspeaks

This is indicative of a helical structure. Further criteria are:

1. $d_{\alpha N}(i,i+1)$ crosspeaks will also be present for these residues but should be of lower intensity than the $d_{NN}(i,i+1)$ and the intraresidue $d_{N\alpha}(i,i)$ crosspeaks.
2. Medium range crosspeaks should be present for these residues, e.g., $d_{\alpha N}(i,i+3)$ and $d_{\alpha\beta}(i,i+3)$ for an α helix and $d_{\alpha N}(i,i+2)$ and $d_{\alpha N}(i,i+3)$ for a 3_{10} helix. These will be weak and may only appear at longer mixing times.
3. These residues should show a nearly uniform $^3J_{HN\alpha} < 5.0$ Hz.
4. The amide protons of the same residues should show slow exchange rates and/or low temperature coefficients. Residues in the first turn of a helix are not involved in helical H-bonding and may therefore have normal values.

3.5.1.3. A Single $d_{\alpha N}(i,i+2)$ or Two Consecutive $d_{NN}(i,i+1)$ Crosspeaks

Such crosspeaks are characteristic of some type of fold in the peptide chain. Table 1 shows that a $d_{NN}(i,i+1)$ is expected between residues 2 and 3 of a Type I β turn and between residues 3 and 4 of both Type I and Type II β turns. In addition, a strong $d_{\alpha N}(i,i+1)$ is expected between residues 2 and 3 of a Type II β turn and a $d_{\alpha N}(i,i+2)$ is expected between residues 2 and 4 of all types of bend.

3.5.1.4. A Run of Sequential $d_{\alpha N}(i,i+1)$ and $d_{NN}(i,i+1)$ Crosspeaks

While $d_{\alpha N}(i,i+1)$ crosspeaks are characteristic of extended structures and $d_{NN}(i,i+1)$ of bends or helices, the presence of both points to a random coil.

This is not a single conformation, but a dynamic situation in which each residue spends most time in extended but also some time in folded conformations. Tabulations of chemical shift data for residues in random coil conformation in DMSO, water, and trifluoroethanol have been published *(2,13,17)*.

3.5.1.5. *Cis* and *Trans* Peptide Bonds to Proline

These can be distinguished using NOE, since $d_{N\delta}$ or $d_{\alpha\delta}$ connections are characteristic of a *trans* isomer and $d_{N\alpha}$ or $d_{\alpha\alpha}$ connections of a *cis* isomer *(13)*.

3.5.1.6. Side-Chain Conformations

Side-chains are more mobile than the backbone, and conformations will not be well defined except in larger peptides. Before attempting a detailed analysis, a minimum requirement is that diastereotopic protons have different δ (and ^3J) values. A rough estimate of both $^3J_{\alpha\beta}$ and NOEs to H^β is often sufficient to identify dominant side-chain conformations *(14,16)*.

3.5.2. Computer Modeling Using NMR Data

Analysis of NMR spectra, as suggested in Section 3.5.1., may identify several elements of secondary structure, but the question remains of how these elements fit together. It is possible to use some form of computation, e.g., distance geometry and/or molecular dynamics (MD), or simulated annealing (SA), using constraints based mainly on NOE and coupling constant data, to define the overall structure. This has worked well for large peptides and small proteins. Many neuropeptides, being small and flexible, will not yield sufficient or precise enough data to define a complete three-dimensional structure. The best option is to build models, consistent with the available data and then to run simulations (MD or SA) to look for all the accessible conformations. An introduction to these techniques is given in Chapter 18.

3.5.3. Conformational Averaging

Many neuropeptides are flexible and exist as ensembles of conformations, undergoing rapid interconversion. The NMR parameters observed will all be averaged values and will not relate to a real conformation *(18)*, the most likely average structure being random coil. Averaging affects the different NMR parameters in different ways and some may still yield useful information.

1. Chemical shift: A contribution from nonrandom structures may be inferred if δ values differ significantly (>0.4 ppm for H^N and >0.2 ppm for other protons) from those listed for random coil *(1,13,17)*, or if δ values from multiple copies of the same amino acid in the peptide or diastereotopic protons (especially Gly H^α) differ significantly.

2. NOE: Even with conformational averaging, distinct NOEs may be observed from individual conformations. Intensities will now depend on the population of each conformation, as well as on the respective interproton distances.
3. Coupling constants/amide temperature coefficients. Extreme values of J ($\leq$5.0 or $\geq$9.0 Hz) and temperature coefficients ($\leq$3.0 or $\geq$6.0 ppb/°C) are fairly easily interpreted, but intermediate values (produced by averaging) are not.

If one conformation cannot account for all the data, the problem then is to decide how many and in what proportions. Efforts are being made to establish computational methodologies for handling multiple conformations *(19,20)*.

3.6. Conclusion

NMR spectroscopy can provide information about the solution conformations of peptides in detail not available from other techniques. Using NMR, not only can elements of structure be identified, as, e.g., with CD and FTIR, but they can also be localized fairly precisely to particular residues. This is because individual resonances in the spectrum can be assigned to individual protons and, in this respect, NMR approaches the resolving power of X-ray crystallography.

This chapter has dealt solely with the use of ^{1}H NMR, though use has been made of other nuclei in studying peptides (*see* Chapter 17). It has also ignored most of the technicalities of acquiring and processing NMR data. Recent developments *(1)*, such as the use of the maximum entropy method in place of the Fourier transform or the use of field gradients for improved resolution and solvent suppression, although possibly yielding better quality spectra, are of secondary interest to the interpreter. If the spectra obtained are disappointing in some way, consult with the spectrometer operator, and explain carefully what you are looking for. He or she may be able to make suggestions that would increase the likelihood of getting the desired result (*see* Note 9). Finally, the reader's attention is drawn to some examples of the application of NMR to neuropeptides *(21–25)*.

4. Notes

1. It follows that the population of the ground state is only slightly greater than that of the excited state (by approx 1 part in 10^5 at 400 MHz). This places an important limitation on the intensity of NMR spectra.
2. When different conformations of small- to medium-sized peptides, representing different sets of torsion angle (Φ, Ψ, and χ) values, are in equilibrium, the rates of interconversion are usually sufficiently fast that only averaged spectra are seen. By contrast, with *cis/trans* isomers with $\omega = \pm 180°$, which are quite common for peptide bonds to Pro or other N-substituted amino acids, the rates of interconversion are usually slow enough to permit resonances from both isomers to be seen.
3. As solvent is present in large excess, if amide and solvent protons undergo fast exchange, the weighted average signal that is observed will be indistinguishable

from that of the solvent. This occurs in water at pH ≥ 6.5 so that studies intending to make use of amide resonances must be carried out at a pH below this value.

4. Certain aspects of the plotted spectrum are under operator control. The spacing between contour lines can be decreased so that more lines are plotted and the intensities of different crosspeaks can be more accurately distinguished. The lowest contour level shown is controlled by a floor parameter, designed to prevent peaks being lost in a sea of noise. Occasionally the routine value used for this parameter will cause a weak crosspeak to be missed. Judicious lowering of the floor may permit observation of such a weak peak. If doubts exist as to whether a weak peak is a real peak or a large noise spike at that position, the spectrum can be acquired over again. Random noise should not give the same peak in two spectra.

   Less under operator control is the appearance of noise in a two-dimensional spectrum. Instrumental instabilities (e.g., instability in the lock signal or minor variations in pulse timings or separations) during the course of a long acquisition mean that successive FIDs may differ by more than the intended variation in t_1. This causes noise, different from thermal noise, which is limited to the f_1 dimension. It is most apparent for strong peaks, and solvent and methyl signals are frequently accompanied by ridges of noise, called t_1 noise, running parallel to the f_1 dimension.

5. With DQF-COSY and TOCSY spectra, whether all possible crosspeaks appear depends critically on the values chosen for some parameters in the pulse sequence. Finding optimum values for these parameters is difficult or even impossible, so missing crosspeaks must be expected from time to time. Commonly, in DQF-COSY, when one proton is coupled to two diastereotopic protons (as with $CHCH_2$ in Phe or $NHCH_2$ in Gly), two crosspeaks are expected, but one may be very weak or even totally missing. In TOCSY, only part of a long spin system (e.g., those connecting the H^N with H^δ in Leu or Lys) may show crosspeaks.

6. The ROESY pulse sequence is related to the TOCSY sequence as both use a so-called spin-locking pulse and TOCSY crosspeaks may appear in the ROESY spectrum. If a genuine ROESY crosspeak appears between protons A and B in two spin systems, false crosspeaks may appear between A and other protons in the spin system of B. The presence of these false peaks depends critically on the position of the spin-locking pulse and if the spectrum is acquired again with this pulse shifted by several hundred Hz, the false peaks should disappear. Generally, experienced users seem to prefer to use NOESY spectra when possible.

7. The most commonly used sample tubes are of 5-mm od (and hold approx 0.5 mL). Tubes (and probes) of 10-mm od are not uncommon, but generally give slightly poorer resolution and so are only to be recommended if sample solubility limits concentration and a larger total sample is needed to compensate. Again, for the highest resolution, high-quality tubes should be used (the NMR center running your sample should either supply them or recommend a suitable supplier). Such a tube merits high-quality treatment. It should not be heated above 50–60°C, to

avoid distorting the glass. An NMR tube eventually needs cleaning because of deposits of peptide or protein. It is best to use a proprietary cleansing solution (e.g., Decon), followed by extensive rinsing. Chromic acid should be avoided to prevent contamination by paramagnetic chromium ions. Afterward, the tube can be dried by a stream of dry gas (e.g., N_2) or gentle heating. Any volatile organic solvent used as a rinse must be free of involatile residues.

8. Some peptides will aggregate at these concentrations and a diluted sample (approx 5–10-fold) should also be run to check that this is not happening, i.e., that the one-dimensional spectrum does not change with concentration.
9. These could include other ways of processing the existing data, altered parameters in the pulse sequences, or conditions (e.g., temperature, concentration, pH, or solvent) for the acquisition of new data.

References

1. Rattle, H. (1995) *An NMR Primer for Life Scientists.* Partnership Press, Fareham, UK.
2. Bundi, A., Grathwohl, C., Hochman, J., Keller, R. M., Wagner, G., and Wüthrich, K. (1975) Proton NMR of the protected tetrapeptides TFA-Gly-Gly-L-X-L-Ala-OCH_3, where X stands for one of the 20 common amino acids. *J. Mag. Res.* **18,** 191–198.
3. Abraham, R. J., Fisher, J., and Loftus, P. (1988) *Introduction to NMR Spectroscopy.* Wiley, Chichester, UK.
4. Pardi, A., Billeter, M., and Wüthrich, K. (1984) Calibration of the angular dependence of the amide proton-Cα proton coupling constant $^3J_{HN\alpha}$ in a globular protein. *J. Mol. Biol.* **180,** 741–761.
5. Neuhaus, D. and Williamson, P. (1989) *The Nuclear Overhauser Effect in Structural and Conformational Analysis.* Verlag Chemie, Weinheim, Germany.
6. Sanders, J. K. M. and Hunter, B. K. (1993) *Modern NMR Spectroscopy, a Guide for Chemists,* 2nd ed. Oxford University Press, Oxford, UK.
7. Derome, A. E. (1987) *Modern NMR Techniques for Chemistry Research.* Pergamon Press, Oxford, UK.
8. Oppenheimer, N. J. and James, T. L., eds. (1989) Nuclear magnetic resonance, Pt A, Spectral techniques and dynamics. *Methods Enzymol.* **176.**
9. Hore P. J. (1989) Solvent suppression. *Methods Enzymol.* **176,** 64–77.
10. Motta, A., Picone, D., Tancredi, T., and Temussi, P. A. (1987) NOE measurements on linear peptides in cryoprotective aqueous mixtures. *J. Mag. Res.* **75,** 364–370.
11. Amodeo, P., Motta, A., Picone, D., Salviano, G., Tancredi, T., and Temussi, P. A. (1991) Viscosity as a conformational sieve. NOE of linear peptides in cryoprotective mixtures. *J. Mag. Res.* **95,** 201–207.
12. Verheyden, P., De Wolf, E., Jaspers, H., and Van Binst, G. (1994) Comparing conformations at low temperature and at high viscosity. *Int. J. Peptide Protein Res.* **44,** 401–409.
13. Wüthrich, K. (1986) *NMR of Proteins and Nucleic Acids.* Wiley, New York.

14. Evans, J. N. S. (1995) *Biomolecular NMR Spectroscopy.* Oxford University Press, Oxford, UK.
15. Dyson, H. J. and Wright, P. E. (1991) Defining solution conformations of small linear peptides. *Ann. Rev. Biophys. Biophys. Chem.* **20,** 519–538.
16. Case, D. A., Dyson, H. J., and Wright, P. E. (1994) Use of chemical shifts and coupling constants in nuclear magnetic resonance structural studies on peptides and proteins. *Methods Enzymol.* **239,** 392–416.
17. Merutka, G., Dyson, H. J., and Wright, P. E. (1995) Random coil H-1 chemical-shifts obtained as a function of temperature and trifluoroethanol concentration for the peptide series GGXGG. *J. Biomolec. NMR* **5,** 14–24.
18. Jardetsky, O. (1980) On the nature of molecular conformations inferred from high-resolution NMR. *Biochim. Biophys. Acta* **621,** 227–232.
19. Mierke, D. F., Kurz, M., and Kessler, H. (1994) Peptide flexibility and calculations of an ensemble of molecules. *J. Am. Chem. Soc.* **116,** 1042–1049.
20. Cicero, D. O., Barbarato, G., and Bazzo, R. (1995) NMR Analysis of molecular flexibility in solution: a new method for the study of complex distributions of rapidly exchanging conformations. *J. Am. Chem. Soc.* **117,** 1027–1033.
21. Guthrie, D. J. S., Geraghty, R. F., Irvine, G. B., and Williams, C. H. (1994) Conformational studies on analogues of the invertebrate neuropeptide pyroGlu. Asp.Pro.Phe.Leu.Arg.Phe.amide, using ^{1}H NMR. J. *Chem. Soc. Perkin Trans* **2,** 1239–1245.
22. Horne, J., Sadek, M., and Craik, D. J. (1993) Determination of the solution structure of neuropeptide-K by high-resolution nuclear-magnetic-resonance spectroscopy. *Biochemistry* **32,** 7406–7412.
23. Morris, M. B., Ralston, G. B., Biden, T. J., Browne, C. L., King, G. F., and Iismaa, T. P. (1995) Structural and biochemical studies of human galanin: NMR evidence for nascent helical structures in aqueous solution. *Biochemistry* **34,** 4538–4545.
24. Mierke, D. F., Durr, H., Kessler, H., and Jung, G. (1992) Neuropeptide-Y-optimized solid-phase synthesis and conformational-analysis in trifluoroethanol. *Eur. J. Biochem.* **206,** 39–48.
25. Breeze, A., Harvey, T. S., Razzo, R., and Campbell, I. D. (1991) Solution structure of human calcitonin gene-related peptide by ^{1}H NMR and distance geometry with restrained molecular dynamics. *Biochemistry* **30,** 575–582.

17

The Study of Membrane- or Receptor-Bound Neuropeptides by NMR

Rickey P. Hicks

1. Introduction

An X-ray crystal structure of a neuropeptide bound to its native receptor protein would provide a wealth of information concerning the structural requirements for ligand-receptor binding. Unfortunately, many neuropeptide-protein complexes are very difficult, if not impossible, to obtain as a single crystal. This difficulty prohibits the use of X-ray crystallography to determine the structure of neuropeptide-receptor complexes. Other than X-ray crystallography, nuclear magnetic resonance spectroscopy (NMR) is the only spectroscopic technique that can provide structural information at atomic resolution *(1)*. The very large size of most neuropeptide receptor proteins (300–600 amino acid residues) makes determination of the three-dimensional structure of the neuropeptide-receptor complex by NMR also very difficult. To date, the use of sequence-specific resonance assignment methods has been limited to uniformly isotope labeled proteins containing 100–150 amino acid residues *(1)*. Owing to overlapping resonances from the receptor, determination of the three-dimensional structure of only the bound ligand may also require uniform isotopic labeling (^{15}N, ^{13}C) of the neuropeptide. In light of the difficulties involved in the direct observation of neuropeptide-receptor complexes, other indirect methods must be employed to obtain information concerning possible biologically active conformations of a neuropeptide. One method extensively used by medicinal chemists involves the preparation and biological evaluation of a large number of conformationally restrained analogs of the neuropeptide. The three-dimensional structures of analogs with good and bad biological activity are then determined by NMR. These structures are then analyzed in terms of the observed biological activity in order to provide insight into the biologically

From: *Methods in Molecular Biology, Neuropeptide Protocols*
Edited by: G. B. Irvine and C. H. Williams Humana Press Inc., Totowa, NJ

active conformations of the neuropeptide. This approach is effective, but it is very time consuming, requiring the syntheses and biological evaluation of a large number of compounds. The conformational analysis of neuropeptides themselves has provided limited structural information. Most neuropeptides are small (<25 amino acid residues) linear polypeptides lacking a defined secondary structure in aqueous environment *(2,3)*.

Spectroscopic investigation of neuropeptides in the presence of membrane model systems has been conducted in order to determine conformations that may exist in the local chemical environment of the receptor site. This approach is based on the two major functions biological membranes are believed to serve in the binding of a neuropeptide to a membrane-embedded receptor *(4,5)*. The first is to increase the concentration of the neuropeptide near the receptor *(4,5)*. The second is to induce a specific conformation onto the polypeptide backbone of the neuropeptide prior to its interacting with its receptor *(4,5)*. This induced conformation may be termed the prebinding conformation. The relationship of the prebinding conformation to the receptor-bound conformation of the neuropeptide is currently unknown. The study of neuropeptides in the presence of membrane model systems has, however, provided a great deal of insight into the conformational requirements for the biological activity of several peptides. Gierasch and coworkers have reported that a series of six peptides related to the *Escherichia coli* ompA signal peptide adopts a high degree of α helical structure in 1:1 trifluoroethanol/water and in the presence of sodium dodecyl sulfate (SDS) micelles *(6)*. The percentage of α-helical content of these peptides could be correlated directly with the observed biological activity within this series of neuropeptides *(6)*.

We have determined that the neuropeptide substance P (Arg-Pro-Lys-Pro-Gln-Gln-Phe-Phe-Gly-Leu-Met-NH_2) in the presence of SDS micelles is involved in a conformational equilibrium between an α and a 3_{10} helix involving the mid region (Pro^4-Gln^5-Gln^6-Phe^7-Phe^8) of the peptide *(3)*. This conformation is consistent with the structure of a NK-1 selective agonist proposed by Convert *(7)*.

Other examples in which the studies of neuropeptides in the presence of membrane models have provided valuable insight into biological activity include: melittin in dimyristoylphosphatidylcholine (DMPC), phosphatidylethanolamine *(8,9)* and dodecylphosphocholine (DPC) micelles *(10,11)*, enkephalin in SDS and lysophosphatidyl choline (LPC) micelles *(12–15)*, and gramicidin A in SDS and DMPC *(16–18)*.

Kyle and coworkers used the conformation adopted by bradykinin in the presence of SDS micelles as a probe molecule to locate the active site of a putative model of the bradykinin B_2 receptor *(19,20)*. A model of the idealized bradykinin B_2 receptor was created based on homology modeling to the known crystallographic structure of bacteriorhodopsin *(21)*. The conformation adopted by bradykinin in the presence of SDS micelles was then tumbled through the model of the bradykinin receptor looking for sites of potential interactions *(19)*.

1.1. Membrane Model Systems

Owing to the amphipilic structure of membranes, peptide-membrane interactions are governed by the free energy transfer of the peptide going from an aqueous to a hydrophobic phase *(22)*. These interactions include: surface interactions owing to head groups or counterions, interfacial adsorption at the micelle-water boundary, partitioning into the hydrophobic region, and comicellizations owing to both hydrophobic and hydrophilic interactions *(22,23)*. The two most common classes (anionic and zwitterionic) of biological membranes may be represented by the anionic membrane model systems SDS and lysophosphatidylglycerol (LPG), as well as the zwitterionic models DPC and LPC *(23)*. All four of these model systems have been used in NMR conformational studies of small peptides *(2,8–18,24,25)*.

The physical and chemical interactions that occur between neuropeptides and membrane model systems are very complex *(3)*. The neuropeptide membrane complex exists as complex equilibrium between the free and bound forms of the neuropeptide (*see* Fig. 1). The observed NMR spectrum represents the time-averaged conformations of all structures that contribute to the equilibrium. In order to properly interpret the NMR data obtained from the investigation of neuropeptides in the presence of membrane model systems, it is necessary to develop an understanding of the interactions taking place between the neuropeptide and the membrane model. The interactions between membrane model systems and neuropeptides, which define the position of the equilibrium, may be broadly characterized as external or internal. An external interaction may be defined as the effect of the membrane model on the properties of the local chemical environment of the solvent, such as viscosity, dielectric constant, pH, and ionic strength *(3)*. An internal interaction requires surface contact between the micelle and the neuropeptide with the hydrophobic side-chains on the neuropeptide penetrating or inserting into the hydrophobic core of the micelle *(3)*. Understanding whether the interaction between a neuropeptide and a membrane is internal or external allows for the determination of the resulting conformational freedom of the neuropeptide. Neuropeptides that interact with a membrane model system internally will be conformationally restrained (owing to the insertion of the hydrophobic side-chains on the neuropeptide into the hydrophobic core of the micelle), whereas neuropeptides that interact externally will exhibit much greater conformational and motional freedom.

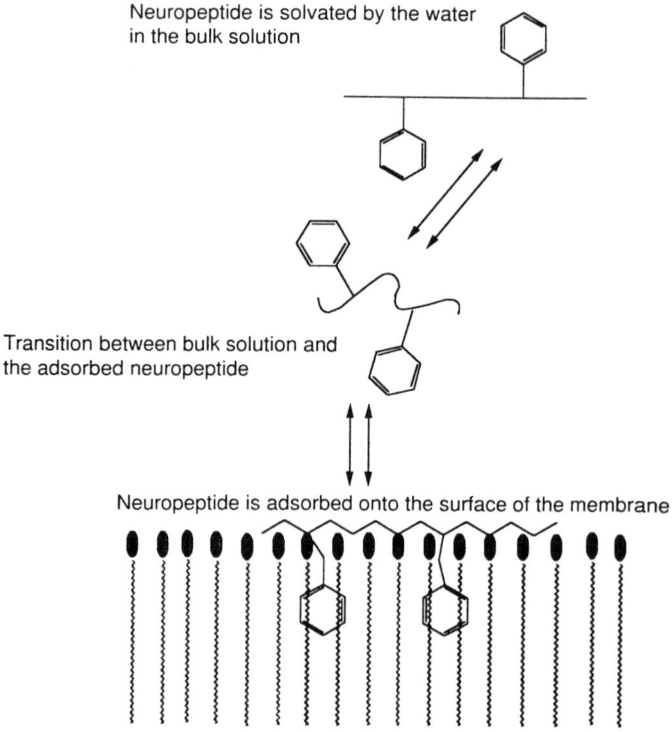

Fig. 1. Cartoon representation of the equilibrium between the free and bound forms of the neuropeptide, in the neuropeptide-membrane model complex.

1.1.1. Characterization of Interactions

Proton longitudinal relaxation studies may be used to make qualitative deductions to help understand the interactions that occur between micelles and neuropeptides *(2)*. For example, proton longitudinal relaxation studies on 30 m*M* SDS (8.3 m*M*, cmc [critical micelle concentration]) have shown that the relaxation times for the protons on SDS increase as one proceeds toward the terminal methyl group *(2)*. This trend is consistent with what has been previously reported and is attributed to a decrease in molecular motions at the center of the micelle *(26,28)*. Neuropeptides that interact internally with SDS decrease the relaxation times for all of the side-chain protons on SDS *(2)*. From the work of Keniry and Smith *(26)*, it is apparent that, if a peptide interacts hydrophobicaly (internally) with a membrane, the longitudinal relaxation time will decrease for the terminal nuclei on the hydrocarbon chain. These results are consistent with what has been reported for several intrinsic membrane proteins that uniformly reduced the relaxation times for lipid bilayers *(26–28)*. Two-dimensional nuclear Overhauser enhancement spectroscopy (NOESY) spectra collected at long mixing times (600 ms) may be used to confirm the insertion of the phenyl rings or alkyl groups on the neuropeptide into the hydrophobic core of the micelles by observation of cross relaxation between the side-chain protons of the membrane model and the side-chain protons on the neuropeptide *(2)*.

1.1.2. Selection of Model Systems

The selection of the appropriate membrane model system for the particular neuropeptide under investigation is a difficult task. The solubility of the neuropeptide in the particular membrane model must be determined. Some neuropeptides that exhibit very good solubility in water will exhibit very poor solubility in the presence of 10 m*M* SDS or DPC, buffered to pH 4.0. Solubility in phospholipid vesicles may also be a problem. For NMR studies of a neuropeptide in the presence of a membrane model system, a concentration of at least 1.0 m*M* is required.

If solubility of the neuropeptide is not a problem, which membrane model is most appropriate? The ideal membrane model systems are derived from phospholipid vesicles. However, the long correlation times associated with these systems limit their application in high-resolution NMR experiments *(3,6)*. The SDS and DPC micelles offer reasonable linewidths and have been used extensively as simple membrane model systems for the investigation of the interactions that occur between polypeptides and membranes *(3,6,17)*. The observed conformations for many polypeptides in SDS are very similar to the conformations observed in phospholipids *(3,6,19,21)*. Lee and coworkers reported that the neuropeptide bradykinin adopts a β turn involving the C terminal residues 6–9 in the presence of 5 molar excess SDS micelles *(29)*, whereas we subsequently reported that bradykinin in the presence of 7.4 m*M* LPC adopts a type IV β turn involving residues 6–9 *(30)*.

1.2. Receptor Models

The use of subunits of a receptor to investigate biochemical questions has been reported by the groups of Reddy *(31)* and Armitage *(32)*. Reddy and coworkers prepared several fragments or domains of the α factor mating pheromone of yeast *Saccharomyces cerevisiae (31)*. The structure of each of these fragments was characterized by CD spectroscopy. The resulting information was used to build a partial putative model of the receptor *(31)*. Armitage and coworkers showed that amino acid residues 173–204 of the α subunit of the nicotinic acetylcholine receptor act as a major recognition site of the native ligand-binding site *(32)*. They then used short, 18–20 amino acid residue polypeptides to model the ligand binding site. Bommakanti and coworkers reported the use of a series of small polypeptides (15–20 amino acid residues) acting as receptor mimetic peptides to map the binding site of the G_{12} and N-formyl peptide receptor of human neutrophils (FPR) *(33)*. Previously, synthetic peptides that corresponded to recognition sites or binding sites on G-proteins were shown to interfere with the physical and functional receptor-G-protein coupling in FPR as well as with other G-protein coupled receptors (GPCR) systems *(34,35)*. Peptides that are effective in blocking GPCR-G protein interactions are believed to mimic interfacial contact sites between the two proteins *(33–35)*. This peptide competition paradigm is now used as a standard method to identify protein interacting pairs and thus recognition sequences *(35)*. The acceptance of the use of short polypeptides corresponding to regions believed to be involved with binding, to locate contact or binding sites, may soon be applied to NMR experiments to model the interaction of recognition sites of neuropeptide receptors.

Most noncovalent complexes of neuropeptides and their corresponding recognition sites would yield NMR spectra that are distinct from the simple sum of the spectra of the disassociated components. The appearance of peaks in the spectrum may be used to determine the rate of exchange. For example, in the slow exchange complex, peaks arising from the unbound form would disappear and new peaks arising from the bound form would appear as the peptide binds to the proteins *(36)*.

1.3. Nuclear Magnetic Resonance Spectroscopy

A review of the technique of sequence-specific resonance assignment and the necessary homonuclear two-dimensional experiments to complete this task are given in Chapter 16 (*see* Note 1). A short discussion of the problems related to the NMR investigation of neuropeptides in the presence of a membrane model will be presented here.

The measurement of nuclear overhauser enhancement (NOE) buildup rates provides a means to measure interproton distances. Several potential problems exist, if one is not careful, in the use of this NOE data. Spin diffusion is a phenomenon associated with T_1 relaxation and may be observed in NOESY studies of macromolecules (*see* Note 2).

The quantification of NOE intensity is a major concern. It must be recognized that in the presence of a membrane model system, there is a possibility that different regions of neuropeptide will experience very different degrees of association with the micelle. Owing to the relatively large size of the micelles (for example, SDS micelles contain 60–100 molecules depending on ionic strength of the solution with a molecular weight range of 18,000–30,000 Dalton) *(3)*, the correlation times in the micellar environment are expected to be much longer than the correlation times observed in an aqueous environment *(2,3,6)*. The net result of the longer correlation times is that the NOEs will be weighted in favor of conformations that are induced by interactions with the micelle, resulting in apparent higher percentage of secondary structure than is actually present *(6)*.

Inverse detection techniques should be used to assign the natural abundance ^{13}C spectrum of a neuropeptide in the presence of a membrane model system by use of the H-C correlation experiments *(37),* heteronuclear multiple quantum coherence (HMQC), and heteronuclear multiple bond coherence (HMBC). Inverse detection experiments are ones in which the chemical shift and coupling information of the insensitive nuclei, in this case ^{13}C, is transferred to the more sensitive nuclei, ^{1}H *(37)*. The advantage of this technique is a great increase in sensitivity (owing to the larger natural abundance of ^{1}H as compared to ^{13}C), meaning fewer scans are required to obtain a signal of sufficient signal to noise. Heteronuclear two-dimensional experiments, unlike homonuclear experiments, do not exhibit a spectrum diagonal—a representation of a standard two-dimensional heteronuclear experiment is given in Fig. 2. The HMQC experiment is used to determine one bond proton-carbon J coupling. The information obtained indicates which protons are directly bonded to which carbons. The HMBC experiment is used to determine two- and three-bond proton-carbon J coupling. The information obtained indicates which protons are two and three bonds away from a particular carbon atom (*see* Note 3). This information is useful in the assignment of nonprotonated carbons. A comparison of data obtained from the HMQC and HMBC experiments is given using the neuropeptide Met-enkephalin in the presence of SDS micelles (*see* Fig. 3).

1.3.1. Analysis of Chemical Shift Information

Wishart and coworkers have reported a simple method for predicting the secondary structures of proteins based on changes in their H^{α} proton chemical shifts *(38)*. In general, the method proceeds as follows. Compare the observed H^{α} proton chemical shift of a particular residue to the random coil chemical shift value ±0.1 ppm for that particular residue. Assign a value of +1 to a residue if its H^{α} proton chemical shift is greater than the range for the random coil value. Assign a value of –1 to a residue if its H^{α} proton chemical shift is less than the range for the random coil value. Assign a value of 0 to a residue if its H^{α} proton chemical shift occurs in the range for the random coil value. A dense grouping of four or more values of –1 that are not interrupted by a value of +1 indicates the presence of an α helix. Any dense group of three or more values of +1 not interrupted by a value of –1 represents a β strand. Even though this method was developed for proteins, it correlates very well with three-dimensional structures determined by NMR for neuropeptides *(3,39)*.

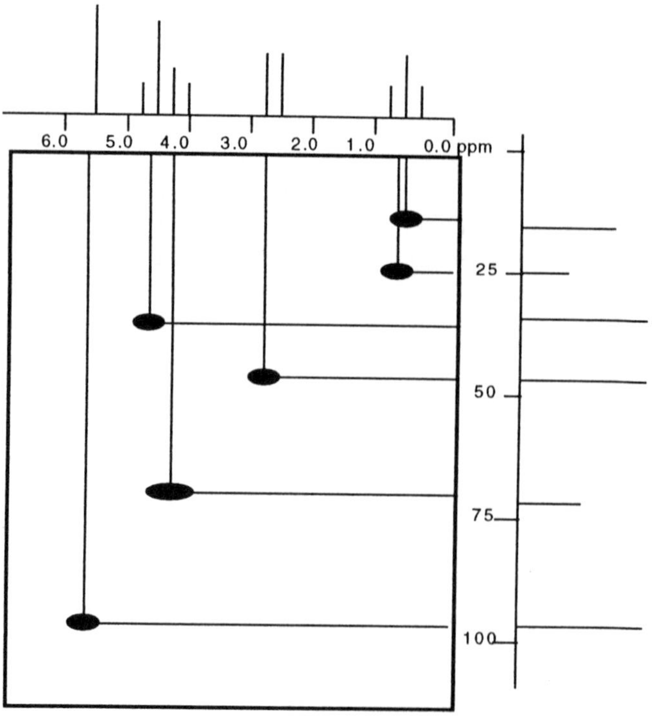

Fig. 2. A general presentation of a two-dimensional heteronuclear experiment. In general, the horizontal axis represents proton chemical shifts, whereas the vertical axis represents carbon-13 chemical shifts. The proton-carbon connectivities are indicated by the lines connecting a resonance on the proton spectrum with a resonance on the carbon spectrum through the cross peak.

Spera and Bax *(40)* have reported a method for predicting the secondary structure of a peptide or protein based on a comparison of the observed α and β carbon-13 chemical shifts to the random coil carbon-13 chemical shift values. Bax reported that, for an α helix, the average secondary shift for C^{α} carbons is 3.09 ± 1.00 ppm and the secondary shift for C^{β} carbons is –0.38 ± 0.85 ppm. For β sheets, the average secondary shift for C^{α} carbons is –1.48 ± 1.23 ppm and the secondary shift for C^{β} carbons is 2.16 ± 1.91 ppm. In general, the secondary shift for an α helix is down field for the C^{α} carbons and up field for the C^{β} carbons; for a β sheet, the secondary shift is up field for the C^{α} carbons and down field for the C^{β} carbons.

Analysis of chemical shift information using the methods discussed above can provide a rapid evaluation of the solution conformation of a neuropeptide in the presence of a membrane model system. The results from these analyses can be used to decide if molecular modeling studies are warranted.

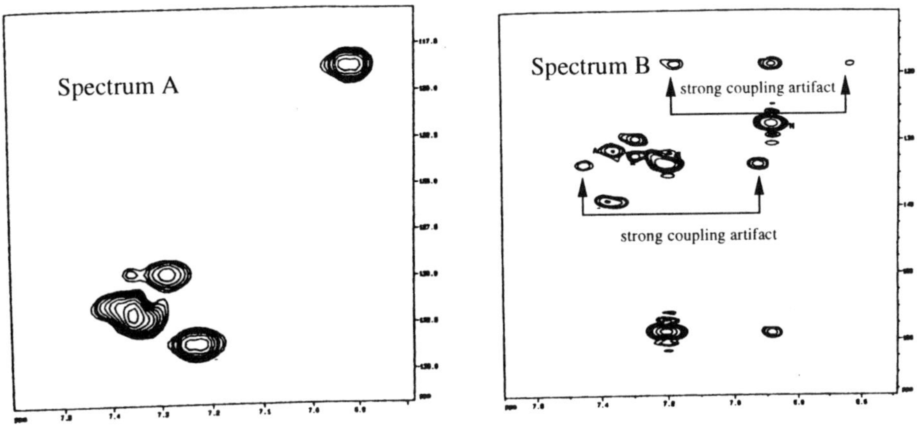

Fig. 3. The aromatic region of **(A)** the ^{1}H-^{13}C HMQC spectrum and **(B)** the ^{1}H-^{13}C HMBC spectrum of 9.0 m*M* methionine-enkephalin in the presence of 50 m*M* SDS at pH 4.10. Notice in spectrum B an increase in the number of signals owing to long range coupling. Also notice in Spectrum B the effect of strong coupling.

1.3.2. Conformational Averaging

Conformational averaging (*see* Chapter 16) is a major problem in NMR investigations of neuropeptides in the presence of membrane model systems. Great care must be taken in the use of interproton distances determined from NOESY and/or rotating frame overhauser enhancement spectroscopy (ROESY) experiments for neuropeptide-micelle complexes as distance constraints in molecular modeling calculations (*see* Note 4).

1.3.3. NMR Investigations of Neuropeptide-Receptor Complexes

Owing to the large size of most neuropeptide-receptor complexes and the resulting severe spectral overlap, normal two-dimensional NMR experiments cannot be employed. One of the most effective NMR methods to determine the three-dimensional structures of ligand receptor complexes involves the use of isotope-editing techniques *(41)*.

Two and three-dimensional ^{15}N and ^{13}C edited homonuclear Hartman-Hahn (HOHAHA) and NOESY experiments are used to assign ^{1}H, ^{13}C, and ^{15}N chemical shifts, and to determine interproton distances *(41–43)*. Three-dimensional NMR experiments require more time for data acquisition than the corresponding two-dimensional experiments, but they yield a dramatic increase in spectral dispersion that makes them the experiments of choice *(41–43)*. The increased spectral dispersion is of particular importance in the investigation of peptides and proteins that adopt an α helix. Owing to the nature of the α helix, the chemical shift ranges for both the amide protons and the $^{\alpha}H$ protons are much smaller than those observed for β sheets *(42)*.

The first step in the assignment of the chemical shifts of interest involves the assignment of the amide proton and nitrogen-15 chemical shifts *(42)*. These assignments are made using the two-dimensional HOHAHA and ^{1}H-^{15}N heteronuclear single quantum coherence (HSQC) experiments on the labeled neuropeptide in water. The second step involves the assignment of the backbone amide protons of the labeled neuropeptide bound to the receptor using the two-dimensional ^{1}H-^{15}N HSQC experiment *(42)*. The three-dimensional NOESY-HSQC and three-dimensional NOESY-TOSCY-HSQC experiments are used to define intraresidue, i to i+1, medium, and long-range NOEs, as well as confirm residue sequence assignments *(42)*.

If the complex formed between the labeled neuropeptide and the receptor is weak and the equilibrium favors the free, instead of the bound, neuropeptide, transfer-NOE *(44)* methods must be used to determine the conformation of the bound neuropeptide. This method is based on the chemical exchange mediated transfer of NOEs, which can easily be observed from the bound form to the free form of the ligand. These techniques have been used to obtain structural information on bound ligands. For example Lippens and coworkers *(44)* employed transfer-NOE experiments to determine the conformation of oxytocin bound to bovine neurophysin I. Normally transfer-NOE experiments are conducted with a ligand-to-protein concentration ratio of 10:1 *(44)*.

1.4. Summary

Two-dimensional homonuclear and heteronuclear NMR experiments may be used to determine the three-dimensional structure of neuropeptides in the presence of membrane model systems with little difficulty. These techniques may also be employed to determine the three-dimensional structure of complexes formed by neuropeptides and small models (15–40 amino acid residues) of their receptor recognition sites. The investigation, however, of neuropeptide-receptor complexes is much more difficult. Uniform isotopic labeling of both the neuropeptide and the receptor are required. In addition, the isotope-edited three-dimensional NMR experiments are often required to obtain sufficient information to determine the three-dimensional structure of the neuropeptide-receptor complex owing to severe spectral overlap.

There are many excellent texts related to the use of NMR for the study of biomolecules; I recommend those by Clore and Gronenborn *(45)*, Evans *(46)*, and Wuthrich *(47)*.

2. Materials

There are hundreds of various modifications of the standard NMR experiments, or pulse sequences, in the literature for the assignment of the ^{1}H, ^{13}C, and ^{15}N spectra, and thus the assignment of the resulting three-dimensional structure of polypeptides and proteins. These experiments range from the simple COSY pulse sequence to isotope-edited, gradient-enhanced sequences. To the scientists interested in using NMR to determine the structure of a neuropeptide in the presence of a membrane model system, the selection of which variations of the standard pulse sequences to use is to the first approximation instrument-dependent. The use of gradient-enhanced or triple resonance experiments requires a sophisticated spectrometer. Described here are the basic experiments needed to determine the three-dimensional structure of neuropeptides (<40 residues) in the presence of a membrane model, using an average NMR spectrometer. The general requirements for these experiments are: for homonuclear proton experiments, the spectrometer must be able to perform spin-lock (HOHAHA, TOCSY, and ROESY) experiments; and for heteronuclear experiments, the spectrometer should equipped to perform inverse detection experiments (HMQC and HMBC). A field strength of 300 MHz or larger is recommended; the larger the field strength, the greater the spectral dispersion of the proton resonances.

For the determination of the receptor-bound conformation of a neuropeptide, a field strength of at least 500 MHz is recommended.

Perdeuterated SDS, DPC, and LPC may be purchased from several suppliers including Cambridge Isotopes (Cambridge, UK) and Merck Isotopes (Rahway, NJ).

3. Methods

3.1. Sample Preparation

A small amount (<0.5 mg) of 2,2-dimethyl-2-silapentane-5-sulfonate (DSS) should be used as an internal chemical shift standard for all samples. All NMR studies should be completed as soon as possible because some neuropeptide-membrane model systems will precipitate out of solution over time.

3.1.1. Micelles

Samples of the neuropeptide to be investigated in the presence of micelles should be prepared in the following manner. A total sample volume of 500–700 μL of 1–2 m*M* neuropeptide in presence of 10–15 m*M* SDS or DPC at pH 4.0 (buffered with 30–50 m*M* sodium acetate) is placed in a 5-mm NMR tube.

1. A buffer of 30–50 m*M* sodium acetate in the pH range of 4.0–5.0, in 75% $^{1}H_2O$/25% $^{2}H_2O$ is best for these studies. Presaturation of the water resonance at a pH >5.0 will result in loss of signal intensity owing to rapid exchange of the amide protons with water.

2. The NMR sample should be prepared by dissolving the necessary amount of neuropeptide in 200 μL of buffer.
3. The necessary amount of perdeuterated SDS or DPC should be dissolved with stirring (a very small magnetic stirring bar in a 2-mL vial on a hot plate stirrer works very well) in 300–500 μL of buffer.
4. After 15 min of stirring, the neuropeptide should be transferred dropwise (a syringe works well) to the vial containing the SDS or DPC micelles. The neuropeptide must be added slowly with stirring to prevent aggregate formation and precipitation. Specific concerns for SDS (*see* Note 5) and DPC (*see* Note 6) are given in Section 4.

3.1.2. Phospholipid Vesicles

Sample preparation is the same as used with micelles except as noted. The necessary amount of perdeuterated phospholipid should be dissolved with sonication at 4°C in 300–500 μL of buffer. After 15 min of sonication, the sample of neuropeptide should be transferred dropwise to the vial containing the phospholipid vesicles. The neuropeptide must be added slowly with stirring to prevent aggregate formation and precipitation.

3.2. NMR Experiments

Basic NMR pulse sequences are given to assign the proton and carbon-13 chemical shifts of a 1–2 m*M* solution of a neuropeptide in the presence of a membrane model system. A 300-MHz proton frequency is used as an example, and sweep widths are given in both Hz and ppm to allow for the conversion to other field strengths.

3.2.2. Proton Spectra

One-dimensional proton spectra should be acquired with the standard presaturation one-pulse experiment shown in Fig. 4. A spectral width of 3623.19 Hz (12 ppm) should be used with 16 K data points. A total of 64–256 scans (depending on concentration and linewidths) should be acquired and then processed with an exponential multiplication, defined by a 1-Hz line broadening (*see* Note 7). Solvent suppression can be accomplished by use of a 2-s presaturation pulse set at the frequency of $^{1}H_2O$ (*see* Note 8).

One should expect in the presence of membrane models, severe line-broadening of particularly the amide and aromatic resonances of the neuropeptide. This line broadening is illustrated in Fig. 5, which shows the amide and aromatic region of substance P in water at pH 4, and in the presence of 10 m*M* SDS at the sample pH.

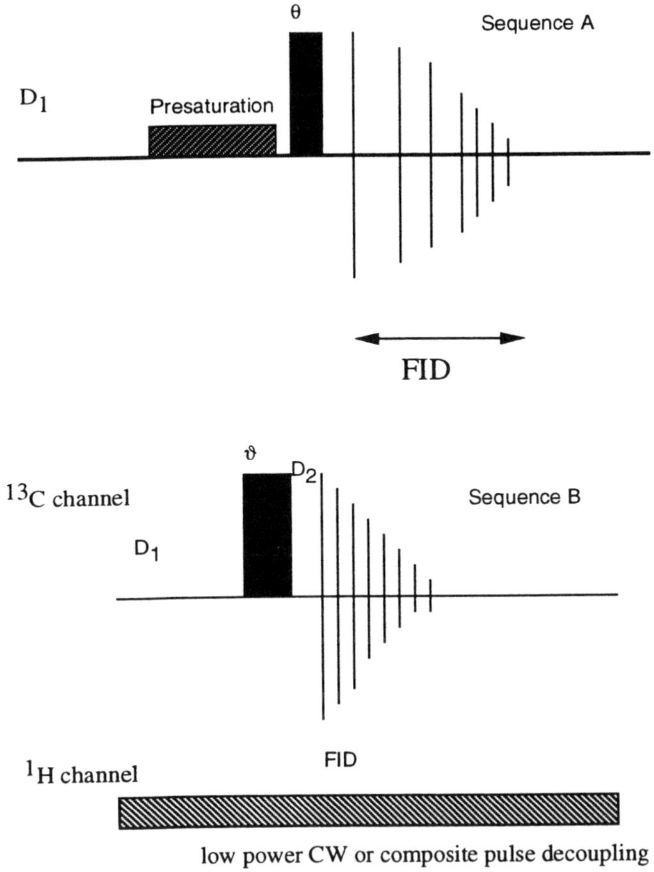

Fig. 4. Standard one-pulse NMR experiments. Sequence A is used to acquire one-dimensional proton spectra, whereas sequence B is used to acquire one-dimensional carbon spectra. The delay period between scans, D1, should be set to 1–2 s. The delay period D2, will be set by the spectrometer.

3.2.2.1. HOHAHA Spectra

Phase-sensitive HOHAHA spectra *(48)* (*see* Fig. 6) should be acquired using an MLEV-17 spin-lock pulse with a total mixing time of 70 ms, including 2.5-ms trim pulses at the beginning and end of the MLEV-17 sequence. Solvent suppression can be accomplished by using a presaturation pulse of 1 s at the frequency of ^1H_2O. A spectral width of 3623.19 Hz (12 ppm) should be used. A total of 1–2 K time-domain data points for 256–512 t_1 values, with 64–256 scans each, should be acquired and then zero-filled to 1 × 1 K (*see* Note 9), followed by processing with a 90° shifted sine squared function in both dimensions.

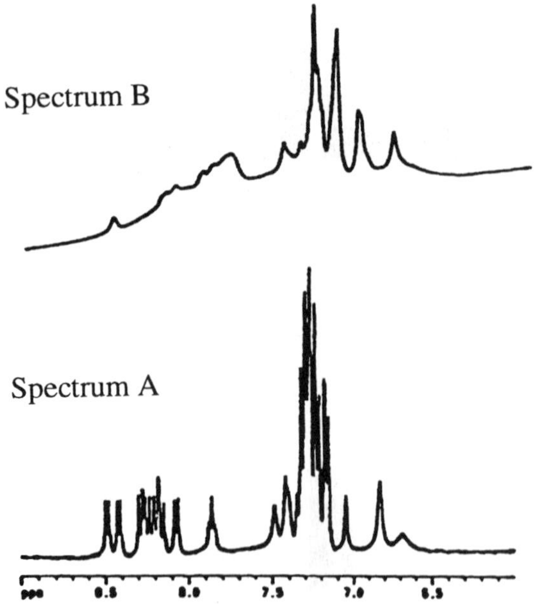

Fig. 5. Illustration of the line-broadening of the amide and aromatic resonances of a neuropeptide caused by SDS micelles. Spectrum **(A)** shows the amide and aromatic region of substance P in water at pH 4, and Spectrum **(B)** shows in the same region in the presence of 30 m*M* SDS at the same pH.

3.2.2.2. ROESY Spectra

A series of phase-sensitive ROESY spectra *(49)* (*see* Fig. 6) should be acquired with mixing pulses of 100, 150, 200, and 300 ms, respectively, to determine ROE buildup rates, as well as to determine at what mixing time spin diffusion becomes a problem. Solvent suppression can be accomplished by using a presaturation pulse of 1.2 s at the frequency of ^1H_2O. A spectral width of 3623.19 (12 ppm) Hz should be used. A total of 1–2 K time-domain data points for 256–512 t_1 values, with 128–256 scans each, should be acquired and then zero-filled to 1 × 1 K followed by processing with a Gaussian multiplication function defined by a –10 to –20 Hz line broadening in both dimensions.

3.2.2.3. NOESY Spectra

A series of phase-sensitive NOESY spectra *(50)* (*see* Fig. 7) should be acquired with mixing pulses of 100, 150, 200, and 300 ms, respectively, to determine NOE buildup rates. Solvent suppression can be accomplished by using a presaturation pulse of 1.5 s at the frequency of ^1H_2O. A spectral width of 3623.19 (12 ppm) Hz should be used. A total of 1–2 K time-domain data

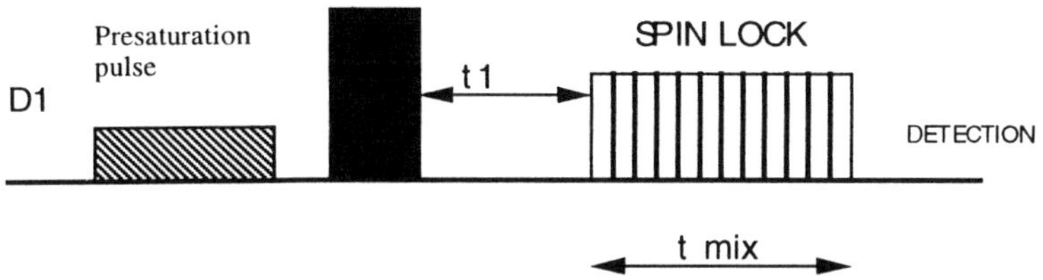

Fig. 6. Standard two-dimensional pulse sequence for the HOHAHA and ROESY spin lock experiments. For both experiments, D1, the delay between scans, should be set to a value between 1.3 × T_1 to 5 × T_1 (T_1 = longitudinal relaxation time) of the slowest relaxing protons in the sample. For experiments using a presaturation pulse, D1 may be set to 500 ms. The low power presaturation pulse is normally set to a duration of 1–2 s. The t_1 period is the increment delay period; for both experiments, t_1 is set to 3 μs in the first block of the experiment and incremented by the proton-dwell time for each following block. For the HOHAHA experiment, higher power for the spin lock pulse is required. A composite pulse sequence such as MLEV-17 is normally used for the spin lock pulse. The power is adjusted to yield a 90° pulse of 20–30 μs to be used in the composite pulse sequence. The total duration of the spin lock pulse for the HOHAHA experiment is 50–80 ms. For the ROESY experiment, a lower power continuous wave pulse is used. The power is adjusted to yield a spin lock field of 4–5 kHz. The duration of the spin lock pulse is from 50 to 700 ms, with the normal duration for the study of neuropeptides in the presence of membrane models being 100–300 ms.

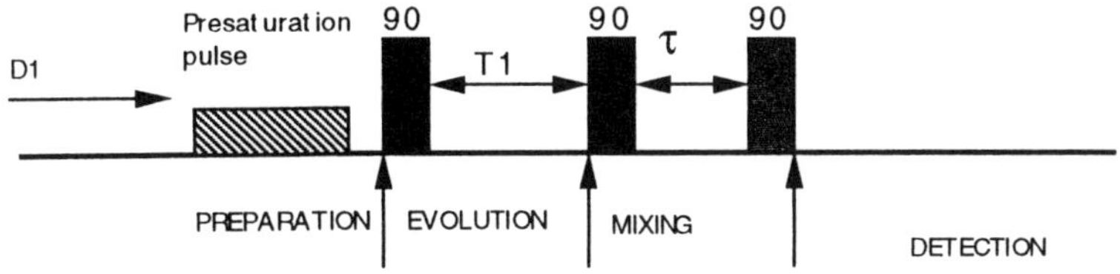

Fig. 7. Standard two-dimensional pulse sequence for the NOESY Experiment. D1, the delay between scans, should be set to a value of 1.3 × T_1 to 5 × T_1 of the slowest relaxing protons in the sample. For experiments using a presaturation pulse, D1 may be set to 500 ms. The low-power presaturation pulse is normally set to a duration of 1–2 s. The t_1 period is the increment delay period; t_1 is set to 3 μs in the first block of the experiment and incremented by the proton-dwell time for each following block. The mixing time τ corresponds to the period where the NOE intensity is built up. The value of τ normally used for the study of neuropeptides in the presence of membrane models is 100–300 ms.

points for 256–512 t_1 values, with 128–256 scans each, should be acquired and then zero-filled to 1 × 1 K followed by processing with a Gaussian multiplication function defined by a –10 to –20 Hz line broadening in both dimensions.

3.2.3. Carbon-13 Spectra

One-dimensional carbon-13 spectra should be acquired using the pulse sequence shown in Fig. 4. A spectral width of 16875 Hz (225 ppm) with 32–64 K data points should be used. A total of 50,000–200,000 scans (this is owing to the very low concentration of the neuropeptide) should be acquired and then processing with exponential multiplication defined by a 1-Hz line broadening.

3.2.3.1. HMQC Spectra

Two-dimensional phase-sensitive ^1H-^{13}C, HMQC (*see* Fig. 8) spectra should be acquired on a new sample dissolved in 100% ^2H_2O, prepared as in Section 3.1.1. This is to reduce the problems associated with presaturation of the large water peak, which can be difficult in the HMQC experiment depending on the model of spectrometer used. The spectrum should be acquired with a proton spectral width of 3623.19 Hz (12 ppm) and a carbon-13 spectral width of 18869.250 Hz (250 ppm). A total of 1–2 K time-domain data points for 512 t_1 values, with 128–256 scans each should be acquired and then zero-filled to 1 × 1 K followed by processing with a Gaussian multiplication function defined by a –5 Hz line broadening, or by a 90° shifted sine squared function, in both dimensions.

3.2.3.2. HMBC Spectra

Two-dimensional magnitude mode ^1H-^{13}C, HMBC spectra (*see* Fig. 9) should also be acquired on samples dissolved in 100% ^2H_2O. Spectra should be acquired with a proton spectral width of 3623.19 Hz (12 ppm) and a carbon spectral width of 18869.250 Hz (250 ppm). A total of 2048 time-domain data points for 512 t_1 values, with 512 scans each, should be acquired and then zero-filled to 1 × 1 K followed by processing with a Gaussian multiplication function defined by a –5 Hz line broadening in both dimensions (*see* Note 10).

3.3. Conclusion

The rapid developments in multidimensional NMR experiments have provided a powerful tool for biochemists to use to determine the three-dimensional structures of neuropeptides bound to membrane models and to models of their receptor proteins. The applications of both homonuclear and heteronuclear two-dimensional NMR experiments to solve structural problems of neuropeptides bound to membrane models are commonplace. These investiga-

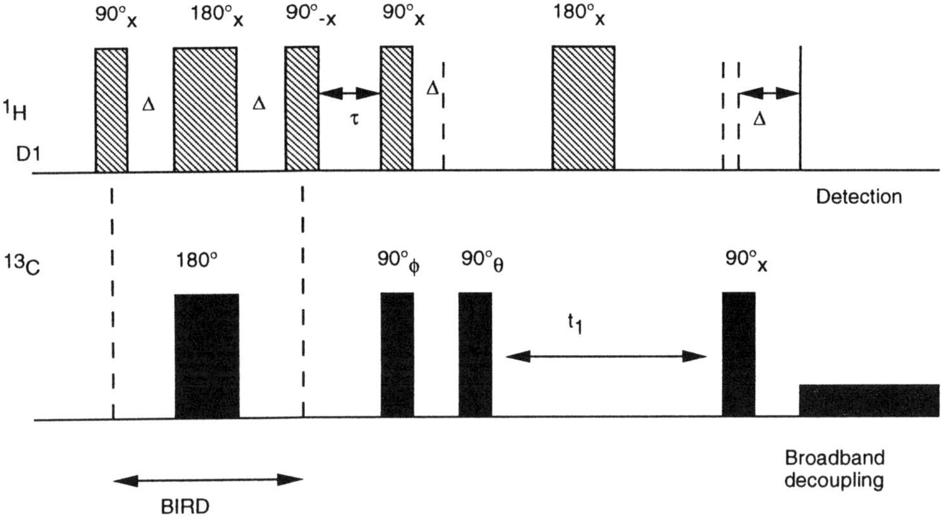

Fig. 8. Standard two-dimensional pulse sequence for the HMQC experiment. The period of the pulse sequence labeled BIRD (bilinear rotational decoupling) is used to achieve broadband homonuclear proton decoupling of protons coupled to ^{12}C. The protons that are not coupled to a ^{13}C are inverted by the BIRD pulse, whereas the protons coupled to ^{13}C are not effected. The BIRD overcomes the dynamic range problem of the standard HMQC sequence by suppressing the signals of the protons that are not coupled to ^{13}C. The period Δ is set to $1/2(^1J_{CH})$ and the interval, t = T/2.7, where T is equal to 1.3 × T_1 of the fastest relaxing proton in the molecule. The delay period between scans, D1, should be set to a value of 1–2 s.

tions require care and careful control of instrumental parameters; however, they will yield useful results. The reader is referred to several examples of the application of these techniques in the literature.

4. Notes

1. In the determination of amide proton temperature coefficients, HOHAHA spectra should be used instead of 1D ^1H spectra to ensure proper assignment of the amide protons. The phase-sensitive double quantum filtered COSY experiment may also be used for this purpose.
2. In an extreme case of spin diffusion, all observed NOEs would be negative and would exhibit equal intensity. In most cases, however, spin-diffusion results in the three-spin effect, or the relay of NOEs. ROESY experiments are less susceptible to spin-diffusion than normal NOESY experiments. A ROESY spectrum at the same mixing time should be collected to ensure that none of the structurally important NOEs were derived from relayed NOEs, or spin-diffusion. Relayed ROEs are antiphase to real ROEs and are thus easy to recognize.

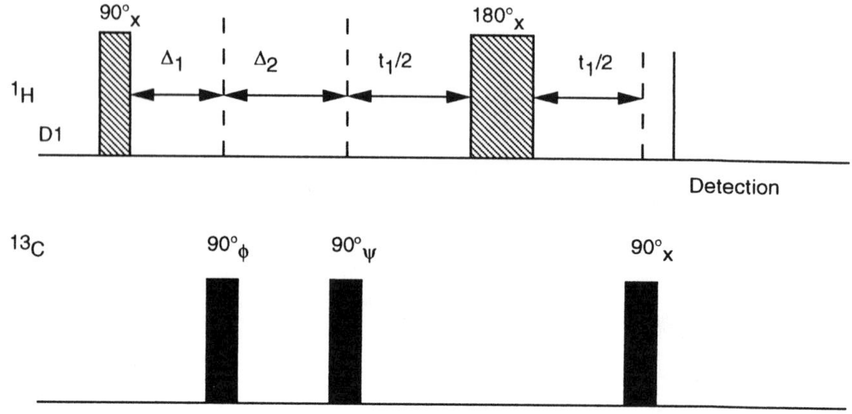

Fig. 9. Standard two-dimensional pulse sequence for the HMBC Experiment. The delay period between scans, D1, should be set to a value of 1–2 s. The delay period Δ1 should be set to $1/2(^1J_{CH})$ and delay period Δ2 should be set to 50–80 ms.

3. Generally, three-bond ^1H-^{13}C coupling constants are larger than two-bond ^1H-^{13}C coupling constants; this should be kept in mind when interpreting HMBC spectra. Generally, two-bond ^1H-^{13}C coupling constants are in the range of 1.0–8.0 Hz, with three-bond ^1H-^{13}C couplings constants in the range of 5.0–15 Hz, depending on the type of carbon atom and other functional groups in the molecule.
4. The problem of conformational averaging may be described in the following manner. Two dipolar coupled protons are separated by a distance r. This distance varies owing to molecular motion and is centered about an average or equilibrium distance r_o. The lower or closer values of r give rise to very large values of r^{-6}. The smaller values of r will dominate the dipolar relaxation of those two protons. Thus, after the averaging of all values of r, the dipolar relaxation will be dominated by the shorter apparent r values given by the equation $(\langle r^{-6}\rangle)^{-1/6}$. In general, the set of averaged r values may not be compatible with any single structure. This is a common factor of all exchange mechanisms that are averaged over the molecule's internal motion. Distance constraints generated from NOE data are, by their nature, necessarily approximate. In an effort to address the issue of conformational averaging, the conservative uniform averaging model developed by Wuthrich may be employed.
5. Because the polar head group of SDS is charged, the ionic strength of the solution has a great effect on its average aggregation number and cmc. The average aggregation number for SDS in water (no salt) is 62 with a cmc of 8.1 m*M*. After increasing the ionic strength to 500 m*M* NaCl, the average aggregation number increases to 141 and the cmc decreases to 0.5 m*M*. In general, the ionic strength should be kept as low as possible to prevent precipitation of the neuropeptide-SDS complex.

6. Perdeuterated DPC is an excellent membrane model system developed specifically for NMR studies. This detergent makes micelles containing approx 56 monomer units with a cmc of 1.1 m*M*. Since the polar head group is zwitterionic, the cmc of DPC is not affected by ionic strength.
7. Apodization: An ideal free induction decay (FID) would smoothly decay to zero given a sufficiently long acquisition time. In practice, however, owing to limited data storage space and acquisition time, the FID is truncated and does not decay to zero. This truncated data set introduces a step function at the end of the FID. Fourier transformation of the truncated FID introduces a ringing or feet around the signal. Apodization is the process of applying a window function (mathematical function) to the raw FID to smooth the FID by causing it to decay to zero. There are several window functions available, such as exponential multiplication, Gaussian multiplication, a shifted sine bell, and so on. These functions can be used to maximize either resolution or sensitivity of the resulting spectrum. Most two-dimensional experiments require apodization, or window functions, to be applied in both dimensions. With each experiment listed, a window function is suggested. However, your data may require different window functions in order to obtain the best results. For two-dimensional data sets, the same or different window functions may be used in each dimension.
8. Zero-Filling: The Fourier transformation of raw data without zero-filling does not use all of the available information in the FID. In the normal collection of a FID, N data points are collected. Fourier transformation of this FID yields two independent data sets, the absorption mode set containing N/2 data points and the dispersion mode set, also containing N/2 data points. Normally, only the absorption mode data is used, with the dispersion mode data being discarded. The process of adding N zeros to the end of the FID retrieves the lost data by yielding an absorption mode data set containing now N, instead of N/2 data points. Normally, the processing of two-dimensional experiments involves one to two zero fillings.
9. Micellular solutions normally yield water resonances that are wider than normal; therefore, shimming of the magnet is critical for effective water suppression. Time spent adjusting the room temperature shims is time well spent. If shimming on the lock signal does not provide a sharp water resonance (and thus good water suppression), try shimming on the FID.
10. HMBC spectra often contain artifacts of strong one-bond ^{1}H-^{13}C couplings. These artifacts yield a doublet, triplet, or quartet on the proton axis, depending on the number of equivalent protons strongly coupled to the carbon atom. These artifacts will normally not line-up with the observed proton chemical shifts. Care must be taken not to mistakenly assign strong coupling artifacts to two- and three-bond ^{1}H-^{13}C couplings.

Acknowledgment

Dedicated in Loving Memory of Richard Adams Hicks, 1913–1995.

References

1. Oschkinat, H., Muller, T., and Dieckmann, T. (1994) Protein structure determination with three- and four-dimensional nmr spectroscopy. *Angew. Chem. Int. Ed. Engl.* **33,** 277–293.
2. Hicks, R. P., Beard, D. J., and Young, J. K. (1992) The interactions of neuropeptides with membrane model systems: a case study. *Biopolymers* **32,** 85–96.
3. Young, J. K., Anklin, C., and Hicks, R. P. (1994) Two-dimensional nmr and molecular modeling investigations of the neuropeptide substance P in the presence of SDS micelles. *Biopolymers* **34,** 1449–1462.
4. Hruby, V. J., Krstenansky, J. L., and Cody, W. L. (1984) Recent progress in the rational design of peptide hormones and neurotransmitters, in *Annual Reports in Medicinal Chemistry-22,* Academic, New York.
5. Surewicz, W. K. and Mantsch, H. H. (1988) Conformational properties of angiotensin II in aqueous solution and in a lipid environment: a Fourier transform infrared spectroscopic investigation. *J. Am. Chem. Soc.* **110,** 4412–4414.
6. Rizo, J., Blanco, F. J., Kobe, B., Bruch, M. D., and Gierasch, L. M. (1993) Conformational behavior of *Escherichia coli* ompA signal peptides in membrane mimetic environments. *Biochemistry* **32,** 4881–4894.
7. Convert, O., Duplaa, H., Lavielle, S., and Chassaing, G. (1991) Influence of the replacement of amino acids by its D-enantiomer in the sequence of substance P. 2. Conformational analysis by nmr and energy calculations. *Neuropeptides* **19,** 259–270.
8. Dempsey, C. E. and Watts, A. (1987) A deuterium and phosphorus-31 nmr study of the interaction of melittin with dimyristoylphosphatidylcholine bilayers and the effects of contaminating phospholipase A_2. *Biochemistry* **26,** 5803–5811.
9. Batenburg, A. M., van Esch, J. H., and de Kruijff, B. (1988) Melittin-induced changes of the microscopic structure of phosphatidylethanolamines. *Biochemistry* **27,** 2324–2331.
10. Lauterwein, J., Bosch, C., Brown, L. R., and Wuthrich, K. (1979) Physicochemical studies of the protein-lipid interactions in melittin containing micelles. *Biochim. Biophys. Acta* **556,** 244–264.
11. Brown, L. R. (1979) Use of fully deuterated micelles for the conformational studies of membrane proteins by high resolution ^{1}H nmr. *Biochim. Biophys. Acta* **557,** 135–148.
12. Zetta, L. and Kaptein, R. (1984) Interaction of β-endorphin with sodium dodecyl sulfate in aqueous solution ^{1}H nmr investigation. *Eur. J. Biochem.* **145,** 181–186.
13. Zetta, L., De Marco, A., and Zannoni, G. (1988) Evidence for a folded structure of met-enkephalin in membrane mimetic systems: ^{1}H-nmr studies in sodium dodecylsulfate, lyso-phosphatidylcholine and mixed lyso-phosphatidylcholine/sulfatide micelles. *Biopolymers* **25,** 2315–2323.
14. Deber, C. M. and Behnam, B. A. (1984) Role of membrane lipids in peptide hormone function: binding of enkephalin to micelles. *Proc. Natl. Acad. Sci. USA* **81,** 61–65.

15. Graham, W. H., Carter, E. S., and Hicks, R. P. (1992) Conformational analysis of met-enkephalin in both aqueous solution and in the presence of sodium dodecyl sulfate micelles using multidimensional nmr and molecular modeling. *Biopolymers* **32,** 1755–1764.
16. Bystrov, V. F., Arseniev, A. S., Barsukov, J. L., and Lomize, A. L. (1986) The 2D nmr of single and double stranded helixes of gramicidin A micelles and solution. *Bull. Magn. Resonance* **8,** 84–94.
17. Killian, J. A., Nicholson, L. K., and Cross, T. A. (1988) Solid state ^{15}N -nmr evidence that gramicidin A can adopt two different backbone conformations in dimyristoylphosphatidylcholine model membrane preparations. *Biochim. Biophys. Acta* **943,** 535–540.
18. Nicholson, L. K., Moll, F., Mixon, T. E., Lograsso, P. V., Lay, J. C., and Cross, T. A. (1987) Solid state ^{15}N nmr of oriented lipid bilayer bound gramicidin A. *Biochemistry* **26,** 6621–6626.
19. Kyle, D. J., Chakravarty, S., Sinsko, J. A., and Stormann, T. M. (1994) A proposed model of bradykinin bound to the rat B2 receptor and its utility for drug design. *J. Med. Chem.* **37,** 1347–1354.
20. Kyle, D. J., Hicks, R. P., Blake, P. R., and Klimkowski, V. J. (1990) Conformational properties of bradykinin and bradykinin antagonists, in *Bradykinin Antagonists: Basic and Clinical Research* (Burch, R. M., ed.), Marcel Dekker, New York, pp. 131–146.
21. Henderson, R., Baldwin, J. M., Ceska, T. A., Zemlin, F., Beckmann, E., and Downing, K. H. (1990) Model for the structure of bacteriorhodopsin based on high resolution electron cryomicroscopy. *J. Mol. Biol.* **213,** 899–929.
22. Deber, C. M. and Behnam, B. A. (1985) Transfer of peptide hormones from aqueous to membrane phases. *Biopolymers* **24,** 105–116.
23. Woolley, G. A. and Deber, C. M. (1987) Peptides in membranes: lipid-induced secondary structure of substance P. *Biopolymers* **26,** S109–S121.
24. Hagler, A. T., Osguthorpe, D. J., Dauber-Osguthorpe, P., and Hempel, J. C. (1985) Dynamics and conformational energetics of a peptide hormone: vasopressin. *Science* **227,** 1309–1315.
25. Wu, C. S. C., Hachimori, A., and Yang, J. T. (1982) Lipid-induced conformation of some peptide hormones and bioactive oligopeptides: predominance of helix over β form. *Biochemistry* **21,** 4556–4562.
26. Keniry, M. A. and Smith, R. (1980) A ^{13}C nmr spin-lattice relaxation study of the interaction of myelin proteins with lipid vesicles. *Biophys. Chem.* **12,** 133–141.
27. Godici, P. E. and Landsberger, F. R. (1974) Dynamic structures of lipid membranes. Carbon-13 nuclear magnetic resonance study using spin labels. *Biochemistry* **13,** 362–368.
28. Gent, M. P. and Prestegard, J. H. (1977) Nuclear magnetic relaxation and molecular motion in phospholipid bilayer membranes. *J. Mag. Res.* **25,** 243–247.
29. Lee, S. C., Russell, A. F., and Laidig, W. D. (1990) Three-dimensional structure of bradykinin in SDS micelles. *Int. J. Pept. Prot. Res.* **35,** 367–377.

30. Young, J. K. and Hicks, R. P. (1994) NMR and molecular modeling investigation of the neuropeptide bradykinin in three different solvents systems: DMSO, 9:1 dioxane/water, and in the presence of 7.4 mM lyso phosphatidylcholine micelles. *Biopolymers* **34,** 611–623.
31. Reddy, A. P., Tallon, M. A., Becker, J. M., and Naider, F. (1994) Biophysical studies on fragments of the α-factor receptor protein. *Biopolymers* **34,** 679–689.
32. Hawrot, E., Colson, K. L., Armitage, I. M., and Song, G.-Q. (1990) in Bungarotoxin binding to acetylcholine receptor-derived synthetic peptides analyzed by nmr in *Frontiers of NMR in Molecular Biology,* New York (Live, D., Armitage, I. M., and Patel, D., ed.), Alan R. Liss, pp. 27–36.
33. Bommakanti, R. K., Dratz, E. A., Sremson, D. W., and Jesaitis, A. J. (1995) Extensive Contact between G_{12} and N-formyl peptide receptor of human neutrophils: mapping of the binding sites using receptor-mimetic peptides. *Biochemistry* **34,** 6720–6728.
34. Hamm, H. E., Deretia, D., Arendt, A., Hargrave, P. A., Konig, B., and Hoffmann, K. D. (1988) Site of protein binding to rhodopsin mapped with synthetic peptides from the alpha subunit. *Science* **241,** 832–835.
35. Smith, G. P. and Scott, J. K. (1993) Libraries of peptides and proteins displayed on filamentous phage. *Methods Enzymol.* **217,** 228–257.
36. Wand, A. J. and Short, J. H. (1994) Nuclear magnetic resonance studies of protein-peptide complexes. *Methods Enzymol.* **239,** 700–717.
37. Bax, A., Marzilli, L. G., and Summers, M. F. (1987) New insights into the solution behavior of cobalamin, studies of the base-off form of coenzyme B_{12} using modern two-dimensional nmr methods. *J. Am. Chem. Soc.* **109,** 566–574.
38. Wishart, D. S., Sykes, B. D., and Richards, F. M. (1992) The chemical shift index: a fast and simple method for the assignment of protein secondary structure through nmr spectroscopy. *Biochemistry* **31,** 1647–1651.
39. Wishart, D. S. and Sykes, B. D. (1994) Chemical shifts as a tool for structure determination. *Methods Enzymol.* **239,** 363–392.
40. Spera, S. and Bax, A. (1991) Empirical correlation between protein backbone conformation and the C^{α} and C^{β} nmr chemical shifts. *J. Am. Chem. Soc.* **113,** 5490–5492.
41. Ikura, M., Kay, L. E., and Bax, A. (1990) A novel approach for sequential assignment of ^{1}H, ^{13}C, and ^{15}N spectra of larger proteins: heteronuclear triple-resonance three-dimensional nmr spectroscopy—application to calmodulin. *Biochemistry* **29,** 4659–4667.
42. Powers, R., Garrett, D. S., March, C. J., Frieden, E. A., Gronenborn, A. M., and Clore, G. M. (1992) ^{1}H, ^{15}N, ^{13}C, and ^{13}CO Assignments of human interleukin-4 using three-dimensional double- and triple-resonance heteronuclear magnetic resonance spectroscopy. *Biochemistry* **31,** 4334–4346.
43. Fesik, S. W., Gampe, R. T., Eaton, H. L., Gemmecker, G., Olejniczak, E. T., Neri, P., Holzman, T. F., Egan, D. A., Edalij, R., Simmer, R., Helfrich, R., Hochlowski, J., and Jackson, M. (1991) NMR studies of [U-^{13}C] cyclosporin A bound to cyclophilin: bound conformation and portions of cyclosporin involved in binding. *Biochemistry* **30,** 6574–6583.

44. Lippens, G., Hallenga, K., Van Belle, D., Wodak, S. J., Nirmala, N. R., Hill, P., Russell, K. C., Smith, D. D., and Hruby, V. J. (1993) Transfer nuclear overhauser effect study of the conformation of oxytocin bound to bovine neurophysin I. *Biochemistry* **32,** 9423–9434.
45. Clore, G. M. and Gronenborn, A. M., eds. (1993) *NMR of Proteins,* CRC, Boca Raton, FL.
46. Evans, J. N. S. (1995) *Biomolecular NMR Spectroscopy,* Oxford University Press, New York.
47. Wuthrich, K. (1986) *NMR of Proteins and Nucleic Acids,* Wiley, New York.
48. Braunschweiler, L. and Ernst, R. R. (1983) Coherence transfer by isotropic mixing: application to proton correlation spectroscopy. *J. Magn. Reson.* **53,** 521–528.
49. Bax, A. and Davis, D. G. (1985) Practical aspects of two-dimensional transverse noe spectroscopy. *J. Magn. Reson.* **63,** 207–213.
50. States, D. J., Haberkorn, R. A., and Rubin, D. J. (1982) A two-dimensional nuclear overhauser experiment with pure absorption phase in four quadrants. *J. Magn. Reson.* **48,** 286–292.

18

Molecular Modeling of Neuropeptides

Sándor Lovas and Richard F. Murphy

1. Introduction

Molecular modeling is the science of the generation, manipulation, and representation of three-dimensional structures of molecules using computational chemistry and high resolution computer graphics. Since peptides of biological interest are large molecules, molecular mechanics (MM) *(1)* is used almost exclusively as a computational tool. Molecular mechanics is a nonquantum method for calculating molecular properties that do not depend on electronic effects. The forces acting on the atoms in a molecule are described in terms of a set of classical potential functions such as harmonic oscillators, Morse potentials, and Lennard-Jones potentials. Parameters of these functions are usually obtained from experimental structural and thermodynamic studies of model molecules. A set of equations together with their parameters is called a force field. Separate potential functions are used to calculate bond stretching, angle bending, bond twisting, and nonbonded interactions such as van der Waals and electrostatic interactions. Molecular mechanics methods reproduce experimental results well if the compound being examined is similar to those used to create the parameters. Consequently, several force fields have been developed for computations on peptides, including CHARMM *(2),* AMBER *(3),* DISCOVER *(4),* GROMOS *(5),* and ECEPP *(6)*. The performance of the different force fields in their application to peptides and proteins, however, should be evaluated on the basis of published data. Commercially available molecular modeling packages implement one of these force fields. The computational procedure described below was carried out using the SYBYL 6.1 *(7)* package in which the AMBER force field was implemented.

Energy minimization optimizes the geometry of a molecule as a function of its energy. Minimization includes two steps. First, the energy for a given con-

From: *Methods in Molecular Biology, Neuropeptide Protocols*
Edited by: G. B. Irvine and C. H. Williams Humana Press Inc., Totowa, NJ

formation is calculated. Second, the coordinates of each of the atoms are adjusted and the energy of the molecule is recomputed. These two steps are repeated until a minimum energy is obtained.

Different algorithms can be used to minimize the energy of a molecule *(8)*. These are generally unable to find the global energy minimum, which is the conformation of a molecule corresponding to the lowest value of the energy; only the local minimum that is closest to the starting coordinates is found. To find the global minimum, it is necessary to explore different sets of starting coordinates or to generate structures based on experimental (NMR or X-ray crystallography) data. The existence of a large number of local minima for each peptide makes it difficult to find the global minimum. It is very important to note that, for all force fields used in calculations of this type, the zero energy is arbitrary. Thus, the total potential energy of different molecules cannot directly be compared. It is acceptable, however, to compare the calculated energies of different configurations of chemically identical systems.

Molecular dynamics (MD) simulation is based on molecular mechanics methods. During the period of simulation, the atoms of a molecule move and the trajectory of the molecular system is generated by integrating Newton's equations of motion for all the atoms in the system. This integration is performed in small time steps, typically 1–10 fs. The trajectory of a simulation can be stored on disk in a history file and, after the simulation has been completed, can be played back to produce a movie of the dynamic nature of the molecule. Static equilibrium quantities can be obtained by averaging over the trajectory, which, therefore, must be of satisfactory length, typically 10–100 ps for peptides, to form a representative ensemble of the state of the system. The main advantage of using MD simulations for molecular modeling is the presence of kinetic energy. Thus, it is possible to search the available conformational space effectively by raising the temperature in the molecular dynamics simulation, since the kinetic energy allows the system to overcome rotational barriers so that many adjacent local minima can be explored. Repeating an MD simulation with different initial velocity distribution is another way to explore the available conformational space of a molecule. It sends the simulations into different regions and thereby increases the probability of finding an energetically stable conformational state. The major limitation to MD simulation is the timescale, since the transitions of interest may take place within microseconds or even milliseconds. Thus, molecular dynamics methods are essentially limited by the computer power available.

In principle, the method of simulated annealing (SA) *(9)* is similar to an experimental annealing technique in which a material is melted and cooled slowly to produce the most stable arrangement (crystalline state). The essential

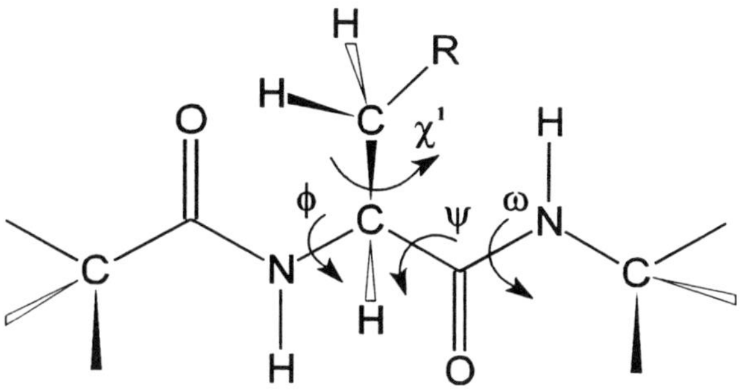

Fig. 1. The conventional torsional angles in a polypeptide unit. The ϕ torsional angle is the angle between two planes formed by the C'-N-Cα and N-Cα-C' atoms. The ψ torsional angle is the angle between two planes formed by the N-Cα-C' and Cα-C'-N atoms and the χ^1 torsional angle is the angle between two planes formed by the N-Cα-Cβ and Cα-Cβ-Cγ atoms.

feature of SA is that it keeps the system for a certain time period at a temperature, typically approx 1000 K, which is high enough to surmount rotational barriers, so that the confirmational space of a molecule can be sampled widely. The temperature is then gradually lowered to near absolute zero and cycled over time. The cooling rate, the average change in temperature per attempted move, is crucially important for the efficiency of SA *(10)*. The best way to lower the temperature is exponential ramping so that the simulation spends most of the time adjusting the lower energy conformation.

Conformations of a peptide chain can be described in conventional (ϕ_i, ψ_i, χ_i^n) torsional angle space (Fig. 1.), which is ideal for comparing local conformations. It is independent of reference frame, but it is not satisfactory for specifying global conformation. Since ϕ and ψ angles give complete description of backbone conformation, a two-dimensional plot of these angles (Ramachandran plot) is used as a representation of a structure of peptides and proteins. The entire ϕ, ψ map is subdivided into nineteen regions, each of which is denoted by single capital letter if $\phi \leq 0°$ and a single capital letter followed by an asterisk if $\phi > 0°$, according to Zimmerman and associates *(11)*. Region A describes the right-handed α helical region (ϕ, ψ) = (–60°, –40°) and region A* the left-handed α helixes (ϕ, ψ) = (60°, 40°). Region C contains the C_7^{eq} conformation, a seven-membered ring closed by a hydrogen bond (ϕ, ψ) = (–80°, 70°), and region E the C_5 extended conformation, a five-membered ring closed by a hydrogen bond (ϕ, ψ) = (–160°, 170°). The central region of the map (ϕ, ψ) ~ (0°, 0°) is the high energy or unallowed conformational region. Side-chain conformations can be classified on the basis of preferred values of χ_i^1 torsional angle. Conformers are denoted as *trans* (t) for $\chi_i^1 = 180 \pm 30°$, *gauche* plus (g+) for $\chi_i^1 = 60 \pm 30°$, and *gauche* minus (g-) for $\chi_i^1 = -60 \pm 30°$.

2. Materials

2.1. Software

Molecular modeling systems should have the following capabilities:

1. Library containing structures of standard amino acids and interface to structural databases such as the Brookhaven X-ray and NMR database.
2. Force field parametrized for peptides.
3. Different minimization methods.
4. MM and MD options.
5. High-resolution graphics that provide interactive viewing and manipulate molecular structures.

For the example below, the SYBYL 6.1 system with modules of Advanced Computation, BIOPOLYMER and DYNAMICS was used.

2.2. Hardware

A broad range of computers are available for molecular modeling, but the minimal requirement is a graphics workstation with at least 32 Mb of RAM and 1 Gb of hard disk space with a minimal speed of 25 MIPS. For the example below, an SGI Power Challenge M computer with R8000 processor, 64 Mb RAM, 2 Gb hard disk, connected to an SGI 4D/210 VGX graphics terminal was used.

3. Methods

3.1. Algorithm

The flow diagram for modeling typical neuropeptides (*see* Note 1) using the simulated annealing technique is shown in Fig. 2.

1. Build up initial structures of target peptides, comparable with the examples, from the SYBYL biopolymer library and set conformation states to random. Side-chain and end functional groups should be charged as at physiological pH. Use the AMBER force-field and all-atom parameter sets. To simulate solvent screening effects on surface charges, use a distance-dependent dielectric function. Use a scaling factor of 0.5 for steric and electrostatic interactions between atoms 1-4 and a nonbonded interaction cutoff of 10 Å (*see* Note 2).
2. Minimize structures in two stages, first in 100 steps by the Powell method and then by the BFGS method, until the norm of the gradient is <0.05 kcal/mol/Å to ensure the stability of systems (*see* Note 3).

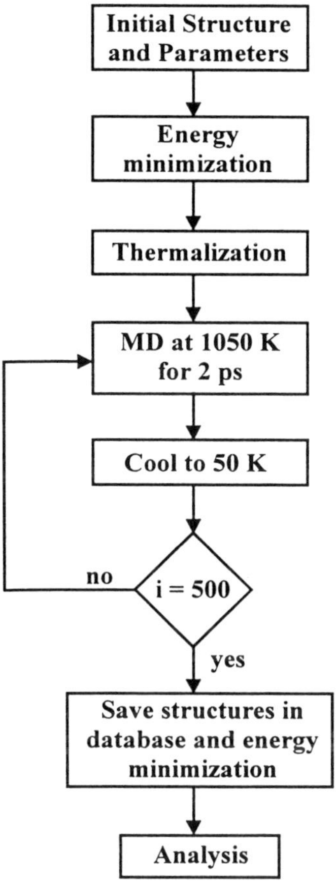

Fig. 2. Flow diagram for molecular modeling of a typical neuropeptide using the simulated annealing technique.

3. Thermalize systems in 8 steps by running NTV MD simulations, with the force field parameters set as above, at 50, 100, 300, 400, 600, 800, 1000, and 1050 K, for 1000 fs at each of the temperatures, using the following parameters: integration step, 0.5 fs; coupling time for temperature regulation, 10 fs; and velocity scale (*see* Note 4). At 50 K, assign random velocities to the atoms and for the rest of the temperatures always use starting velocities from the previous step to start the motion in a direction determined by the strain in the molecules.
4. After thermalization, start the SA procedure by running the NTV MD simulation at 1050 K for 2 ps using initial velocities from the previous step. Use the same parameters as above except for the coupling time for temperature regulation, which should be 2 fs. Cool the system to 50 K over 2 ps at a cooling rate of

0.5 K/fs using the exponential annealing function and repeating the steps 500 times (*see* Note 5).
5. Save the 500 low-energy structures in a database and conduct energy minimization in 1000 steps by the BFGS method.
6. After the energy minimization, analyze the conformational properties of structures using the molecular spreadsheet and graphical interface of SYBYL (*see* Note 6).

4. Notes

1. It is impossible to provide a universal recipe for modeling peptides, since the method being applied depends on the objectives of a calculation. Nevertheless, the procedure described in this paper could be used to study conformational properties of any neuropeptide and limiting factors are the CPU time and the memory of the computer being used. For simulations of the two peptide examples, GnRH (pGlu-His-Trp-Ser-Tyr-Gly-Leu-Arg-Pro-Gly-NH_2) and GnRH-III (pGlu-His-Trp-Ser-His-Asp -Trp -Lys-Pro-Gly-NH_2), 110 h of computer time were required.
2. SYBYL provides two interpretations of the AMBER (version 3.0A) force field, the united-atom and the all-atom force fields. The united-atom force field replaces CH_3, CH_2, and CH groups with a single spherical united atom that has the same radius and point charge as the parent groups. The reduced number of atoms allows minimization of larger peptides. The all-atom force field, by contrast, interprets each atom explicitly. The effect of solvents can be treated with explicit solvent and without explicit solvent using a distance-dependent dielectric constant (for the example, $\varepsilon = 4.5\ R$ was used, where R is the distance in Å between atoms interacting by a pairwise Coulomb potential). For explicit treatment of aqueous solvation, use either the SPC *(12)* or the TIP3P *(13)* water model and constant dielectric ($\varepsilon = 1.0$).
3. For initial minimization, use the Powell method, which is a conjugate gradient method to shake off any strains present in the system. The BFGS method, which is a quasi-Newton method, is superior in convergence to conjugate gradient methods, though the best performance can be obtained if the system is close to either a global or a local minimum. The specific value of the norm of a gradient depends on the objective of the minimization. Usually, 1.0–0.05 kcal/mol/Å is sufficient to relax a system before MD simulation. However, if quantitative comparisons of molecular structures are to be done, it may be necessary to allow the minimization to converge to the order of 0.001 kca/mol/Å.
4. The most common type of MD simulation at constant temperature corresponds to the canonical or NVT ensemble (constant number of particles, N, constant volume, V and constant temperature, T). The velocities of the atoms have to be scaled so that the kinetic energy of the molecule conforms to a preset temperature. The temperature of a simulation has to be reached by raising the temperature in steps from near 0 K. A 1000-fs simulation time at each temperature is long enough to relax the system.

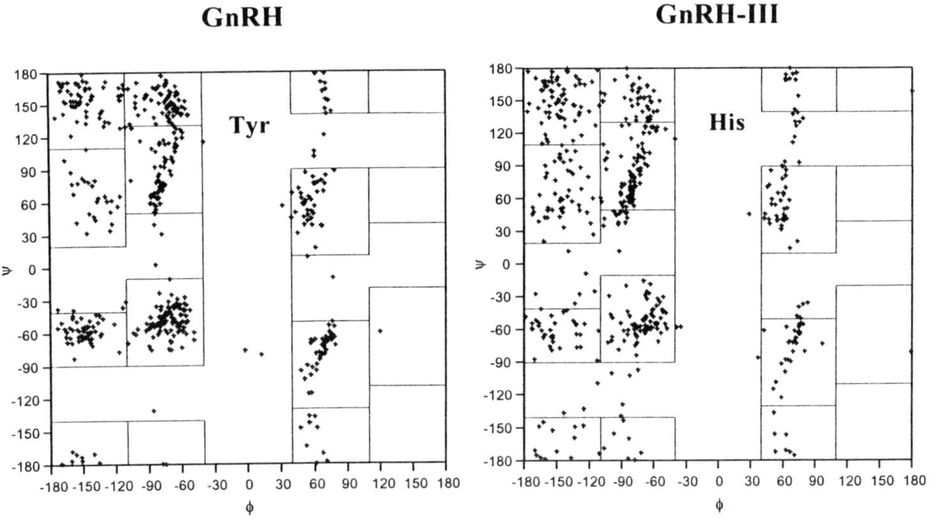

Fig. 3. Ramachandran plot of the 500 low-energy structures for residues Tyr[5] and His[5] in GnRH and GnRH-III, respectively. Boxes designate the regions defined by Zimmerman *(11)*.

5. Cooling rates should not generally be higher than 2 K/fs, so that local relaxation of the system is facilitated. A shorter coupling time for temperature regulation was needed for the GnRH peptides because of the slow cooling rate. Usually, 2 ps for high temperature simulation is long enough to relax the system, although longer time increases the possibility of visiting different conformational states. The value 50 K was selected as the lowest temperature since the description of the motion of a molecular system as a system of point masses is not adequate at 0–10 K according to the laws of classical mechanics. The number of cycles determines the number of configurations of the system to be generated. Typically, at least 500 or more cycles should be run to observe where final structures are clustering. The number of cycles is limited only by the computer power available.
6. Backbone conformations were compared by generation of ϕ, ψ maps of the 500 low-energy structures for each of the residues in both examples, GnRH and GnRH-III. Results were in good agreement with the experimental data showing that the GnRH analogs are flexible molecules in solution. Conformational space was comparable for residues in both GnRH and GnRH-III. Figure 3 shows the ϕ, ψ maps of the Tyr[5] in GnRH and His[5] in GnRH-III. Topology analysis showed that the aromatic side-chains in both GnRH and GnRH-III prefer either *trans* or *gauche* (-) positions except for Trp[3] (Fig. 4). In GnRH, the side-chain topology for Trp[3] is either *trans* or *gauche* (-), whereas in GnRH-III, topomers are normally distributed around the *syn* position.

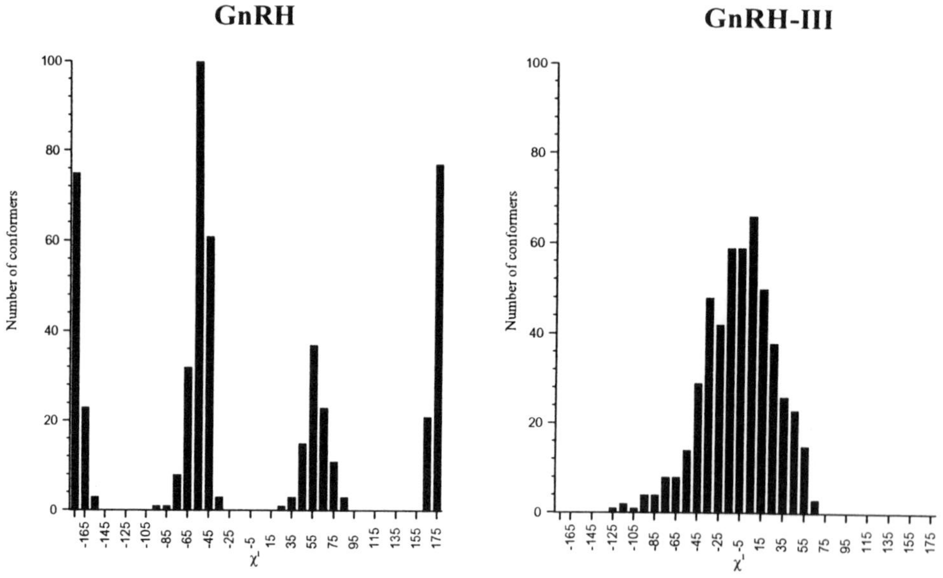

Fig. 4. Distribution of the χ^1 torsional angle for residue Trp^3 in the 500 low-energy structures of both GnRH analogs.

References

1. Burkert, U. and Allinger, N. J. (1982) *Molecular Mechanics.* American Chemical Society, Washington, DC.
2. Brooks, B. R., Bruccoleri, R. E., Olafson, B. D., States, D. J., Swaminathan, S., and Karplus, M. (1983) A program for macromolecular energy minimization and dynamics calculations. *J. Comput. Chem.* **4,** 187–217.
3. Weiner, S. J., Kollman, P. A., Case, D. A., Singh, U. C., Ghio, C., Alagona, G., Profeta, S., Jr., and Weiner, P. (1984) A new force field for molecular mechanical simulation of nucleic acids and proteins. *J. Am. Chem. Soc.* **106,** 765–784.
4. Maple, J. R., Hwang, M.-J., Stockfish,T. P., Dinur, U., Waldman, M., Ewig, C. S., and Hagler, A. T. (1994) Derivation of class II force fields. I. Methodology and quantum force field for the alkyl functional group and alkene molecules. *J. Comput. Chem.* **15,** 162–182.
5. van Gunsteren, W. F. and Berendsen, H. J. C. (1987) *Groningen Molecular Simulation (GROMOS) Library Manual.* Biomos, Groningen.
6. Némethy, G., Gibson, K. D., Palmer, K. A., Yoon, C. N., Paterlini, G., Zagari, A., Rumsey, S., and Cheraga, H. A. (1992) Energy parameters in polypeptides. 10. Improved geometrical parameters and nonbonded interactions for use in the ECEPP/3 algorithm, with application to proline-containing peptides. *J. Phys. Chem.* **96,** 6472–6484.
7. SYBYL 6.1 (1994) Tripos, St. Louis, MI 63144–2913.

8. Shlick, T. (1992) Optimization methods in computational chemistry, in *Reviews in Computational Chemistry III* (Lipkowitz, K. B. and Boyd, D. B., eds.), VCH, New York, pp. 1–71.
9. Kirkpatrick, S., Gelatt, C. D., Jr., and Vecchi, M. P. (1983) Optimization by simulated annealing. *Science* **220,** 671–680.
10. Collins, N. E., Eglese, R. W., and Golden, B. L. (1988) Simulated annealing—an annotated bibliography. *Am. J. Math. Management Sci.* **8,** 209–307.
11. Zimmerman, S. S., Pottle, M. S., Némethy, G., and Scheraga, H. A. (1977) Conformational analysis of the 20 naturally occurring amino acid residues using ECEPP. *Macromolecules* **10,** 1–9.
12. Berendsen, H. J. C., Grigera, J. R., and Straatsma, T. P. (1987) The missing term in effective pair potentials. *J. Phys. Chem.* **91,** 6269–6271.
13. Jorgensen, W. L. (1981) Transferable intermolecular potential functions for water, alcohols and ethers. Application to liquid water. *J. Am. Chem. Soc.* **103,** 335–340.

19

Tritium Labeling of Neuropeptides

Géza Tóth, Sándor Lovas, and Ferenc Ötvös

1. Introduction

Investigation of neuropeptide receptors requires biologically active peptide analogs containing radioactive, fluorescent, chemiluminescent, and/or (photo)-affinity labels or other chemically reactive marker groups. Radioisotopes are the most frequently used type of marker for peptides. Depending on the purpose of radiolabeled peptide usage, different radionuclides can be selected for peptide labeling. For in vivo receptor imaging in medical practice, radioisotopes with relatively high energy and a short half-life are required (^{123}I, ^{131}I, ^{99m}Tc). Receptor studies in vitro, biochemical receptor analyses, and hormone assays are usually based on peptide labeled with either ^{3}H or ^{125}I. In studies where high potency of the ligand is crucially important, ^{3}H, ^{14}C, and ^{35}S labeling are preferred over replacement of H with ^{125}I since the latter generally alters the physicochemical properties and frequently reduces the potency and selectivity of neuropeptides.

Radioactive peptides can be prepared by:

1. *De novo* synthesis using labeled amino acids (^{3}H, ^{14}C, ^{35}S);
2. Catalytic isotope exchange reactions using $^{3}H_2$;
3. Substitution of a functional group by radioactive acylation reagent; and
4. Insertion of a radioactive metal ion into a chelator chemically bound to peptides.

Tritium is a low energy beta-emitting radionuclide with a half-life of 12.47 yr. One milliatom of tritium represents a radioactivity of 28.6 Ci (1080 GBq). Thus, tritium incorporation is suitable in studying subnanomolar concentrations of neuropeptides.

This chapter describes a general method for tritiation of peptides by exchange reactions using tritium gas. It does not describe the preparation of the

From: *Methods in Molecular Biology, Neuropeptide Protocols*
Edited by: G. B. Irvine and C. H. Williams Humana Press Inc., Totowa, NJ

various amino acid or peptide precursors that are used for such exchange reactions, but refers the reader to appropriate literature sources.

1.1. Methods for Tritium Labeling of Neuropeptides

Synthetic methods available for tritium labeling of peptides are summarized in Fig. 1.

1.1.1. Isotope Exchange Reactions

The $^3H/H$ isotope exchange methods were developed from the technique of Wilzbach *(1)*. Among them, one of the most promising approaches is the heterogeneous catalytic exchange reaction in solution/solid systems. The most frequently used catalyst is $PdO/BaSO_4$. A disadvantage of this method is that peptides are randomly labeled. Peptides containing histidine are most suitable for heterogeneous catalytic $^3H/H$ exchange reactions since high specific radioactivity can be achieved *(2)*.

High temperature solid phase catalytic tritiation was recently developed *(3)* in which tritium is activated on the surface of the catalyst covered previously by substrates to be tritiated and the reaction proceeds in the organic compound layer. Reactions are usually performed at 80–160°C and 5–50 kPa tritium pressure. The distribution of tritium label in product molecules depends on the structure of the labeled compound and the reaction temperature.

Although $^3H/H$ isotope exchange reactions synthetically are not demanding, the high percent of impurities formed during radiolytic side reactions requires great efforts of purification to obtain the final product.

1.1.2. Derivatization of Neuropeptides by $[^3H]CH_3I$ or Reductive Methylation Using Tritiated Metal Hydrides

Peptides can be tritiated by methylation of side-chain functional groups using either $[^3H]CH_3I$ or a tritiated metal hydride and formaldehyde. Introduction of tritiated methyl groups to N-methyl, O-methyl, or S-methyl positions is generally accepted. N-methylation can be performed either by using $[^3H]CH_3I$ under basic condition or by reductive methylation using formaldehyde. The aldehyde reacts with the free amine resulting in a Schiff base that is reduced with tritiated sodium borohydride to a secondary amine *(4)*. Applicability of this procedure is limited since it may also reduce disulfide bridges if the pH of the medium is not maintained close to neutrality.

S-methylation can be carried out with $[^3H]CH_3I$ starting from peptides containing a homocysteine residue to obtain peptide tritiated at methionine residues *(5)*.

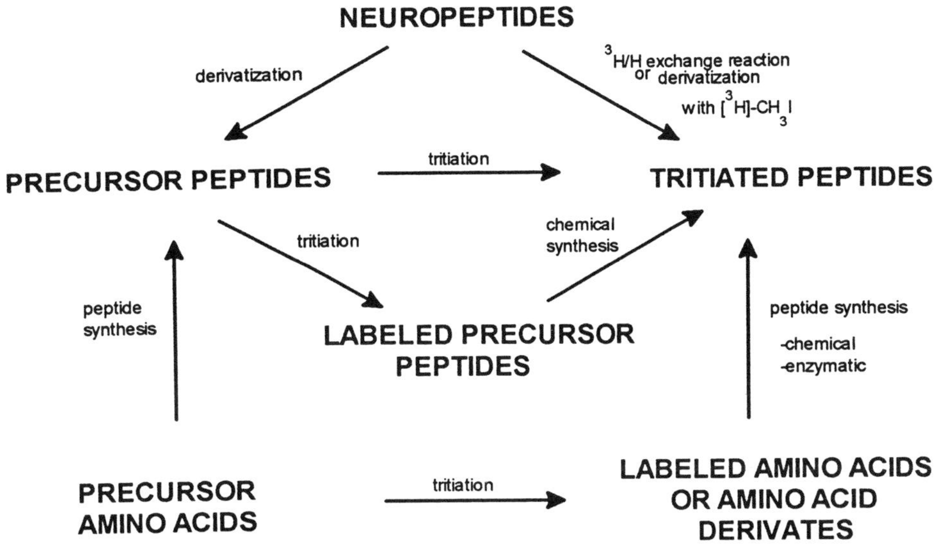

Fig. 1. Different synthetic routes for tritiated neuropeptides.

Modified side-chain groups may alter the biological activity of peptides. However, in many cases the effect of the modification on the biological activity is negligible and use of this derivative procedure is reasonable.

1.1.3. Synthesis of Peptides from Labeled Amino Acids

It is possible to synthesize labeled peptides using tritiated amino acids with high specific radioactivity by either solution or solid-phase synthetic methodology. It is very advantageous that tritiated amino acids used for peptide synthesis are already characterized, that specific activity and position of tritium atoms incorporated into amino acids are known. Specific radioactivity and position of the tritium labels in peptides could be decided before the synthesis and, by using more than one tritiated amino acid, the specific radioactivity would be over 100 Ci/mmol (3.7 TBq/mmol). Nevertheless, work with very high doses of radioactivity could limit applications of this method. Further disadvantage is the autoradiolytic decomposition during storage of tritiated amino acids and peptides of very high specific activity.

Tritiated peptides using labeled amino acids can also be synthesized by enzymatic techniques that offer a number of advantages over the classical peptide synthetic methods:

1. Enzymatic coupling reactions take place in mild conditions;
2. No protecting and deprotecting steps are required; and
3. No racemization is expected.

Hellio and associates *(6)* described a microscale procedure of enzymatic coupling that was applied to the synthesis of fully tritiated Leu-enkephalin. The enzyme used for the synthesis was carboxypeptidase Y (CP-Y) from *Saccharomyces cerevisiae*. The specific radioactivity of the peptide was 139 Ci/mmol (5.14 TBq/mmol), in agreement with the summation of the specific radioactivity of the constituent amino acid residues.

1.1.4. Synthesis of ^{3}H-Neuropeptides Using Precursor Peptides

1.1.4.1. Postsynthetically Modified Derivatives

Precursor peptides for tritiation can be prepared by a number of chemical modifications resulting in derivatives that can be reduced by tritium gas to obtain ^{3}H-labeled peptides. These modifications have been extensively reviewed earlier by Morgat and Fromageot *(7)* and Teplán et al. *(8)*.

The most important chemical modification is the iodination of peptides. Tyrosine and histidine residues in the peptides can be iodinated using different methods, for example using I_2 solution in methanol, ICl, *in situ* generated iodine by the reaction of HI and HIO_3 under strong acidic conditions, reaction of chloramine T with iodide, electrolysis of iodide, enzymatic iodination with peroxidases, and reaction of iodogen and iodobeads *(9)*.

In the iodination of tyrosine and histidine, mono- and diiodinated analogs are obtained depending on the ratio of iodine and substrate, temperature, and reaction time. Reaction mixture containing mono-, diiodinated, and non-iodinated peptides should be purified by HPLC. For tritiation, the diiodo analog is the favorable derivative. After tritiation of the iodinated precursors, tyrosine and histidine carry the radiolabel in positions 3', 5' and 2', 4' of the aromatic rings, respectively.

Tryptophan, cysteine, and methionine are sensitive to the oxidative conditions of direct halogenation. To avoid such difficulties, Brundish and Wade *(9)* prepared α-ACTH(1-24) containing Trp in position 9 by fragment condensation using diiodinated C-terminal tetradecapeptide and N-terminal decapeptide. Bienert and associates labeled Met-enkephalin using the ICl method to prepare precursor peptide *(10)*. Besides the iodination of Tyr residue by ICl, a complete transformation of methionine thioether group into sulfoxide took place. To prevent S-oxidation, S-t-butylsulfonium intermediates were synthesized. After iodination, the thioether group was regenerated by heating the peptide in solution at 60°C for 2 h.

Another modification is the reaction between *o*-nitrophenylsulfenyl (NPS) chloride and the indole ring of tryptophan in the peptide. Pentagastrin, LH-RH, and somatostatin were labeled with this method, and the NPS-tryptophan residue was tritiated to produce specifically labeled peptide *(11)*.

In summary, the method of modified precursor peptide for peptide tritiation is a widely used procedure because of its simplicity and the known position of the radiolabel in a molecule. In addition, it is not necessary to perform a separate, time-consuming peptide synthesis to get a precursor. Nevertheless, owing to side reactions as mentioned in this section, an effective purification step using HPLC is essential.

1.1.4.2. Synthetic Precursors for Tritiation

Precursor neuropeptides for tritiation can be obtained by peptide synthesis using amino acid derivatives containing halogens, double or triple bonds. Amino acids used for precursor synthesis are shown in Fig. 2. Boc and Fmoc chemistry can equally be used although, in the presence of double bonds, usage of Fmoc chemistry is more suitable.

The most frequently used amino acids are 3',5'-diiodotyrosine and 3',5'-dibromotyrosine, although tritium labels at *ortho* positions to OH group are more labile than at *meta* positions. Opioid peptides such as enkephalins, endorphins, casomorphins, dermorphins, and deltorphins were labeled using these amino acids for the synthesis of the precursor *(12–15)*. For enkephalins, to introduce radiolabels to more stable positions, 2',6'-dibromotyrosine was used in the synthesis of the precursor *(16)*. Using *p*-iodophenylalanine or other *para*-halogenated Phe for precursor synthesis, the specific activity will be less in the tritiated peptide, but the label is more stable *(17)*. Using diiodotyrosine and *p*-iodophenylalanine at the same time, the specific activity can be increased. Tritium atoms can be incorporated into histidine or tryptophan using 2',4'-diiodohistidine-*(18)* or 5',7'-dibromotryptophan-containing peptides *(19)*, respectively.

Peptides with a disulfide bridge can cause many problems during tritiation. The bridge can cleave, and in some cases the product can have alanine instead of cysteine. Janáky et al. *(20)* tritiated vasopressin analogs, using 3',5'-di-I-Tyr^2-vasopressin analogs, and they proved the presence of alanine in the product by amino acid analysis. We have also proved this reaction in the labeling of somatostatin analogs *(21)* by mass spectrometry.

Peptides containing dehydroproline, dehydroleucine, dehydroisoleucine, propargyl, or allyl-glycine are also frequently used as precursors for tritiation. Hasegawa and associates prepared [^{3}H]Leu-enkephalin with a specific radioactivity of 4.39 TBq/mmol starting with [4,5-dehydro-Leu]enkephalin *(16)*. In addition to saturation of the double bond, tritium incorporated to positions 3 and 6, possible by ^{3}H/H isotope exchange, reaction resulted in a specific radioactivity higher than was expected by introducing two tritium atoms. 4,5-dehydroleucine was used to tritiate sub-

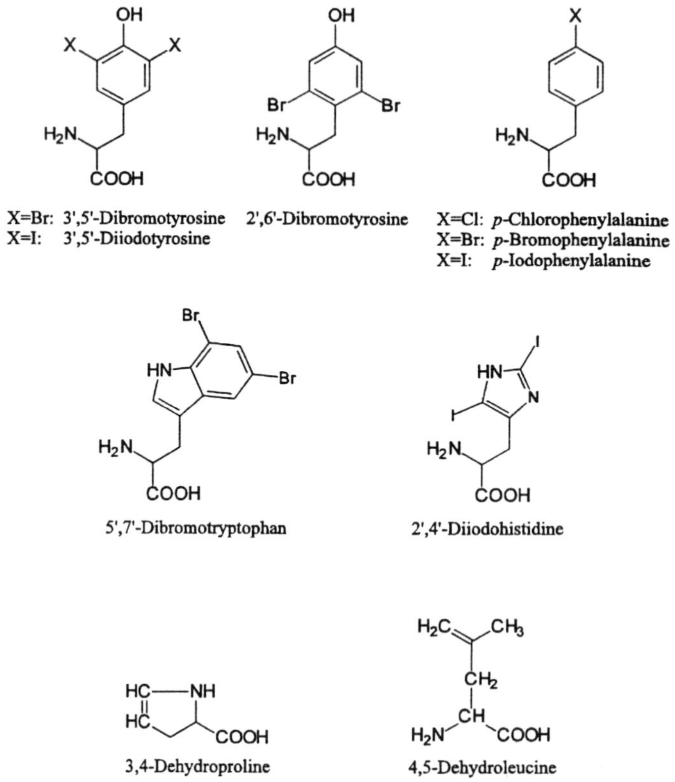

Fig. 2. Amino acid derivatives used for synthesis of precursor neuropeptides for tritiation.

stance P *(22)* and neurokinin A *(23)* with over 100 Ci/mmol (3.7 TBq/mmol) of specific radioactivity. Eberle and associates used propargylglycine to synthesize an α-MSH precursor and in the course of tritiation, propargylglycine was reduced to norvaline *(24)*. Replacement of valine to norvaline in the sequence of α-MSH did not alter binding properties of the peptide. Precursors containing dehydroproline were used to label ACTH, GnRH *(25)*, TRH, and substance P.

2. Materials

1. Tritium gas is available from Technabexport, Russia and contained at least 98% $^3H_2$.
2. Dimethylformamide, purified by vacuum distillation and dried over molecular sieves (5Å) prior to use.

3. Triethylamine, purified by vacuum distillation and dried over molecular sieves (5Å) prior to use.
4. Catalysts: PdO, PdO/$BaSO_4$, Pd/C, Rh, and uranium wire are available from Merck (Rahway, NJ).
5. Pyrophoric uranium (*see* Note 1).
6. Spectrophotometer (e.g., Shimadzu-160, Kyoto, Japan) for measurement of the amount of tritiated peptide by UV detection.
7. Liquidfluor scintillant (BDH, London, UK).
8. Liquid scintillation counter (e.g., Searle Delta 300) for counting of tritiated samples (Searle Analytic, Des Plaines, IL).
9. Berthold Radiochromatogram Tracemaster to check radiochemical purity of peptides (Laboratorium Prof. Dr. Berthold, Wildbad, Germany).
10. Thin layer chromatography (TLC) plates (silica gel $60F_{254}$/Merck).
11. TLC solvents: n-butanol:acetic acid:water (4:1:1 by volume); butanol:acetic acid:ethyl-acetate:water (1:1:1:1 by volume); butanol:pyridine:acetic acid:water (13:12:3:10 by volume).
12. HPLC facility (e.g., Vydac 218TP54 C_{18} column, eluent: 0.1% TFA/acetonitrile and 0.1% TFA/water. Detection, UV at 220 nm on a JASCO HPLC).

3. Methods

Synthesis of tritiated peptides from labeled amino acids or using [^{3}H]CH_3I can be performed in laboratories qualified to handle tracer amounts of tritium. Laboratories handling multicurie quantities of tritium should be specially equipped. Recently, a number of different glass or metal equipments have become available to perform tritium labeling. In all cases, however, closed vacuum manifolds should be used for multicurie purposes and operated under a fume hood.

3.1. Handling of Tritium Gas in a Closed Vacuum Manifold

In-house designed equipment *(26)* shown in Fig. 3. was built for multicurie applications of tritium gas. The transfer of tritium gas from/to pyrophoric uranium (Note 1) in containers (U_1, U_2) and the reaction vessel (R) is carried out in a vacuum system using valves (1-22) and a Toepler pump (P). Uranium containers and reaction vessel are connected to the manifold via ground glass joint. A vacuum pump is connected to the manifold through the T vacuum trap frozen with liquid N_2. If higher vacuum (10^{-2}–10^{-4} Pa) is needed, a diffusion vacuum pump can be used through valve 3. The pressure is measured by an open-end mercury vacuum gauge (G) and a McLeod manometer (M). An additional gas container (C) is suitable to store vapors of a volatile solvent to be condensed in the reaction vessel or hydrogen for preparing pyrophoric uranium to store tritium.

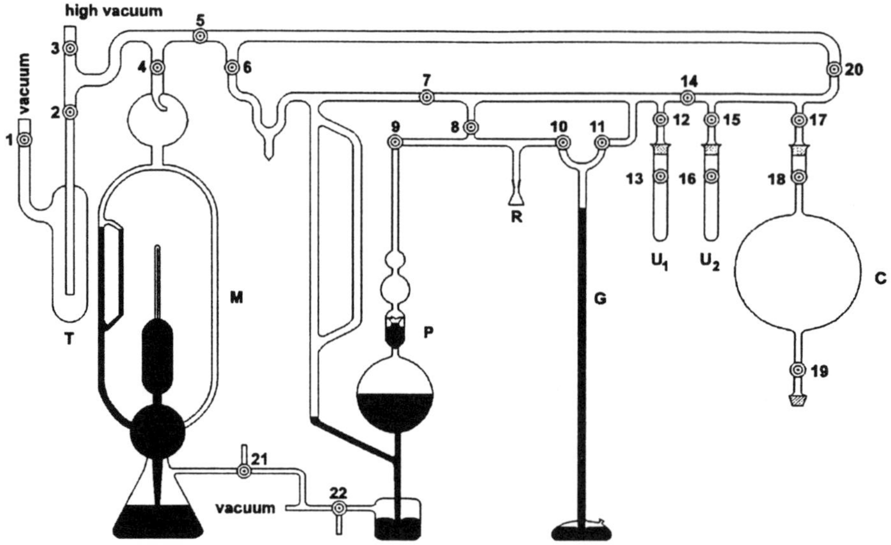

Fig. 3. Vacuum manifold for tritiation using multicurie quantities of tritium.

3.2. General Method for Tritium Labeling of Peptides

1. Dissolve 1–3 µmol of the peptide derivative in 0.5–1 mL of solvent (*see* Note 2).
2. Add 5–50 mg of Pd catalyst and 2–6 µmol of triethylamine (i.e., twice the molar amount of peptide at step 1) to the solution (*see* Note 3) and use a magnetic stirrer to stir the reaction mixture.
3. Connect the reaction vessel with the reaction mixture to the tritium manifold.
4. Freeze the solution using liquid N_2 and evacuate the air from the apparatus by vacuum opening valves 8, 14, 20.
5. Transfer 10–20 Ci (370–740 GBq) of tritium gas from container U_1, by heating to 300–400°C in an electric oven, to the reaction vessel and opening of valves 8 and 10.
6. Close valve 8 and remove the liquid N_2 to start the reaction. Remove heating oven and allow container U_1 to cool to room temperature to absorb excess tritium from the tubes onto the pyrophoric uranium. Close the valve 10 after 10 min.
7. Stir the reaction mixture, following the tritium pressure by observation of manometer G at room temperature (*see* Note 4). Reaction is complete when the tritium pressure becomes constant (30–240 min).
8. Freeze the reaction mixture with liquid N_2 so as to absorb the unreacted tritium onto the pyrophoric uranium in container U_2 through valves 8,14,15, and 16.

9. Close all the valves when the absorption is completed (as indicated by no further change in pressure on mercury vacuum gage G). Remove the liquid N_2 and take off the reaction vessel from the tritium manifold to work up the crude mixture.
10. Filter the catalyst through Whatman GF/C filter using a filter holder and syringe and wash three times with ethanol:water (1:1). Work under a fume hood.
11. Evaporate the solvent by rotary evaporator in a fume hood and remove the labile tritium by repeated evaporation (three times is usually sufficient) either with water or other protic solvent (*see* Note 5).
12. Purify the crude tritiated peptide by TLC and/or HPLC.
13. Check the chemical and radiochemical purity of the final product by chromatography (*see* Note 6).
14. Determine the amount of the labeled peptide by UV spectrometry using unlabeled peptide as standard.
15. Measure the radioactivity of the peptide by liquid scintillation counting.
16. Calculate the specific radioactivity from the total radioactivity and the amount of tritiated peptide.

4. Notes

1. Pyrophoric uranium absorbs tritium gas as uranium tritide (UH_3) at room temperature and releases it at 375°C under vacuum. Generally two reservoirs are used, one for the fresh and the other one for the used tritium gas. The used tritium should be collected separately, because it is mixed with hydrogen from the solvent and other components in the reaction mixture because of isotope exchange reactions. In the preparation of pyrophoric uranium: 1–2 g of uranium wires are washed with hexane, 50% nitric acid, water (three times), and acetone (three times) and are taken into container U_1 or U_2 and connected to the tritium manifold. Evacuate the air from it by vacuum. Add hydrogen gas from the gas container C and heat the container U_1 to 275°C by a temperature-controlled oven. The pressure of hydrogen can be followed by manometer G and occasionally more hydrogen may be added. The UH_3 is a powder. Elevate the temperature to 300–400°C and remove the liberated hydrogen from the container by vacuum. The remaining solid is the pyrophoric uranium that is used to absorb the tritium gas at room temperature.
2. Aprotic solvents such as DMF or *N,N*-dimethylacetamide are generally used. Dehalogenation reaction can occur in buffer solution in controlled pH (pH > 7). In special cases other solvents can also be used.
3. Heterogen catalysts (Pd/C, $Pd/BaSO_4$, Pd/Al_2O_3, PdO, Rh) are most popular, but other catalysts (homogenous: Wilkinson) can also be used. If the precursor peptide is halogenated, triethylamine is used to neutralize the hydrogenhalide (HI, HBr, or HCl) appearing in the reaction.
4. Room temperature is frequently applied; however, in some cases, elevated temperature may enhance the random exchange in the molecule and could increase the specific radioactivity.

5. It is an important step to remove tritium from exchangeable positions such as $-NH_2$, -NHR, -OH, -SH, because in bioassays labile tritium results in incorrect counting. In some cases, tritium label introduced by exchange reaction is not stable under acidic or basic conditions. In all cases, it is more advantageous to check the stability of the label during assay and storage.
6. Conventional detection methods are used to determine the components and the impurities. Methods include HPLC, using UV absorbance or fluorescence detectors; and TLC on silica gel plates (followed by use of suitable spray reagents or, if plates impregnated with a fluorescent dye are used, by examination of UV-absorbing spots). Radioactivity in the mass peaks of HPLC chromatograms can be measured with continuous flow radioactivity detectors or by collecting fractions and measuring their activity separately by liquid scintillation counting. Distribution of the radioactivity on a TLC plate can be determined by autoradiography or using radiochromatogram scanners.

Acknowledgments

Thanks are owed to Zsuzsa Zámbori and Éva Gulyás for technical assistance.

References

1. Wilzbach, K. E. (1957) Tritium-labelling by exposure of organic compounds to tritium gas. *J. Am. Chem. Soc.* **79,** 1013.
2. Oehlke, J., Mittag, E., Tóth, G., Bienert, M., and Niedrich, H. (1987) Enhanced incorporation of nonhydrolyzable tritium in GnRH and TRF by catalytic exchange labeling. *J. Labelled Compd. Radiopharm.* **24,** 1483–1491.
3. Zolotarev, Y. A., Kozik, V. S., Dorokhova, E. M., and Myasoedov, N. F. (1991) The investigation of solid-state catalytic hydrogenation of organic compounds. II. The synthesis of tritium-labelled valine. *J. Labelled Compd. Radiopharm.* **24,** 997–1007.
4. Cooper, D. and Reich, E. (1972) Neurotoxin from venom of cobra, *Naja siamensis. J. Biol. Chem.* **247,** 3008–3013.
5. Nägren, K., Franzén, H. M., Ragnarsson, U., and Längström, B. (1988) Synthesis of two S-(methyl-^{3}H)-labelled enkephalins and S-(methyl-^{14}C) substance P. *J. Labelled Compd. Radiopharm.* **25,** 141–148.
6. Hellio, F., Lecocq, G., Morgat, J. L., and Gueguen, P. (1990) Total synthesis of fully tritiated leu-enkephalin by enzymatic coupling. *J. Labelled Compd. Radiopharm.* **28,** 991–999.
7. Morgat, J. L. and Fromageot, P. (1973) Preparation of tritium-labelled peptide hormones of high specific radioactivity, in *Radiopharmaceuticals and Labelled Compounds II* IAEA-SM-171/60, Vienna, 109–119.
8. Teplán, I., Mezô, I., Nikolics, K., Seprôdi, J., Kéri, Gy., Bienert, M., and Klauschenz, E. (1982) Tritium Labeling of Brain peptides, in *Hormonally Active Brain Peptides* (McKerns, K. W. and Pantic, V., eds.), Plenum, New York, pp. 599–618.

9. Fraker, P. J. and Speck, J. C. (1978) Protein and cell membrane iodinations with soluble chloroamide 1,3,4,6-tetrachloro-3a,6a-diphenilglycolouril. *Biochem. Biophysical Res. Comm.* **80,** 849–857.
10. Brundish, D. E. and Wade, R. (1973) Synthesis of [3,5-$^3H_2$-Tyr^{23}]-α-corticotrophin-(1-24)-tetracosapeptide. *J. Chem. Soc. Perkin Trans* **1,** 2875–2879.
11. Bienert, M., Klauschenz, E., Nikolics, K., and Niedrich, H. (1981) Protection of methionine in peptides during iodination by sulfonium salt formation. *Int. J. Pept. Protein Res.* **19,** 310–314.
12. Morgat, J. L., Girma, J. P., and Fromageot, P. (1977) Tritium labelling of peptidic hormones *First International Symposium on Hormonal Receptors in Digestive Tract Physiology*, INSERM Symp. No. 3. p. 43, Elsevier, Amsterdam.
13. Fellion, E., Gacel, G., Roques, B. P., Roy, R., and Morgat, J. L. (1990) Tritium labelling of highly selective probes for δ-opioid receptors. *J. Labelled Compd. Radiopharm.* **28,** 867–876.
14. Hartrodt, B., Tóth, G., Neubert, K., Sirokmán, F., Baláspiri, L., and Schulz, H. (1983) Synthesis of ^{14}C- and ^{3}H-labelled β-casomorphin-5. *J. Labelled Compd. Radiopharm.* **20,** 39–52.
15. Nevin, S. T., Kabasakal, L., Ötvös, F., Tóth, G., and Borsodi, A. (1994) Binding characteristics of the novel highly selective delta agonist, [^{3}H-Ile^{5,6}]deltorphin II. *Neuropeptides* **26,** 261–265.
16. Hasegawa, H., Shinohara, Y., and Baba, S. (1991) Synthesis of ^{3}H-labelled enkephalins for metabolic and pharmacokinetic studies, in *Synthesis and Applications of Isotopically Labelled Compounds* (Buncel, E. and Kabalka, G. W., eds.), Elsevier, Amsterdam, pp. 486–489.
17. Búzás, B., Tóth, G., Cavagnero, S., Hruby, V. J., and Borsodi, A. (1992) Synthesis and binding characteristics of the highly delta-specific new tritiated opioid peptide [^{3}H]deltorphin II. *Life Sciences* **50,** PL 75–78.
18. Shu, A. Y. L. and Heys, R. (1991) Synthesis of a tritiated growth hormone releasing peptide, in *Synthesis and Application of Isotopically Labelled Compounds* (Buncel, E. and Kabalka, G. W., eds.), Elsevier, Amsterdam, pp. 85–88.
19. Hsi, R. S. P., Stolle, W. T., and Bundy, G. L. (1994) Synthesis of tritium labeled renin inhibitor ditekiren. *J. Labelled Compd. Radiopharm.* **24,** 1175–1186.
20. Janáky, T., László, F. A., and Morgat, J.-L. (1984) Tritiation of vasopressin analogues and their metabolic fate after intravenous injection in the rat. *Acta Medica Hungarica* **41(1),** 43–54.
21. Tóth, G., Gulyás, É., and Kéri, Gy., unpublished results.
22. Aharony, D., Catanese, C. A., and Woodhouse, D. P. (1991) Binding of the novel ligand [4,5-^{3}H-Leu^{10}]substance P to high-affinity NK-1 receptors on guinea pig lung membranes: modulation by GTP analogs and sulfhydryl modifying agents. *J. Pharm. Exp. Ther.* **229,** 146–155.
23. Aharony, D., Conner, G. E., and Woodhouse, D. P. (1992) Pharmacologic characterization of the novel ligand [4,5-^{3}H-Leu^9]neurokinin-A binding to NK-2 receptors on hamster urinary bladder membranes. *Neuropeptides* **23,** 121–130.

24. Eberle, A. N., Siegrist, W., Drozdz, R., Verin, V. J., Bagutti, C., Solca, F., Girard, J., and Zeller, A. (1991) Radiolabelled melanotropic peptides (α-MSH and MCH) for receptor identification, in *Synthesis and Application of Isotopically Labelled Compounds* (Buncel, E. and Kabalka, G. W., eds.), Elsevier, Amsterdam, pp. 110–115.
25. Felix, A. M., Wang, C. T., Liebman, A. A., Delaney, C. M., Mowles, T., Burghardt, B., Charnecki, A. M., and Meienhofer, J. (1977) Synthesis, biological activity and tritiation of L-3,4-dehydroproline-containing peptides. *Int. J. Pept. Protein Res.* **10,** 299.
26. Tóth, G. and Sirokmán, F. (1981) Különféle ^{3}H-jelölési módszerek, in *Izotóptechnika* (Budapest) **24**, 259–267.

20

The Use of IODO-GEN for Preparing ^{125}I-Labeled Peptides and Their Purification by Reversed-Phase High Performance Liquid Chromatography

J. Michael Conlon

1. Introduction

Several methods have been developed for introducing an ^{125}I atom into peptides and proteins. These may be divided into oxidative techniques involving generation from Na^+ $^{125}I^-$ of a reactive intermediate ($I^+/I_3{}^+$) that participates in the electrophilic attack on primarily tyrosine residues as well as on histidine; and nonoxidative techniques involving the use of conjugation reagents containing an ^{125}I atom to modify α-amino groups and/or the ε-amino group of lysine. Among the commonly used oxidative reagents, chloramine T enables a high degree of incorporation of radioactivity but results in oxidation of tryptophan to oxindole and oxidation of methionine to its sulfoxide derivative. The use of lactoperoxidase/hydrogen peroxide restricts oxidative damage, but incorporation of radioactivity is generally low owing to self-iodination of the enzyme and yields are irreproducible. Among the conjugation reagents, the use of ^{125}I-labeled Bolton-Hunter reagent (N-succinimidyl-3[4-hydroxyphenyl] propionate) provides a gentle method of labeling, but the reagent is expensive, the reaction is slow and technically quite difficult to perform (owing to the instability of the reagent towards moisture), and incorporations are often low and variable.

Since its introduction by Fraker and Speck in 1978 *(1)*, IODO-GEN (1,3,4,6-tetrachloro-3α,6α-diphenylglycoluril) (Fig. 1) has become the reagent of choice for radioiodination of neuroendocrine peptides *(2–5)*. The reaction is simple to perform and gives reproducibly high levels of incorporation of radioactivity. Because the reagent is virtually insoluble in water and is used as a film

From: *Methods in Molecular Biology, Neuropeptide Protocols*
Edited by: G. B. Irvine and C. H. Williams Humana Press Inc., Totowa, NJ

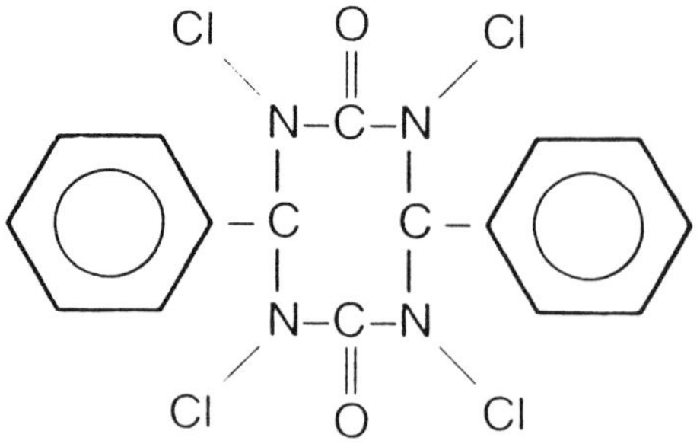

Fig. 1. The structure of IODO-GEN (1,3,4,6-tetrachloro-3α,6α-diphenylglycoluril).

on the walls of reaction vessels, oxidative damage to sensitive residues in the peptide is minimal.

Irrespective of the method of iodination employed, it is necessary to separate as completely as possible the radiolabeled peptide from the unreacted starting material in order to obtain a tracer of sufficiently high specific activity to be of use in radioimmunoassay, radioreceptor assay, or autoradiography. Techniques such as selective adsorption to diatomaceous materials (e.g., talc or microfine silica and gel permeation chromatography) give tracers of low specific activity; ion-exchange chromatography is time-consuming and results in a sample dilution that may be unacceptable. Reversed-phase high performance liquid chromatography (RP-HPLC) combines rapidity and ease of operation with optimum separation of labeled and unlabeled peptide. The availability of wide-pore C_{18} and C_4 columns permits good recoveries of labeled peptides with molecular mass (M_r) >6000.

The strategy proposed for the preparation of undamaged tracers of high specific activity involves trace labeling (approx 10% of the molecules are radioiodinated) using IODO-GEN followed by complete separation of labeled and unlabeled peptide by reverse-phase HPLC. Trace-labeling has the advantage of minimizing oxidative damage and production of the diiodotyrosyl derivative of the peptide.

2. Materials

2.1. Apparatus

No specialized equipment is required to carry out the iodination, but the reaction should be carried out in an efficient fume hood. Reaction takes place in a 1.5-mL natural-colored polypropylene microcentrifuge (Eppendorf tube), e.g., Fisherbrand Cat. No. 05-407-5 (Fisher Scientific, Pittsburgh, PA) (*see* Note 1), immersed in an ice bath. A nitrogen or argon cylinder is required for removal of solvent. Liquids are dispensed with 10- and 100-µL pipets (e.g., Gilson Pipetman) that, because of inevitable contamination by radioactivity, should be dedicated to the reaction and stored in a designated area.

For HPLC, a system capable of generating reproducible linear two solvent gradients is required. Again, because of contamination by radioactivity, an injector, e.g., Rheodyne Model 7125 with 1-mL sample loop; a 1-mL leak-free injection syringe, e.g., Hamilton Gastight #1001 (25 × 0.46 cm) analytical reversed-phase column (*see* Note 2 on column selection); and fraction collector capable of collecting a minimum of 70 samples, e.g., Frac 100 (Pharmacia, Uppsala, Sweden) should be dedicated to the radioiodination procedure. A UV-detection system and chart recorder/integrator are not necessary.

2.2. Chemicals

2.2.1. Iodination Reagents

1. IODO-GEN (Pierce, Rockford, IL): The reagent should be stored in a desiccator and the bottle protected from light.
2. Dichloromethane (stabilized, HPLC grade, Fisher): Redistillation is not required.
3. Carrier-free Na^+ $^{125}I^-$ in 0.1*M* NaOH (3.7 GBq/mL; 100 mCi/mL, Amersham, Arlington Heights, IL).
4. 0.2*M* disodium hydrogen phosphate/sodium dihydrogen phosphate buffer, pH 7.5.

2.2.2. Chromatography

1. Solvent A: Add 1 mL trifluoroacetic acid (Pierce HPSC/spectro grade) to 1000 mL water (*see* Note 3).
2. Solvent B: Mix 700 mL acetonitrile (Fisher optima grade) with 300 mL water and add 1 mL trifluoroacetic acid. The solvents should be degassed, preferably by sparging with helium for 1 min, but passage through a filter is unnecessary.

3. Methods

3.1. Radioiodination

1. Dissolve 1.5 mg of IODO-GEN in 2 mL of dichloromethane.
2. Pipet 20 mL of the solution into a polypropylene tube and remove the solvent in a gentle stream of nitrogen or argon at room temperature. The aim is to produce a film of IODO-GEN on the wall of the tube. If the reagent has formed a visible clump, the tube should be discarded and a new tube prepared. According to the manufacturer's instructions (Pierce), the tubes can be stored in a vacuum desiccator for up to 2 mo, but it is recommended that a tube is prepared freshly for each reaction. The tube is set in an ice bath for 10 min prior to the iodination.
3. Dissolve 10 nmol of the peptide (*see* Note 4) in 0.2*M* sodium phosphate buffer (100 µL) and pipet the solution into the chilled IODO-GEN-coated tube.

4. Add the Na^+ $^{125}I^-$ solution (5 μL; 0.5 mCi or 10 μL; 1 mCi depending on the quantity of tracer required).
5. Allow the reaction to proceed for between 1 and 20 min (*see* Note 5). The contents of the tube should be gently agitated by periodically tapping the side of the tube with a gloved finger.
6. Reaction is stopped by aspirating the contents of the tube into a solution of 0.1% (v/v) trifluoroacetic acid/water (500 μL) contained in a second polypropylene tube.

3.2. Chromatography

1. Prior to carrying out the iodination, the column is equilibrated with 100% solvent A at a flow rate of 1.5 mL/min for at least 1 h.
2. The instrument is programmed to increase the proportion of solvent B from 0–70% over 60 min using a linear gradient (*see* Note 6).
3. The reaction mixture is injected onto the column (*see* Note 7). The fraction collector, programmed to collect 1-min fractions, is started and the linear elution gradient is begun. A total of 60 fractions are collected.
4. At the end of the chromatography, the column is irrigated with 100% solvent B for 30 min. The column can be stored in this solvent.
5. The radioactivity in aliquots (2 μL) of each fraction is counted in a gamma scintillation counter (*see* Note 8 for optimum storage conditions of tracer).

The results of a typical radioiodination are illustrated in Fig. 2. The reaction mixture comprised 10 nmol [Tyr^0]bradykinin and 0.5 mCi Na^+ $^{125}I^-$ and the reaction time was 1.5 min. Unreacted free iodide was eluted at the void volume of the column (fractions 3–5). The fraction denoted by the bar (tube 40) was of high specific activity (approx 74 TBq/mmol; 2000 Ci/mmol). The earlier eluting minor peak of radioactivity probably represented the diiodotyrosyl derivative. Before use in radioimmunoassay or radioreceptor studies, the quality of the tracer is assessed by incubating an aliquot (approx 20,000 cpm) with an excess of an antiserum raised against bradykinin (1:1000 dilution) in 0.1*M* sodium phosphate buffer, pH 7.4 (final volume 300 μL) for 12 h at 4°C. Free and bound radioactivity are separated by addition of 100 μL of a 10 mg/mL solution of bovine γ-globulin (Sigma, St. Louis, MO) and 1 mL of 20% (w/v) solution of polyethylene glycol 6000 (Sigma, approx M_r 8000) in water followed by centrifugation (1600*g* for 30 min at 4°C). Under these conditions, >90% of the radioactivity is bound to antibody.

4. Notes

1. Irreversible binding of most peptides to the walls of polypropylene tubes is much less than to glass or polystyrene.
2. The choice of column is dictated by the nature of the radiolabeled peptide to be purified. For relatively small (M_r < 3000) peptides, good resolution and

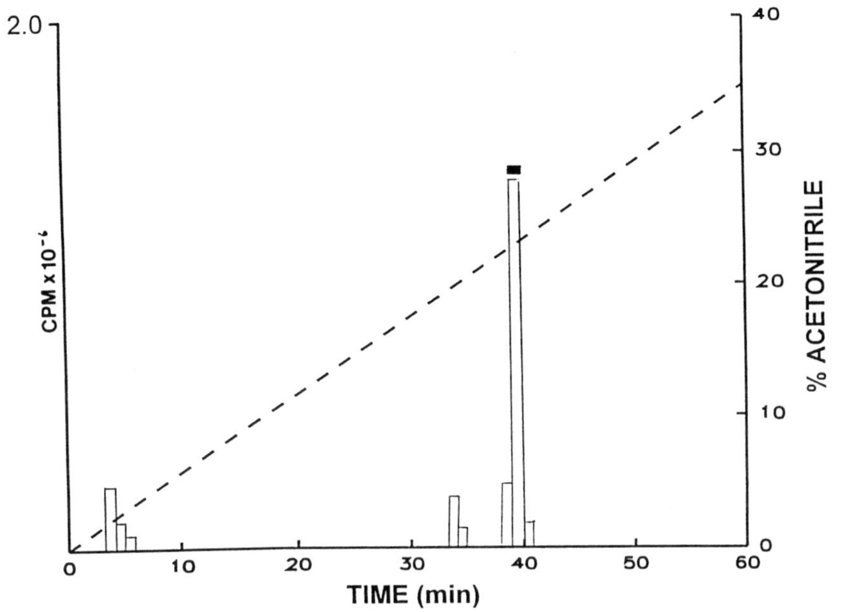

Fig. 2. RP-HPLC on a (0.46 × 25 cm) Vydac 218TP54 (C_{18}) column of the reaction mixture following incubation of 10 nmol [Tyr^0]bradykinin with 0.5 mCi $Na^{125}I$ in an IODO-GEN-coated tube for 1.5 min. Fractions (1 min) were collected and the fraction denoted by the bar contained tracer of high specific activity (74 TBq/mmol). The dashed line shows the concentration of acetonitrile in the eluting solvent.

recoveries are generally obtained with (0.46 × 25 cm) narrow pore (80 Angstrom), 5-μm particle size octadecylsilane (C_{18}) columns such as Supelcosil LC-18-DB (Supelco, Bellefonte, PA), Ultrasphere ODS (Beckman, Duarte, CA), or Spheri-5 RP-18 (Brownlee/Applied Biosystems, Foster City, CA). For purification of radiolabeled tracers of higher molecular mass (M_r > 3000), the use of columns containing wide-pore (300 Angstrom) 5-μm particle size C_{18} packing materials is recommended. Suitable columns include Vydac 218TP54 (Separations Group, Hesperia, CA), Spherisorb wide-pore C18 (Phase Separations; U.K.), Waters Delta-Pak C18 (Millipore, Milford, MA), and Ultrapore C18 (Beckman). For purification of tracers of molecular mass >6000, such as insulin/proinsulin and the pituitary glycoprotein hormones, sharper peaks and better recoveries of radioactivity may be obtained using wide-pore silica with C_3 (e.g., Beckman Ultrapore C3) or C_4 (e.g., Vydac 214TP54) columns.

3. Suitable water can be obtained using a Milli-Q purification system (Millipore) supplied with water that has been partially purified by single distillation or with a deionization resin.

4. As many peptides are relatively insoluble in buffers of neutral pH, it is recommended that the peptide first be dissolved in a minimum volume (approx 5 µL) of 0.1% (v/v) trifluoroacetic acid/water and the volume made up to 100 µL with 0.2*M* sodium phosphate buffer, pH 7.5.
5. The optimum reaction time must be determined for each peptide, but some general guidelines can be given. For small peptide (<15 amino acid residues) containing a tyrosine residue in a sterically unhindered region of the molecule e.g., the N- or C-terminal residue, the reaction proceeds rapidly and reaction times of between 0.5 and 2 min are generally sufficient. For larger peptides and proteins, it is often necessary to prolong the reaction time to between 10 and 15 min. Neurohormonal peptides that do not contain a tyrosine residue but possess a sterically unhindered histidine (e.g., neurokinin B *[6]* and secretin) may be iodinated using IODO-GEN, but longer reaction times (up to 20 min) may be required.
6. The relatively steep gradient (0 → 70% solvent B, equivalent to 0 → 49% acetonitrile over 60 min) is recommended as the initial elution conditions when preparing a radiolabel for the first time. Better separation of the tracer and the unlabeled peptide will be obtained using a shallower gradient. For example, relatively hydrophilic peptides such as [Tyr^0]bradykinin (Fig. 2) and [Tyr^8]substance P may be purified using a gradient of 0 → 35% acetonitrile over 60 min, whereas hydrophobic peptides such as corticotropin-releasing hormone and neurokinin B may be purified using a gradient of 21 → 49% acetonitrile over 60 min.
7. In some published protocols, radiolabeled peptide and unreacted $^{125}I^-$ are separated prior to RP-HPLC, e.g., by adsorption on Sep-Pak C_{18} cartridges (Waters) or by gel-permeation chromatography on a Sephadex G-10 desalting column (Pharmacia). This procedure is not necessary and it is recommended that the reaction mixture is injected directly onto the HPLC column.
8. The stability of radiolabeled peptides varies dramatically, with useful lives ranging from a few days to more than 2 mo. Repeated freezing and thawing of the tracer is not recommended and so the HPLC fraction(s) containing the radiolabel should be aliquoted immediately, diluted with one volume ethanol or methanol, and stored at as low a temperature as possible (–70°C is preferred). The volume of the aliquot should be related to the size of a typical assay.
9. Although IODO-GEN is almost insoluble in water, its solubility in buffers containing detergent increases appreciably. Under these circumstances, oxidative damage to the peptide may occur and the use of an alternative reagent IODO-BEADS (Pierce) should be considered. IODO-BEADS comprise the sodium salt of N-chloro-benzenesulfonamide immobilized on nonporous polystyrene beads *(7)*. The reaction conditions and purification protocol using IODO-BEADS are the same as using IODO-GEN except that one or more of the beads are substituted for the film of IODO-GEN. High incorporations of radioactivity are observed even in the presence of detergents or chaotropic reagents, e.g., urea.

References

1. Fraker, P. J. and Speck, J. C. (1978) Protein and cell membrane iodination with a sparingly soluble chloroamide 1,3,4,6-tetrachloro-3α,6α-diphenylglycoluril. *Biochem. Biophys. Res. Commun.* **80,** 849–857.
2. Salacinski, P. R. P., McLean, C., Sykes, J. E. C., Clement-Jones, V. V., and Lowry, P. J. (1981) Iodination of proteins, glycoproteins and peptides using a solid-phase oxidizing agent, 1,3,4,6-tetrachloro-3α,6α-diphenyl glycoluril (Iodogen). *Anal. Biochem.* **117,** 136–146.
3. Conlon, J. M., Whittaker, J., Hammond V., and Alberti, K. G. M. M. (1981) Metabolism of somatostatin and its analogues by the liver. *Biochim. Biophys. Acta* **677,** 234–242.
4. O'Harte, F., Smith, D. D., Lanspa, S. J., and Conlon, J. M. (1992) Measurement of T-kinin in rat plasma using a specific radioimmunoassay. *Regul. Peptides* **41,** 139–148.
5. Conlon, J. M. (1991) Regionally-specific antisera to human β-preprotachykinin, in *Methods in Neurosciences,* vol. 6 (Conn, P. M., ed.), Academic, San Diego, pp. 207–221.
6. Conlon, J. M. (1991) Measurement of neurokinin B by radioimmunoassay, in *Methods in Neurosciences,* vol. 6 (Conn, P. M., ed.), Academic, San Diego, pp. 221–231.
7. Markwell, M. A. K. (1982) A new solid-state reagent to iodinate proteins. *Anal. Biochem.* **125,** 427–432.

21

Production of Antisera Using Peptide Conjugates

Thomas E. Adrian

1. Introduction

Since an immunogen requires both an antigenic site and a T-cell receptor binding site, there is a minimum size necessary *(1)*. Natural immunogens have a molecular weight >5000. Small molecules such as neuropeptides may be able to bind to the surface of B-cells, but do not stimulate an immune response. Such molecules are known as haptens. A hapten is an incomplete immunogen but can be made immunogenic by coupling to a suitable carrier molecule. There are a variety of different crosslinking agents utilized for the coupling of peptides to carrier proteins; examples of each type are covered in this chapter. In the case of larger neuropeptides, such as calcitonin gene-related peptide (CGRP), it is possible to stimulate an immune response by presenting the peptide together with a carrier such as polyvinyl-pyrrolidone without the need for conjugation. This method, described in Section 3.5., has proven to be useful for the author for a number of different peptides. Unfortunately, however, the success of this method compared with the responses to conjugated peptides is largely a matter of trial and error.

2. Materials

2.1. Protein Carriers

Factors governing the choice of the carrier include immunogenicity, solubility, and availability of functional groups for crosslinking. Whereas substances such as mucopolysaccharides, poly-L-lysine, and polyvinyl-pyrrolidone have been used as carriers, proteins are more widely used. Common protein carriers include serum albumin, ovalbumin, hemocyanin, and thyroglobulin. To find the very best immunogen, it would be ideal to prepare conjugates with several different carriers with a range of hapten to carrier coupling ratios (*see*

From: *Methods in Molecular Biology, Neuropeptide Protocols*
Edited by: G. B. Irvine and C. H. Williams Humana Press Inc., Totowa, NJ

Note 1). The cost and time involved will usually make this impractical, however, and it is therefore necessary to carefully select the carrier most suitable for a particular antigen. In the classical hapten carrier system, T-lymphocytes recognize processed carrier determinates and cooperate with B-cells that produce hapten-specific antibody response. Note that the amounts of carrier and peptide for conjugation are given in molar terms in the following conjugation protocols. This is necessary because of the wide variation in the molecular weights of potential carriers and neuropeptides to be coupled. The carrier protein represented by 100 nmol is approx 7 mg of bovine albumin, 4.5 mg of ovalbumin, 15 mg of gamma globulin, and 70 mg of thyroglobulin.

2.1.1. Bovine Serum Albumin

Because of its wide availability, high solubility, and relatively high number of coupling sites, albumin is a popular choice as a carrier for weakly antigenic compounds. Albumin has a molecular weight of 67,000 and has 59 lysine residues providing primary amines useful for conjugation.

2.1.2. Ovalbumin

Ovalbumin (egg albumin) also has wide availability since it is the primary protein constituent of egg white. This protein is smaller than serum ovalbumin with a molecular weight of 45,000, but contains 20 lysine residues, 14 aspartic acid, and 33 glutamic acid residues for conjugation *(2)*. Ovalbumin exists as a single polypeptide chain with an isoelectric point of 4.6. Half of its 400 residues are hydrophobic. Caution should be exercised in handling of ovalbumin since it is denatured at temperatures above 56°C or even by vigorous shaking.

2.1.3. Hemocyanin

Keyhole limpid hemocyanin (KLH) is a copper-containing protein that belongs to a family of non-heme proteins found in arthropods and mollusca. The KLH exists in five different aggregate states at neutral pH that will dissociate into subunits above pH 9.0 *(3)*. KLH is a valuable carrier protein because of its large molecular mass (approx 1×10^6 to 1×10^7) and numerous lysine groups for coupling. This property of dissociation at high pH can be utilized because it increases the availability of angiogenic sites and this can produce improved antigenic responses *(3)*. The disadvantage of using KLH as a carrier protein is its poor water solubility. Whereas this makes the protein difficult to handle, it does not impair its immunogenicity.

2.1.4. Thyroglobulin

Thyroglobulin is another large molecular-weight protein with a limited solubility. The advantage of thyroglobulin as a carrier comes from its large content of tyrosine residues that can be used for conjugation using the diazo reaction. The molecular weight of thyroglobulin is 670,000.

2.2. Coupling Agents

Chemical coupling agents or crosslinkers are used to conjugate small peptide haptens to large protein carriers. The most commonly used crosslinking agents have functional groups that couple to amino acid side-chains of peptides (*see* Table 1). There are several things that need to be taken into consideration when selecting a bifunctional coupling reagent. First is the selection of functional groups; this can be used to produce a specific type of conjugate. For example, if the only primary amine available is in the N-terminal end of a peptide, then a coupling agent can be selected to specifically couple in this position, leaving the carboxyl terminal end of the peptide free and available as an antigenic site. Second is the length of the cross bridge; the presence of a spacer arm may make the hapten more available and therefore produce a better immune response. Third is whether the crosslinking groups are the same (homobifunctional) or different (heterobifunctional). Once again this can alter the specificity of the coupling reaction. Last is whether the coupling reaction is chemical or photochemical. For a good antigenic response it is necessary to maintain the native structure of the protein complex, and this can be achieved only using mild buffer conditions and near neutral pH. The reactive groups that can be targeted using crosslinkers include primary amines, sulfhydryls, carbonyl, and carboxylic acids (*see* Note 2). It is difficult to predict the proximity of protein–peptide interactions. The use of bifunctional reagents with spacer arms can prevent steric hindrance and make the hapten more available for producing a good immune response.

2.2.1. Carbodiimide

Carbodiimide condenses any free carboxyl group (nonamidated C-terminal aspartate, or glutamate residue) or primary amino group (N-terminal or lysyl residue), to form a peptide bond (CO-NH). The most commonly used water-soluble carbodiimide is 1-ethyl-3-(3-dimethylaminopropyl)-carbodiimide hydrochloride (CDI or EDC). This coupling agent is very efficient, easy to use, and usually couples at several alternative points on a peptide, giving rise to a variety of antigenic responses *(4)*. Because of the unpredictable nature of the antibody responses, the process using this bifunctional agent has been termed "shotgun-immunization." If it is necessary to raise antisera to a particular region of the peptide, then this is not the method of choice. Furthermore, because the peptide bond linking the hapten to the carrier cannot rotate and holds the hapten physically close, steric hindrance is considerable.

Table 1
Bifunctional Crosslinking Agents Useful for Conjugation of Peptide Haptens to Protein Carriers for Immunization

Coupling method	Chemical name and formula	Reactive toward	Comments
CDI or EDC	1-Ethyl-3-[3-dimethylaminopropyl] -carbodiimide hydrochloride 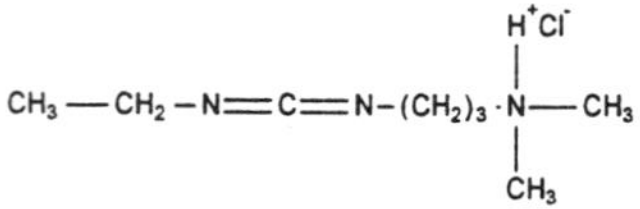	Carboxyls or primary amines	Condenses any free carboxyl or primary amino groups (C- or N-terminus or side-chain), rapid, efficient, considerable steric hindrance, nonspecific coupling ("shotgun-immunization").
Glutaraldehyde	Glutaraldehyde $CHO—(CH_2)_3—CHO$	Primary amines	Forms bridge between primary amino groups (N-terminus or lysine residue), rapid, efficient, allows free rotation of hapten and thereby avoids steric hindrance.
Diazo	Bis-diazotized benzidine 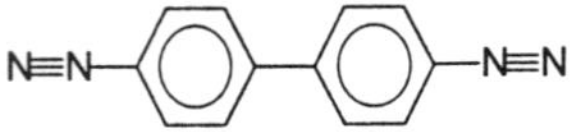	Tyrosyl or histidyls	Bis-diazonium salts bridge residues between hapten and carrier, very specific, overnight procedure, spacer arm holds hapten away from carrier, usually results in an excellent antigenic response.
Sulfo-SMCC	Sulfo-succinimidyl 4-[N-maleimidomethyl] -cyclohexane-l-carboxylate 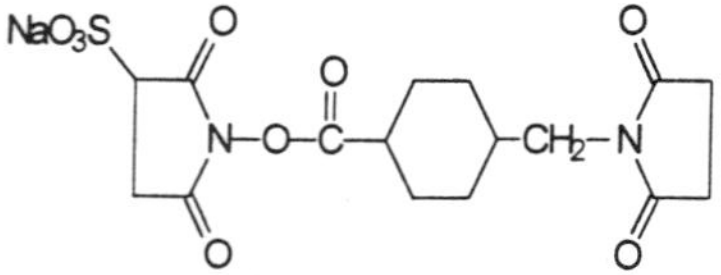	Primary amine and sulfhydryl	Contains a maleimide to react with a free sulfhydryl group and an NHS-ester group to react with a primary amino group, efficient and stable coupling, highly specific (e.g., coupling a synthetic peptide with a terminal cysteine residue).

Carbodiimide reacts with available carboxyl groups to form an active O-acylurea intermediate, which is unstable in aqueous solution, making it ineffective in two-step conjugation procedures. This intermediate then reacts with the primary amine to form an amide derivative. Failure to immediately react with an amine results in hydrolysis of the intermediate. Furthermore, hydrolysis of CDI itself is a competing reaction during the coupling and is dependent on temperature. 4-Morphylinoethansulfonic acid (MES) can be used as an effective carbodiimide reaction buffer in place of water. Phosphate buffers reduce the efficiency of the CDI reaction, although this can be overcome by increasing the amount of CDI used to compensate for the reduction in efficiency. Loss of efficiency of the CDI reaction is even greater with Tris, glycine, and acetate buffers and, therefore, use of these should be avoided.

2.2.2. Glutaraldehyde

Glutaraldehyde links primary amino groups (either the N-terminal or lysyl residues) on both the peptide hapten and the carrier. This linkage allows free rotation of the hapten, which reduces possible steric hindrance that may otherwise block access to the immune system by the large carrier molecule.

2.2.3. Sulfo-Succinimidyl 4-[N-Maleimidomethyl] Cyclohexane-1-Carboxylate (Sulfo-SMCC)

A peptide with a free sulfhydryl group, such as a synthetic peptide with a terminal cysteine residue, provides a highly specific conjugation site for reacting with sulfo-SMCC (Pierce, Rockford, IL). This crosslinker contains a maleimide group that reacts with free sulfhydryl groups, along with an N-hydroxysuccinimidyl ester group that reacts with primary amines *(5)*. All peptide molecules coupled using this chemistry will display the same basic antigenic conformation. They will have a known and predictable orientation, leaving the molecule free to interact with the immune system. This method can preserve the major epitopes on a peptide while enhancing the immune response. The water solubility of sulfo-SMCC, along with its enhanced maleimide stability, makes it a favorite for hapten carrier conjugation.

2.2.4. Bis-Diazotized Benzidine

Bis-diazonium salts bridge tyrosyl or hystidyl residues between the hapten and carrier. Overnight treatment of benzidine at 4°C with nitrous acid (hydrochloric acid and sodium nitrite) results in the two amino groups being diazotized. These two diazonium groups allow coupling at both ends of the molecule. Although limited in coupling points, diazotized benzidine provides a spacer arm holding the hapten away from the carrier and usually results in an excellent antigenic response.

2.3. Adjuvants

The adjuvant is important for inducing an inflammatory response. The author has had continued success over two decades using Freund's adjuvant. Various synthetic adjuvants are available such as AdjuPrime (Pierce), and Ribi Adjuvant System (Ribi Immunochemicals, Hamilton, MT) *(6)*. The author has had very limited success with the latter adjuvant when used to immunize with several different small peptides. In contrast, responses with Freund's adjuvant run in parallel have always been good.

The conjugate is injected in the form of an emulsion made with Freund's adjuvant, which is a mixture of one part of detergent (Arlacel A, Sigma, St. Louis, MO) with four parts of n-hexadecane. This permits slow release of the coupled hapten into the circulation and may serve to protect labile antigens from degradation. Freund's adjuvant alone ("incomplete") causes an inflammatory response that stimulates antibody formation, and when made "complete" by addition of 1 mg/mL heat-killed *Mycobacterium butyricum,* this response is further enhanced. It is convenient to purchase complete and incomplete Freund's adjuvant ready mixed (Sigma or Calbiochem, San Diego, CA).

2.4. Synthetic Peptides as Haptens

There is considerable advantage to be gained by using synthetic peptides as haptens. First, it is possible to raise region-specific antibodies directed perhaps to one end of a neuropeptide molecule. Second, it is possible to insert particular amino acids with specific side-chains for coupling. For example, a peptide can be synthesized with an extra cysteine residue at one end of the molecule to enable coupling using sulfo-SMCC *(5)*. Alternatively, a tyrosine residue can be inserted that will enable specific coupling through the bis-diazotized benzidine reaction *(7)*. This latter approach is valuable, since the same synthetic peptide can then serve as a radioligand in the radioimmunoassay, with the assurance that the antibodies raised will not be directed toward the tyrosine residue that will be iodinated *(7)*.

2.5. Choice of Animal for Immunization

There are several factors that need to be considered when choosing animal species for an immunization program, including cost, ease of handling, and the volumes of antisera required (*see* Note 3). Small animals (such as rats and mice) have low blood volumes and present difficulties with bleeding. Large animals such as sheep or goats are expensive to house, particularly over long periods. Rabbits or guinea pigs provide a near optimal solution, since they are relatively cheap to house and bleeding an ear vein or cardiac puncture in guinea pigs can provide between 10 and 30 mL of plasma from each bleed. For production of MAb immunization of mice is required.

3. Methods

3.1. Carbodiimide Procedure

1. Dissolve the peptide to be coupled (400 nmol) and protein carrier (100 nmol) in a small volume of water (<1 mL if possible) (*see* Note 4).
2. Add carbodiimide (200 µmol, 50 mg) to this solution.
3. Incubate the mixture overnight at 4°C (*see* Notes 5 and 6).

3.2. Glutaraldehyde Procedure

1. Dissolve the peptide (400 nmol) and carrier protein (100 nmol) in 1 mL of 0.1*M* phosphate buffer pH 7.5 (*see* Note 4).
2. Add glutaraldehyde (30 µmol, 1.5 mL of a 0.02*M* solution) dropwise over 15 min.
3. Incubate the mixture overnight at 4°C (*see* Notes 5 and 6).

3.3. Sulfo-SMCC Procedure

1. Activate the carrier (100 nmol) by conjugating the active ester of sulfo-SMCC (2 µmol) via an amino group, in PBS pH 7.2 (with 5 m*M* EDTA) for 60 min at room temperature. This reaction results in the formation of an amide bond between the protein and the crosslinker with the release of sulfo-N-hydroxysuccinimide as a byproduct.
2. If desired, the carrier protein can then be isolated by gel filtration to remove excess reagent using a gel such as Sephadex G25 (*see* Note 5). Desalt by eluting the column with PBS, collecting 0.5-mL fractions. Locate the protein peak using a protein assay (BioRad [Hercules, CA] micro method).
3. At this stage, the purified carrier possesses modifications generated by the crosslinker, resulting in a number of maleimide groups projecting from its surface. The maleimide group of sulpho-SMCC is stable for several hours in solution at physiological pH. Therefore, even after activation and purification, the greatest possible activity will still be left for conjugation with the peptide.
4. The maleimide group of sulpho-SMCC reacts at pH 7.0 with free sulfhydryls on the peptide to form a stable thioether bond. The peptide (100 nmol) with a free sulfhydryl group is incubated with the maleimide-activated carrier in 1 mL PBS (with 5 m*M* EDTA) for 2 h (or overnight if more convenient) at 4°C (*see* Note 6).
5. Keep sulfo-SMCC away from moisture since it is subject to hydrolysis.

3.4. Diazo Procedure

1. Freshly prepare bis-diazotized benzidine on each occasion in the following manner: dissolve benzidine hydrochloride (80 µmol, 20.5 mg) in 10 mL of 0.18*M* HCl and gently mix overnight with 1 mL of 0.16*M* $NaNO_2$ (11 mg). This reaction should take place in an ice bath inside a cold room, and the temperature should never be allowed to rise above 4°C.
2. Dissolve the peptide (400 nmol) in 130 µL (1 µmol) of fresh bis-diazotized benzidine solution.

3. Add $NaHCO_3$ (40 µmol, 3.4 mg) followed immediately by the addition of the carrier predissolved in a minimal volume of aqueous solution (carrier protein 100 nmol) at 4°C.
4. Adjust the pH to 9.8 with NaOH using a microelectrode.
5. Incubate the mixture overnight at 4°C (*see* Notes 5 and 6).

3.5. Immunization Without Conjugation to Carrier Protein

1. This procedure, using polyvinyl-pyrrolidone as a noncovalent carrier, can be very valuable for peptide antigens with more than 40 amino acid residues. Good results have been obtained with 30–40 amino acid peptides, but it is unlikely to produce useful antisera with smaller antigens. As well as being quick and easy, it has the added advantage of retaining tertiary peptide structure.
2. Dissolve the peptide in aqueous solution containing a 100 molar excess of polyvinyl pyrrolidone (with respect to peptide concentration).
3. Emulsify the solution in Freund's adjuvant in the manner described in Section 3.6.

3.6. Making an Emulsion in Freund's Adjuvant

1. Dissolve conjugate in water (between 10 and 100 nmol of conjugated peptide/mL, 1 mL for each rabbit being immunized).
2. Make Freund's adjuvant by mixing one part of Arlacel A with four parts of n-hexadecane (allowing a little more than 1 mL per rabbit).
3. For primary injections only complete Freund's adjuvant is used. This is made by addition of 1 mg/mL heat-killed mycobacteria. (Boosts are given in incomplete Freund's.)
4. When preparing the emulsion, care should be taken to ensure that the oil remains in the continuous phase. Injection of aliquots of the aqueous conjugate solution into the oil via a fine bore needle, followed by repeated aspiration and ejection of the crude emulsion, will produce the required result.
5. A simple test for the success of the preparation is to add a drop of the emulsion to the surface of water in a tube. The emulsion should stay in a single droplet without dispersing; confirming that it is immiscible and thus oil-phase continuous.

3.7. Immunization Procedure

The emulsified conjugate can be administered in a variety of ways. For rabbits, the most frequently used are multiple (30–50) id injections in the neck or back region, or by four sc injections into each groin and axilla *(8)*. The latter is the method we have successfully adopted for more than 20 yr. Injection into the foot-pads, which was at one time commonly employed, provides no advantage in terms of antibody response and should be avoided to prevent distress to the animals.

The procedure is as follows:

1. Bleed the animals and collect preimmune serum for later comparison with antisera produced by the immunization procedure.
2. Primary inoculations are given in complete Freund's adjuvant, 0.5 mL of emulsion into each groin and axilla.
3. Booster injections are given at 2–4 wk intervals, in the same manner but with incomplete Freund's adjuvant. The optimum is probably about 4 wk, but time constraints and cost may necessitate a shorter immunization schedule. With small synthetic haptens, 3–5 or more boosts may be required to produce the desired high titer or high avidity antibody (*see* Notes 7 and 8).
4. After the first and subsequent boosts, blood should be collected from an ear vein to test for the antibody titer and avidity.

3.8. Antibody Characterization for Radioimmunoassay

1. Serial dilutions of antisera are incubated with radiolabeled peptide under routine assay conditions in order to determine a working dilution.
2. The maximum displacement of radioactively labeled hormone from the antibody by the minimum amount of unlabeled peptide (the maximum displacement slope) is one of the main criteria for radioimmunoassay sensitivity.
3. Rapid screening for slope can be achieved by the addition of small amounts (usually approx 10 fmol) of standard peptide to one set of a series of replicate antiserum dilutions set up to determine the antiserum titer. The amount of standard used should reflect the useful range (e.g., the concentration at which a hormone circulates).
4. Antibody heterogeneity may be owing to use of nonhomogenous antigens for immunization, polymerization or degradation of the hapten or carrier after coupling, or individual differences in the lymphocytic response to the antigen *(9)*.
5. Existence of heterogeneity can be revealed by Scatchard analysis. However, for high titer antisera there is frequently effectively only a single class of antibody that predominates in the reaction. Other populations of lower concentrations and avidities make insignificant contributions.
6. Specificity should be tested using related peptides.

4. Notes

1. The ratio of hapten reactive with protein is usually arranged to be in excess of 4:1 to achieve better antigenicity with respect to the hapten. Some authorities prefer ratios as high as 40:1, but in our experience this gives a lower affinity antibody response. This is presumably owing to conjugation between hapten molecules rather than between hapten and carrier. The hapten and protein carriers should both be present in high concentration to increase the efficiency of the crosslinking between the molecules.
2. Coupling of a hapten at a specific site gives more chance of governing which part of the peptide becomes the antigenic determinant for the antibody, since the particular area of peptide where coupling occurs is likely to be hidden from immune surveyance.

3. Success in raising antisera is to some extent a hit-or-miss affair. Some workers have needed to immunize many animals to produce useful antisera, whereas three or four rabbits may produce the desired product in another immunization program. This depends, in part, on the antigenicity of the peptide and on the goals set for sensitivity and specificity. Some peptides are particularly susceptible to proteolysis (such as members of the vasoactive intestinal polypeptide family) or oxidation (such as cholecystokinin). In general, these less stable peptides make relatively poor antigens.
4. Usually the coupling agent is added after the hapten and carrier have been mixed together in order to minimize self-polymerization of either component.
5. Although not usually necessary, excess unreacted hapten and toxic byproducts may be removed by dialysis or gel permeation chromatography.
6. Quantification of the success of the coupling reaction is conveniently obtained by addition of a small amount of radioactively-labeled hapten to the mixture prior to adding the coupling agent. Thus, before and after the reaction, small aliquots are removed and chromatographed. Small disposable columns containing Sephadex G-25 (Pharmacia, Uppsala, Sweden) are ideal for this purpose. Proportion of radiation diluting in the high molecular weight position, together with the carrier, indicates the amount of coupling achieved.
7. If a particular animal has been boosted three or four times without producing detectable antibody, then the likelihood of subsequent antibody production is small and further effort is unprofitable.
8. On other occasions when animals do show a response but further boosting results in little improvement in avidity or titer, then variation in the coupling method for subsequent boosts can help. Changes including a different carrier protein or crosslinking agent, or both, may result in production of a higher titer or more avid antisera.

References

1. Germain, R. (1986) The ins and outs of antigen processing and presentation. *Nature* **322,** 687–689.
2. Harlow, E. and Lane, D. (1988) *Antibodies: A Laboratory Manual.* Cold Spring Harbor Laboratory, Cold Spring Harbor, NY, pp. 56–100.
3. Bartel, A. and Campbell, D. (1959) Some immunochemical differences between associated and dissociated hemocyanin. *Arch. Biochem. Biophys.* **82,** 2332–2336.
4. Bauminger, S. and Wilchek, D. (1980) The use of carbodiimides in the preparation of immunizing conjugates. *Methods Enzymol.* **70,** 151–159.
5. Samoszuk, M. K., Petersen, A., Lo-Hsueh, M., and Rietveld, C. (1989) A peroxide-generating immunoconjugate directed to eosinophil peroxidase is cytotoxic to Hodgkin's disease cells *in vitro. Antibody Immunocon. Radiopharm.* **2,** 37–46.
6. Chedid, L. and Lederer, E. (1978) Past, present and future of the synthetic immunoadjuvant MDP and its analogs. *Biochem. Pharmacol.* **27,** 2183–2186.

7. Adrian, T. E., Bacarese-Hamilton, A. J., and Bloom, S. R. (1985) Measurement of cholecystokinin octapeptide using a new specific radioimmunoassay. *Peptides* **6,** 11–16.
8. Vaitukaitis, J., Robbins, J. B., Nieschlag, E., and Ross, G. T. (1971) A method for producing specific antisera with small doses of immunogen. *J. Clin. Endocrinol. Metab.* **33,** 988–991.
9. Parker, C. W. (1976) *Radioimmunoassay of Biologically Active Compounds.* Prentice Hall, Englewood Cliffs, NJ, pp. 36–67.

22

Radioimmunoassay

Thomas E. Adrian

1. Introduction

There is frequently a need to measure concentrations of neuropeptides in tissue perfusates, tissue extracts, chromatographic column fractions, and so on. Since the concentrations of neuropeptides encountered are often low (usually the low fmol/mL range in perfusates and pmol/g range in tissue extracts), it is necessary to employ sensitive methods for their measurement (*see* Note 1). Adequate sensitivity usually equates with the use of immunoassays, although sensitive bioassays are available for some neuropeptides, and these can be made specific by the judicious use of high-affinity receptor antagonists *(1)*. Since the advent of the liquid-phase radioimmunoassay, several modifications to the general concept have been employed, including the use of radiolabeled antibodies instead of peptide ligand (immunoradiometric assay), nonradioactive markers (enzyme immunoassays), and "sandwich assays" in which the ligand is trapped between one antibody coupled to a solid phase such as a microplate and a second antibody attached to some detection molecule. Refinements such as signal enhancement and the sandwich technique can improve both the sensitivity and the useful range of an assay, but at the expense of considerable method development. As a general rule, such investment is worthwhile in the development of assays with a clinical or other commercial need, but not for research. Indeed, for assays employing a single antiserum, the sensitivity accomplished is dependent on the characteristics of that antiserum. In such a system, it is hard to improve on a monoiodinated ligand molecule for sensitivity and convenience. This chapter describes the radioimmunoassay technique as it could be applied for the measurement of a newly isolated neuropeptide.

The radioimmunoassay of a hormone depends on the competition between radioactively-labeled and unlabeled hormone for the specific binding site of an

From: *Methods in Molecular Biology, Neuropeptide Protocols*
Edited by: G. B. Irvine and C. H. Williams Humana Press Inc., Totowa, NJ

antibody (*see* Notes 2 and 3). The amounts of antibody and labeled hormone are fixed, the only variable being the unlabeled hormone concentration. The higher the concentration of unlabeled hormone, the less radioactively labeled hormone will be bound to antibody. After separation of the bound from the free hormone, the amount of labeled hormone bound to antibody can be assessed by counting the radioactivity. By using known quantities of pure hormone, a standard curve can be prepared with which unknown samples can be compared. The amount of antibody in the assay is adjusted to bind about 50% of the label added (in the absence of unlabeled hormone). This point, being midway between the nonspecific binding of the label and the maximal binding by excess antibody, gives the steepest standard curve slope and also optimal precision in counting.

2. Materials

2.1. Antisera

Antisera are available commercially for a variety of neuropeptides, but for new or novel peptides it may be necessary for an investigator to raise his own antibodies as described in Chapter 21. Selection of an appropriate antiserum is essential as this is a critical part of the radioimmunoassay. No substantial improvement can be made to an assay to compensate for an antibody with poor slope, whereas improving the quality of a label will usually increase the sensitivity of an assay with an antibody with a steep slope. Small aliquots (to avoid repeated freezing and thawing) of precious antibodies can be safely stored lyophilized in vacuum-sealed vials at –20°C for several years.

2.2. Radiolabeled Neuropeptide Ligand

The neuropeptide can be iodinated on a convenient tyrosine residue or with harsher oxidizing conditions on a histidine residue. Oxidizing agents include chloramine T (N-chloro-*p*-methyl benzenesulfonamide), lactoperoxidase, or the insoluble compound iodogen as described in Chapter 20 *(2,3)*. Selection of the oxidizing agent is largely a matter of personal choice and convenience. Oxidation produces damage to peptides, particularly by sulfoxidation of methionine residues and cleavage of tryptophans, often resulting in a marked loss of immunoreactivity. There is little evidence that one oxidizing agent causes more damage than another, although lactoperoxidase may be more reproducible because of the longer time involved (minutes rather that 10–20 s with chloramine T). To reduce oxidative damage, a useful approach is that of "trace iodination," the aim of which is to iodinate only a small proportion of available peptide using relatively mild oxidation *(4)*. It is essential to purify trace-iodinated products using a high-resolution technique such as reversed-phase (RP) or ion exchange HPLC. If a tyrosine is not available, or if labeling a tyrosine will inhibit antibody binding, then the nonoxidative (Bolton and Hunter reagent) labeling procedure can be used *(5)*. Following iodination, the peptide should be purified using RP-HPLC to produce pure monoiodo-peptide (*see* Chapter 20).

Many radiolabeled neuropeptides are commercially available (Amersham, Arlington Heights, IL or New England Nuclear, Boston, MA). Whereas these commercial preparations are expensive, they are generally of high quality and for small quantity or infrequent usage they may prove to be most convenient.

The useful shelf-life of a radiolabeled neuropeptide is usually 2–4 mo, but some labeled peptides appear to be more stable than others (*see* Note 4). Labels should be divided into single-use aliquots and either frozen at –20°C or lyophilized.

2.3. Neuropeptide Standards

In order to readily detect any error in any single standard point, 8–10 concentrations of standard are preferred for the preparation of the standard curve, which should cover the full range of expected unknown values. Samples that are off-scale are dealt with by dilution prior to assay. Standards should ideally be prepared using the peptide sequence of the species for which the assay is to be used. However, in contrast to peptide hormones, most neuropeptides are highly conserved and differences in amino acid sequence do not usually occur between mammalian species. Because of the lability of peptide hormones, standards should be very carefully prepared, freeze-dried in aliquots, and then stored below –20°C. The composition of a suitable freeze-drying solution is given in Table 1.

The lactose helps to produce a large soluble pellet so that the material dissolves readily, the albumin prevents peptide binding to surfaces, the cysteine helps prevent peptide oxidation damage, aprotinin inhibits the action of any proteolytic enzymes, and the citric acid keeps the pH low for better peptide stability during freeze-drying. The citric and formic acids should be omitted and the pH of the solution neutralized for lyophilizing acidic peptides. Similarly, the cysteine hydrochloride should be omitted when freeze-drying cyclic peptides to prevent reduction of disulfide bonds.

The use of neuropeptide-containing control samples is essential to evaluate within assay precision and also as a check on standards between assays. Although the World Health Organization has undertaken to provide international reference standard preparations of some hormones, research standards are not available for neuropeptides.

Table 1
A Solution for Freeze Drying Neuropeptide Standards

Lactose	140 m*M* (50 g/L)
Bovine albumin	40 μ*M* (2.5 g/L)
Citric acid	10 m*M* (2 g/L)
L-Cysteine hydrochloride	6 m*M* (1 g/L)
Aprotinin (Trasylol)	800,000 Kallikrein inhibitor U/L
Formic acid	100 m*M* (4.25 mL 90% formic acid/L)

2.4. Buffers

Optimal assay conditions vary with different antisera, and the best buffer type, molarity, pH, and incubation time must therefore be determined for each system in order to achieve the greatest sensitivity. Phosphate and veronal buffers, with a molarity of about 0.05 and pH of 7–8 are, however, suitable for most assays. The composition of a suitable phosphate buffer is given in Table 2.

Bovine serum albumin is usually added (0.3–1.0% w/v as needed) to radioimmunoassay buffers to reduce adherence of peptides to surfaces. This is particularly important for basic peptides that bind to negatively charged glass and plastic surfaces.

2.5. Separation of Bound from Free Ligand

There are a multitude of possible procedures for the separation of bound from free peptide fraction in the radioimmunoassay. These include second antibody precipitation, organic solvent precipitation, chromatography, and adsorption onto agents such as talc or dextran-coated charcoal. For simplicity, reliability, and low cost, the author has standardized all of his procedures using precipitation with polyethylene glycol (PEG 6000) with crude gamma globulin as a carrier to produce a visible precipitate. Dextran-coated charcoal is equally efficient and cheap, but has the disadvantage of being messy. Second antibody procedures are very reliable, but each batch of antibody requires optimization and the reagents are costly.

2.6. Equipment for Radioimmunoassay

The equipment for setting up radioimmunoassays can be broadly categorized, by degree of sophistication, into hand pipets, automatic dilutors, and pipeting stations. Large numbers of micropipeting operations are fatiguing to perform, particularly where precision is of prime importance. The computing of standard curves and calculation of results is another stage that is both

Table 2
Sorenson's 60 m*M* Phosphate Buffer Solution pH 7.4, Suitable for Radioimmunoassay

Buffer volume	1 L, g	2 L, g	5 L, g
KH_2PO_4	1.82	3.64	9.1
Na_2HPO_4	9.5	19.0	47.5
EDTA	3.72	7.44	18.6
NaN_3	0.5	1.0	2.5

Dissolve salts in deionized water, adjust pH to 7.4, and store at 4°C.

time-consuming and prone to error. Automation can provide the benefits of a greater sample throughout, greater precision and reliability, and can help to reduce errors in calculation. The system chosen depends on the number of samples assayed, the stage of development of the assay and, above all, on cost.

2.6.1. Hand Pipeting

The addition of small volumes of samples and standards and the repeated dispensing of antibodies and labels require considerable precision. The addition of standards can be made, for example, by disposable precision glass capillaries that come in a large range of volumes (e.g., Drummond [Broomall, PA] Microcaps). Though these are expensive, it is reassuring to see the correct volumes actually being added to the all important tubes of the standard curve. Individual sample additions can be made with a gas-tight syringe (e.g., a Hamilton [Reno, NV] syringe with a Chaney adaptor) or by one of the many displacement pipets with disposable individual sample tips. Although widely used, displacement pipets often become quite inaccurate after continued usage. The volume delivered is simple to check gravimetrically, however. The precise repetitive addition of labels and antisera is conveniently performed by the use of a repeating gas-tight dispenser (e.g., Hamilton repeating dispenser ratchet syringe or Eppendorf [Madison, WI] repeating dispenser).

2.6.2. Autodilutors

The repetitive nature of plasma and reagent addition can be overcome by the use of a more expensive automatic dispenser. Several such machines are marketed; examples include Tecan (Tecan US, Research Triangle Park, NC), and the digital diluter (Hamilton). However, these can be slower to use than simple hand-held pipets.

2.6.3. Automatic Preparation Units

The entire assay procedure may be effectively automated on systems that feature automatic sample handling and linked centrifuge separation and counting systems. Labeling of individual test tubes can be avoided as they remain in the same rack throughout with consequent saving of time. The Tomtec (Hamden, CT) microplate system is an excellent example. Of course, the high cost of such systems, which can equal two years' salary for an experienced technician, may more than balance out the advantages, even for a large laboratory. It should not be forgotten that all such systems entail a certain loss of time for cleaning and maintenance.

2.6.4. Gamma Counting

Because large numbers of samples are handled in hormone radioimmunoassay, a γ-counter that can automatically load hundreds of samples has been an essential item. Manufacturers include Wallac (Gaithersburg, MD), and Packard Instruments (Meriden, CT). Machines that count up to 10 tubes simultaneously are available for rapid throughput. An alternative is the use of a low-cost, multiple-well counter (Wallac). These machines have to be hand-loaded, but make short work of high-count assays with less risk of mechanical breakdown. Automated data reduction is essential for high throughput and several good radioimmunoassay software packages are available such as RIACALC from Wallac.

In addition to the automatic counters, an independent single-well counter is very useful for label checks and counting the purification fractions after iodination, as this prevents radioactive contamination of the larger machines.

3. Methods

3.1. Assay Incubation Conditions

1. It is not always possible for radioimmunoassays to be set up in a theoretically optimal manner, but the assay of a neuropeptide should be as precise (reproducible), as sensitive (able to measure small amounts), and as specific (showing no cross-reaction with other substances) as possible (*see* Notes 5–7).
2. The sensitivity of an assay is governed by the product of two factors—the standard curve slope (steepness) and the precision or lack of variability of each value. A sensitive assay will thus have a steep standard curve and little error between replicates, allowing small differences between unknowns to be measured confidently.
3. As a general rule, a reduction in the amount of labeled peptide added to an assay will produce a standard curve with a steeper slope as the unknown competes for antibody binding with fewer labeled molecules.

4. The preparation of labeled peptide of high specific activity, i.e., approximately one radioactive iodine atom per hormone molecule (monoiodinated ligand), which is the practical maximum, will enable the amount of label added to each tube to be reduced to about 1 fmol without a very long counting time being necessary. This allows detection of a comparable quantity of unknown hormone, which is particularly appropriate for neuropeptides, which exist at low concentrations because of their inherent biological potency.
5. In practice, the steepness of the standard curve, and thus the sensitivity of assay, increases with incubation time, reaching a maximum after about 5–6 d. Thus, for assays in which sensitivity is important because of low neuropeptide concentrations, a long incubation period is used, whereas for assays of tissue extracts in which higher peptide concentrations may be encountered, an incubation period of 1–2 d may be sufficient.
6. A low temperature (4°C) and the addition of a bacteriostat such as sodium azide (0.5 g/L) to prevent bacterial contamination are important. This temperature, as low as possible without the risk of actually freezing the medium, also helps to minimize proteolytic degradation and evaporation. Furthermore, the avidity of antibodies increases with reduction in temperature and therefore greater sensitivity is achieved at 4°C than at room temperature.
7. The total volume in the assay tube must be commensurate with the volume of sample available (often limited) and the cost of reagents. If the volume of sample and amount of antiserum used in the assay are of no consequence, the counting time can be halved if all the assay volumes (and thus the total amount of label) are doubled. Volumes have to be scaled down, and counting times increased, to cope when available sample volumes are low.

3.2. Assay Format

The scheme used in our laboratory for a typical neuropeptide radioimmunoassay is shown in Table 3. Note that the total volume in each tube is identical at 300 μL. The sample volume shown is 10 μL but may vary between 0.1 and 100 μL depending on the anticipated concentration of peptide. The buffer volume would be adjusted accordingly.

Several inclusions found to be useful are (*see* Note 4):

1. Blank: A tube with all assay reagents except antibody, to evaluate the nonspecific binding of labeled hormone, and so on.
2. Half concentration of labeled hormone: A tube that contains half the amount of label added to the rest of the tubes in the assay. This is useful for assessing whether greater sensitivity could have been achieved in an assay by adding a smaller amount of label, and may also be useful in determining whether there is a danger of a "hump" if less label was used (*see* Chapter 21).
3. Twice concentration of labeled hormone: A tube that contains twice the amount of label added to the other assay tubes, which is useful for assessing the sensitiv-

Table 3
Standard Format for a Neuropeptide Radioimmunoassay

Constituent	Blank	½× Label	2× Label	Zero	Standard	Sample
Buffer	240 µL	215 µL	140 µL	190 µL	90 µL	190 µL
Standard	None	None	None	None	100 µL	None
Sample	None	None	None	None	None	10 µL
Vehicle	10 µL	10 µL	10 µL	10 µL	10 µL	None
Label	50 µL	25 µL	100 µL	50 µL	50 µL	50 µL
Antibody	None	50 µL	50 µL	50 µL	50 µL	50 µL

Vehicle: solution with the same characteristics as the sample. For example, most neuropeptides (except acidic ones, such as CCK-8) are extracted in boiling 0.05*M* acetic acid; the same acid solution is then used as vehicle in the assay. Label: counts adjusted according to specific activity of label and required sensitivity of assay, usually 2000–4000 cpm/tube (equivalent to 1–2 fmol of a monoiodo peptide). The antiserum should be diluted to bind approx 50% of the label in the absence of any standard ("cold") hormone (*see* Chapter 21, Section 3.8. for antibody characterization).

ity of the assay with respect to the amount of label added. This tube may be used for calculating the specific activity of the labeled hormone.

4. Zero hormone tube: Tubes containing zero unlabeled hormone (but that contain hormone label and antibody, and so on) should be run at frequent intervals throughout the assay to detect and assess the degree of any drift, e.g., owing to reagent deterioration or syringe fatigue.
5. Excess antibody: A tube containing excess antibody that will bind all of the bindable hormone (to avoid wasting antiserum, this may be an antiserum of high titer but not of sufficiently high avidity to be useful for assays). This tube is a measure of the immunological integrity of the labeled hormone, as all should be bound, and in conjunction with the blank tube, it can be used to assess the quality of the label.
6. In general, as a label becomes older and starts to deteriorate, the percentage of bound radioactivity in the blank tube rises and the percentage of bound radioactivity in the excess antibody tube falls. However, high excess binding is not always associated with good label, as iodinated fragments of hormones may occasionally bind well to this excess antibody, but are not suitable for an assay. All tubes, including standards, controls, and test samples should be set up in duplicate in order to increase precision and detect random errors (*see* Note 8).

3.3. Selection of Antisera

1. Antisera vary widely in terms of their specificity, avidity or slope, titer, and susceptibility to nonspecific interference. Great care should be taken in selecting an antiserum for a particular radioimmunoassay procedure. The ideal antiserum has a steep slope (the sensitivity will then enable detection of small changes

between adjacent tubes), a high titer (so that a small amount of antiserum can be used to measure several thousand samples), and its binding to the peptide being measured is not affected by nonspecific factors. High titer and lack of nonspecific effects often, but not always, go together.

2. Selection of an antibody with a particular regional specificity may provide an assay that, for example, only measures biologically active peptides and not inactive precursors or degradation products. Similarly, antisera should be selected to avoid undesirable crossreactivity with related neuropeptides or hormones.
3. Procedures for testing titer, and slope of antisera are outlined in Chapter 21.

3.4. Trace Iodination Using Chloramine T—Typical Procedure

1. Dissolve the peptide (2 nmol) in 30 µL of 0.4*M* phosphate buffer pH 7.4. The high molarity buffer is needed to neutralize the NaOH vehicle in which the $Na^{125}I$ is received.
2. Add 0.2 nmol of $Na^{125}I$ (5 µL Amersham IMS 30, 0.2 nmol = 0.5 mCi).
3. Add 40 nmol (11 µg) chloramine T in 5 µL of 0.04*M* phosphate buffer pH 7.4.
4. Mix by aspiration for 10 s.
5. Add 100 nmol (20 µg) sodium metabisulfite in 10 µL of 0.04*M* phosphate buffer pH 7.4. This reducing agent should be omitted when iodinating peptides with disulfide bonds to prevent reduction of these bonds. Instead rapidly proceed to step 6.
6. Dilute fivefold in starting column eluent and inject on HPLC.

3.5. Trace Iodination Using Lactoperoxidase—Typical Procedure

1. Dissolve the peptide (2 nmol) in 30 µL of 0.2*M* sodium acetate buffer pH 5.0.
2. Add 0.2 nmol of $Na^{125}I$ (5 µL Amersham IMS 30, 0.2 nmol = 0.5 mCi).
3. Add 0.002 nmol lactoperoxidase (80 IU = 1 mg = 13 nmol) in 5 µL buffer as in Section 3.5.1.
4. Add 10 nmol hydrogen peroxide (7 µL of a 1:1000 dilution of 20 vol H_2O_2) in buffer as above.
5. Mix by aspiration for 15 min.
6. Dilute fivefold in starting column eluent and inject on HPLC.

3.6. Ligand Purification by RP-HPLC

1. For purification of trace-iodinated radioligands, it is essential to choose conditions that will separate labeled from nonlabeled peptide, as well as removing damaged (oxidized) products. Therefore, the conditions chosen usually include either isocratic elution of the column or a very shallow gradient. Such conditions will even allow separation of peptides with different tyrosines labeled when the peptide contains more than one tyrosine residue.
2. Reverse C-18 or phenyl columns are usually suitable for purification of neuropeptides.
3. A suitable low-viscosity solvent is acetonitrile usually with 0.1% trifluoroacetic acid (TFA) as counterion. Most neuropeptides elute from C-18 columns at

acetonitrile concentrations between 20 and 40%, but conditions should be tested for individual peptides.

4. Further details on reversed-phase chromatography of radiolabeled peptides can be found in Chapter 20.

3.7. Testing of the Radioligand

1. Following iodination, it is wise to test the integrity of each labeled peak eluting from the HPLC.
2. We have found that overnight testing of blank (nonspecific binding), zero binding (no added cold peptide), binding in the presence of a single standard concentration (usually 10 fmol/tube), and excess binding are essential.
3. Measurement of binding in the presence of half and twice label concentrations gives further information on the integrity of the label (*see* Table 4) and the latter allows calculation of the specific activity of the label.
4. The good label fraction will have low blank, high excess, and a substantial fall from zero binding in the presence of cold standard hormone. This good label fraction will also have high specific activity, signified by small differences between zero, half, and twice label concentrations.
5. If the percentage of half-label binding is lower than zero (and perhaps even the zero binding lower than the twice label), then this indicates that the assay is susceptible to the hump of hook discussed in Chapter 21. If zero binding is much greater than 50%, this can be overcome by increasing the antibody dilution. If it is 50% or lower, then increasing the label concentration (higher counts) will usually correct the problem.
6. It is a good idea to test new label fractions in parallel with the previous label. The indication of the degree of improvement will be valuable in assessing the usefulness of the new label before precious samples are wasted.

3.8. Calculation of Specific Activity

1. The specific activity of a labeled peptide is the amount of radioactivity per mole of peptide. The unit of specific activity is based on the SI unit of radioactivity, the Bequerel (Bq), which is equivalent to 1 disintegration per second.

   Specific activity in Bq/fmol = (disintegrations/s)/fmol

   Notes: *1 Bq = 0.27×10^{-10} Curies. Pure ^{125}I has a specific activity of 81 Bq/fmol, so the specific activity of monoiodinated peptides should be close to this.*
2. Calculate the amount of radioactive hormone added to each tube by reading off the standard curve a tube to which precisely twice the usual concentration of label has been added.
3. Obtain absolute disintegrations/s by correcting the count for the efficiency with which the particular γ-counter performs. This may be of the order of 60–70% (note that the manufacturer's relative "efficiency" is much higher and is based on different criteria).

4. Specific activity in Bq/fmol = $\frac{\text{additional counts in twice label tube} \times 100/\text{efficiency \%}}{\text{fmol of hormone producing the additional counts}}$

   Example: assuming counter is 67% efficient, total count in standard tube is 102 counts/s, count in twice-labeled tube is 202 counts/s (i.e., additional count = 100 counts), twice-labeled tube reads 2.2 fmol/tube (not per mL) on standard curve.

   $$\text{Specific activity} = [(202 - 102) \times 100/67]/2.2 = 67.8 \text{ Bq/fmol}$$

3.9. Separation Methods

1. Many methods of separating the free labeled hormone from antibody-bound labeled hormone are available, the most widely used being immunoprecipitation (e.g., by addition of goat antirabbit antibody), chemical precipitation (e.g., with alcohol, polyethylene glycol, or ammonium sulfate) and adsorption (e.g., with cellulose, talc, or charcoal). The amounts of each of these agents required will vary with the assay and conditions always need to be tested. It is convenient to test blank (nonspecific binding), zero (binding of antisera in the absence of unlabeled cold peptide), and binding in the presence of an excess (50–100-fold) amount of antibody. The most commonly employed methods are polyethylene glycol precipitation, second antibody precipitation, or adsorption to dextran-coated charcoal.
2. For simplicity's sake, chemical precipitation using polyethylene glycol is employed for all assays in the author's laboratory. In addition to providing ease of handling for large numbers of samples, the polyethylene glycol method is rapid and cheap. The optimal amount of polyethylene glycol and carrier γ-globulin added should be assessed for each assay. Generally, for assay tubes containing 300 µL, addition of 1 mg of gamma globulin (50 µL of a 20 mg/mL solution) followed by 0.5 mL of a 20% solution of polyethylene glycol (PEG 6000), a final PEG concentration of 12% will suffice. The tube contents are then thoroughly mixed, centrifuged for 20 min at 4°C, and the supernatant removed using a Pasteur pipet attached to a vacuum pump. The entire separation procedure is carried out at 4°C.
3. Typically, second antibody dilutions will be of the order of 1:500 to 1:5000 final, and carrier serum will need to be added at a final dilution of 1:100 to 1:1000.
4. Charcoal precipitation, another commonly used procedure, will usually require 4–24 mg norit A charcoal per tube. The charcoal is coated with dextran (70T dextran at 10% of the amount of charcoal), prepared as a constantly mixed slurry with a final volume of 250–500 µL being added to each tube prior to centrifugation and separation.

3.10. Scintillation Counting and Expression of Results

1. The precipitated antibody-bound peptide fractions are counted and the percentage of total counts bound is plotted against standard concentration.
2. Results are best expressed in molar concentrations; this has three advantages. First, it conforms with the Systeme Internationale (SI units). Second, several

Table 4
Specific Problems Encountered in Radioimmunoassay Using PEG Separation and Their Remedies

Problem	Possible cause	Other observations	Remedy
High blank (nonspecific binding)	Contaminated label (nonpeptide radioactivity)	Associated with low zero and excess binding, shallow or flat standard curve.	Purify label or reiodinate.
	Damaged label (degraded)	Associated with low zero and excess binding, initially good label may deteriorate with age, difference between zero and twice label binding small (apparent high specific activity), shallow standard curve.	Reiodinate.
	Endogenous antibodies in sample	Very high zero, unknowns give negative values.	Measurement not possible without careful extraction.
Low zero binding	Contaminated label	High blank, low excess binding, shallow standard curve.	Purify label or reiodinate.
	Damaged label (degraded)	Associated with low zero and excess binding, initially good label may deteriorate with age, difference between zero and twice label binding small (apparent high specific activity), shallow standard curve.	Reiodinate.

	Low specific activity label	Low blank, high excess, shallow standard curve, twice label large fall from zero, ½× label well above zero.	Add less label (calculate specific activity) repurify label, or reiodinate.
	Inadequate γ-globulin	Low excess binding, shallow standard curve.	Retest procedure.
	Incorrect conditions (buffer, pH, incubation time)	Shallow standard curve.	Retest procedure.
	Inadequate antibody	Dilution wrong or deteriorated.	Recheck with fresh antibody.
Low excess binding	Damaged label	High blank, low zero, shallow standard curve.	Reiodinate
	Contaminated label	High blank, low zero, shallow standard curve.	Repurify or reiodinate.
Shallow standard curve, all other parameters normal	Weak standards	Quality control samples read high.	Check for errors in preparation, storage, damage, loss by adsorption to glassware; freeze-dry with protein; use vial only once.
Shallow standard curve, with high zero	Excessive antibody, too little label	First standard and twice label tube may be higher than zero (hump)	Retest conditions.
"Negative" sample concentration	Standard curve conditions not identical to samples	The majority of samples may have binding greater than zero.	Make standard curve conditions identical to those of samples.

(continued)

Table 4 *(continued)*

Problem	Possible cause	Other observations	Remedy
Inappropriate peptide concentration in unknowns	Presence of agents in samples that may interfere with binding (protein, urea, bacterial proteolytic enzymes)		Use bacteriostat and enzyme inhibitors, test for serial dilution parallelism of unknown to standards.
	Inaccurate standards	Unexpected standard curve, when standards too strong, slope is too steep and quality controls too low, when standards too weak slope is too shallow and quality controls too high.	Recheck standards.
	Damaged label	Very shallow standard curve giving inaccurate readings.	Reiodinate.
	Antibody with inappropriate specificity	If specificity too broad, then biologically inactive peptides with similar amino acid sequence will also be measured;	Test reaction with related peptides.
		if specificity too narrow, then biologically active fragments of peptide may be ignored.	Test ability of peptide fragments to displace label from antibody.

neuropeptides have separate molecular species of different sizes with different molecular weights (e.g., somatostatin and cholecystokinin) and antisera may be used that fully crossreact with these different molecules. In this situation, results quoted in absolute gravimetric units are inappropriate and confusing. Finally, the concentrations of different neuropeptides and metabolites can be directly compared when all are considered on a molar basis. Presumably, also, the hormone receptors on a cell react in molar (and not weight) terms.

4. Notes

1. Compared with the bioassay, the radioimmunoassay offers several advantages. It is more sensitive, often capable of detecting a few fmol of hormone, which is well beyond the capabilities of most bioassays. Because of this sensitivity, the radioimmunoassay allows the assay of small volumes. Second, radioimmunoassay is potentially extremely specific (especially with a carefully chosen region-specific antiserum). Third, the simplicity of the system permits the simultaneous estimation of a large number of samples.
2. In spite of its inherent advantages, the radioimmunoassay is a structural assay recognizing a particular molecular configuration, rather than the biological activity of a peptide hormone. The antigenic determinant, or site on a peptide recognized by the binding site of an antibody molecule, usually comprises only three or four amino acids and, if the peptide has a folded structure, these need not even be adjacent residues. As the antibody recognizes only a particular structure, it may detect inactive precursors or fragments, as well as unrelated proteins that happen to exhibit the amino acid sequence of the antigenic determinant. It should be remembered, therefore, that results from this type of assay should not be directly equated with the degree of biological activity. An additional problem is that any external agent that nonspecifically alters the binding of antigen to antibody will cause interference in the assay and give rise to error.
3. It is clear that any substance that interferes in any way with the precise quantitative binding of the antibody will cause erroneous results. In practice this is all too common. An interfering substance may either be related to the antigen or be dissimilar (i.e., the interference is nonspecific). Related substances include other peptides with a similar amino acid sequence or biologically inactive molecules such as inactive neuropeptide precursors or degradation fragments. The specificity of each antibody should be tested against synthetic fragments and other related peptides when possible. The former will give precise information on the regional specificity of the antiserum. Nonspecific interference in the assay can be produced by many factors, but proteins are frequently involved. This can be circumvented by extraction of the peptide from the crude sample by precipitating the proteins using organic solvents or, better still, by the use of disposable reverse-phase cartridges (i.e., Sep-Paks, Millipore, Bedford, MA) (*see* Chapter 1). This latter method, although expensive, has the added advantage of allowing concentration of the sample by lyophilization. Furthermore, it has been our experience that samples eluted from Sep-Paks in 50%

acetonitrile with 0.1% TFA can be assayed directly without loss of sensitivity. Indeed the titer and slope of some antisera are actually improved in the presence of this solvent. Susceptibility to nonspecific interference may be sufficient cause to exclude one antiserum in favor of another when characterizing a radioimmunoassay.

4. For a sensitive assay, a high specific activity tracer is essential. It is therefore essential that the state of the radioactive antigen be constantly monitored. This can be done quite easily by including some control tubes in each run. These are the blank, excess antibody, zero peptide, and half and twice label concentration tubes mentioned above in assay format (*see* Section 3.2.). The blank tube contains only buffer and tracer but no antibody. If the label is damaged or contains traces of free ^{125}I, the blank will be high. This condition is usually associated with low excess and low zero-binding tube values. When the tracer has a low specific activity, this will be indicated by a reduced zero, normal excess binding, and large binding differences between the twice label concentration tube, the zero, and the half label concentration tube.
5. The sensitivity of the assay depends on the slope of the standard curve and the error between replicates. The parameters that affect the slope include the antibody avidity, incubation time, nature of the buffer, specific activity of the labeled antigen, and the rate of degradation. A robust assay will function with no loss of sensitivity, precision, or accuracy over a wide range of conditions. However, not all assays are robust, so it is very important that the ideal conditions be found for each antibody.
6. Reproducibility is a very important aspect of radioimmunoassay. Nonrobust assays are susceptible to a lack of reproducibility. With such assays, small variations in incubation time, buffer, pH, and so on, may have profound effects on apparent neuropeptide concentrations. Thus, it is critical not to diverge from established optimal conditions. Intra-assay variation should be monitored using control samples covering the useful range of the assay that are included in each run.
7. It is common knowledge that laboratories often disagree over true concentrations of a peptide. One reason for discrepancies is the difference in standard curve environments. Some laboratories will erroneously employ buffer standard curves and use these to calculate peptide concentrations in samples such as plasma. In a robust assay, the consequences of this may be minimal. However, nonspecific protein interference is commonly encountered in nonrobust assays. It is thus essential that the environment of the standard curve is identical to that of the samples. If measurements are to be made in crude plasma samples, then the standard curve should contain hormone-free normal plasma produced by affinity chromatography using antisera to the peptide being measured. Better still, the samples should be extracted on Sep-Paks and the eluting vehicle added to the standard curve and zeros (*see* Chapter 1).
8. Some specific problems encountered in radioimmunoassays and possible ways of overcoming them are summarized in Table 4.

References

1. Adrian, T. E., Zucker, K. A., Bilchik, A. J., and Modlin, I. M. (1990) A novel micro-method for pancreatic acinar secretion. *Int. J. Pancreatol.* **6,** 61–70.
2. Hunter, W. H. and Greenwood, F. C. (1962) Preparation of iodine-131 labelled human growth hormone of high specific activity. *Nature* **194,** 495,496.
3. Holohan, K. N., Murphy, R. F., Flanagan, R. W. J., Buchanan, K. D., and Elmore, D. T. (1973) Enzymic iodination of the histidyl residue of secretin:a radioimmunoassay of the hormone. *Biochim. Biophys. Acta* **322,** 178–180.
4. McFarlane, A. S. (1958) Efficient trace-labelling of proteins with iodine. *Nature* **182,** 53,54.
5. Bolton, A. E. and Hunter, W. M. (1973) The labelling of proteins to high specific radioactivities by conjugation to a 125I-containing acylating agent. *Biochem. J.* **133,** 529–539.
6. Adrian, T. E., Bacarese-Hamilton, A. J., and Bloom, S. R. (1985) Measurement of cholecystokinin octapeptide using a new specific radioimmunoassay. *Peptides* **6,** 11–16.

23

Enzyme-Linked Immunosorbent Assay of Peptides

Laszlo Otvos, Jr. and Gyorgyi I. Szendrei

1. Introduction

Enzyme-linked immunosorbent assay (ELISA) is a member of the solid-phase immunoassay family that detects specific antigen–antibody binding reactions *(1,2)*. A great many variables of antigen or antibody presentation, treatment, and detection are considered when performing ELISA *(3,4)*. This chapter will describe the simplest form, a direct assay, in which the peptide antigen is immobilized on a polystyrene carrier, and a specific primary antibody in solution (detector) is added that recognizes the plate-bound antigen. A secondary, horseradish peroxidase (HRP)-coupled polyclonal antibody (reporter) binds to the primary antibody. The enzymatic activation of hydrogen peroxide releases a chromophore from *o*-phenylenediamine that is detected and quantified.

A general ELISA consists of two consecutive assay steps: First, the active dilution range of an antibody preparation to a given peptide antigen is determined by varying the antibody concentration in one direction on the ELISA plate, and the peptide concentration in the other. A typical antibody dilution curve is found in Fig. 1. The second step is to determine the selectivity of a given dilution of antibody toward various peptide antigens. Figure 2 shows the selective recognition of a serine-phosphorylated human τ peptide by an antipaired helical filament MAb *(5)*.

Protein-specific monoclonal and polyclonal antibodies recognize antigenic fragments or sites that are interconnected (linear epitopes) or fragments that are, by chance, located in a spatial arrangement complementary to an antigen binding site (discontinuous epitopes) *(6,7)*. In most instances, the biologically active protein sites can be modeled by an uninterrupted short peptide segment. It is currently not established what forces peptides and protein antigens use to adhere to plastic carriers, but hydrophobic interaction is suggested to be the

From: *Methods in Molecular Biology, Neuropeptide Protocols*
Edited by: G. B. Irvine and C. H. Williams Humana Press Inc., Totowa, NJ

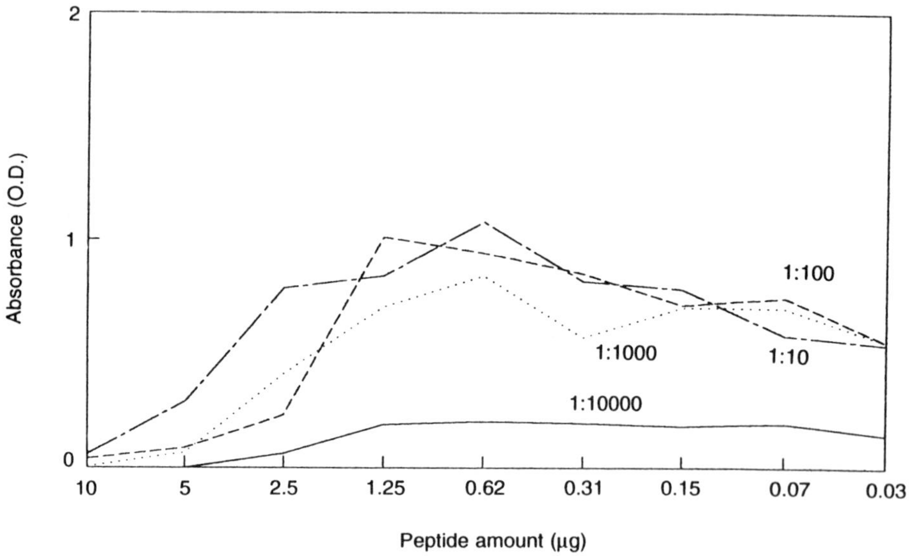

Fig. 1. Binding of MAb Tau-1 to peptide T1NM (corresponding to amino acids 192-204 of human τ protein) *(10)*. The antibody dilutions are as follows: solid line (1:10,000), dots (1:1000), dashes (1:100), dots and dashes (1:10).

strongest one among them *(8)*. Proteins can bind with one of their many hydrophobic surfaces without masking antigenic sites. In contrast, when peptides bind to the plastic, the elimination of one of the few hydrophobic domains may considerably affect their biologically active conformation. We developed an ELISA protocol in which the peptide antigens are plated in the structure-inducing solvent trifluoroethanol (or other aqueous organic solvents that conserve secondary structure), and have found significantly increased recognition by MAbs *(9)* (Fig. 3) (*see* Note 1).

2. Materials

2.1. Equipment and Plate

1. Reversed-phase high performance liquid chromatography (RP-HPLC) system or UV spectrophotometer are necessary to determine the actual concentration of the peptide antigen in any solution to be tested. Alternatively, amino acid analysis can be used.
2. Linbro 96-well tissue culture plates with cover (Flow Laboratories, McLean, VA).
3. Computer-controlled MR4000 microplate reader (Dynatech, Chantilly, VA), equipped with a 450-nm filter and online data analysis software. This instrument can be omitted when results are analyzed only qualitatively.

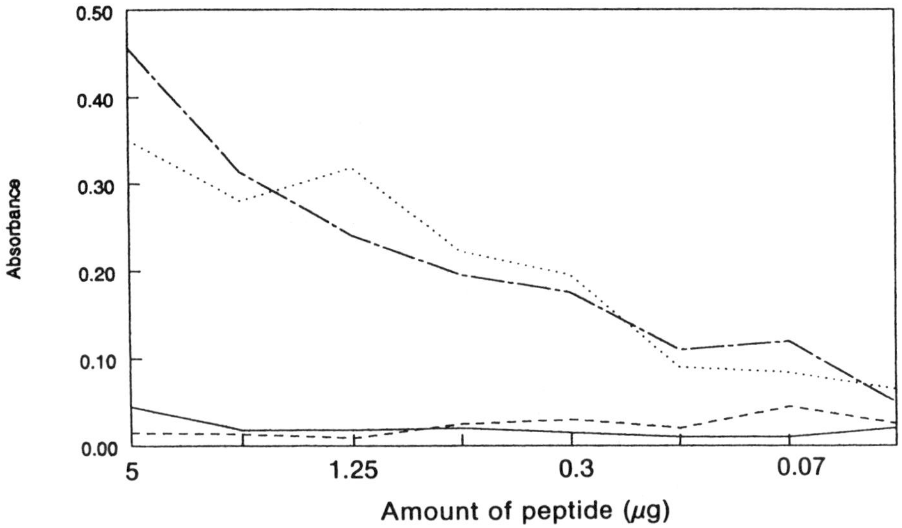

Fig. 2. Binding of a phosphoserine-specific, antipaired helical filaments MAb PHF-1 (dilution = 1:1000) to synthetic τ peptides. Solid trace: a 14 amino acid-long peptide phosphorylated on the tyrosine residue 394 (T3YP); dots: the same peptide phosphorylated on a serine residue 396 (T3P); dashes: the previous peptide elongated by an additional 9 amino acid stretch without any phosphate groups (T3+9); dots and dashes: this longer peptide phosphorylated on serine 396 (T3P+9).

2.2. Chemicals

1. Double-distilled and ion-exchanged water.
2. Phosphate-buffered saline (PBS), pH 7.2: 150 m*M* sodium chloride and 150 m*M* sodium phosphate (Sigma, St. Louis, MO).
3. GG (gamma-globulin) free horse serum (Gibco, Gaithersburg, MD) (*see* Note 7).
4. Secondary antibody (antirat, antimouse, antirabbit IgG) conjugated to HRP (Cappel, West Chester, PA) (*see* Note 2).
5. 0.1*M* Citrate buffer, pH 4.5 (Sigma) (*see* Note 3).
6. *o*-phenylenediamine hydrochloride (Sigma) (*see* Note 3).
7. Hydrogen peroxide (Sigma).
8. 0.1*M* Sodium fluoride (Sigma).
9. 2,2,2-trifluoroethanol, 99.5%, NMR grade (Aldrich, Milwaukee, WI).
10. Acetonitrile, 99.5%, spectrophotometric grade (Aldrich).
11. Ethanol, spectrophotometric grade (Aldrich).

3. Methods (*see* Note 4)

1. Prepare 100 μg/mL peptide antigen solutions. Use water and trifluoroethanol-water, alcohol-water, or acetonitrile-water mixtures as solvents.

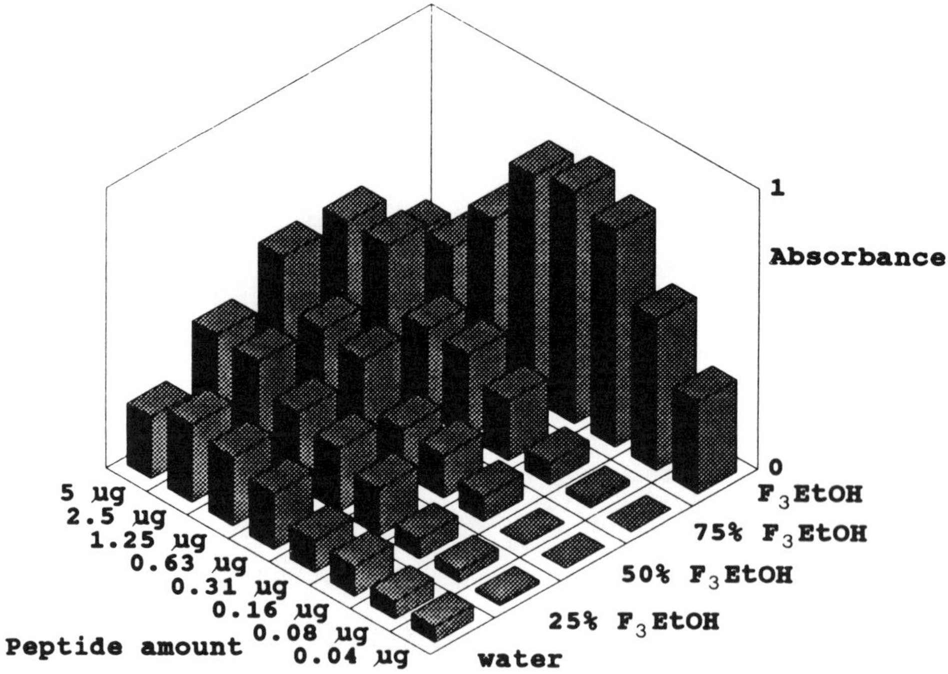

Fig. 3. Binding of MAb FNP 7 to a human neurofilament heptadecapeptide immobilized at different aqueous trifluoroethanol concentrations. Reprinted from ref. *9* with permission from Elsevier Science B. V. Compare with the circular dichroism spectra in Chapter 15.

2. Add 50 µL of antigen per well to the first row of Linbro tissue culture plates. Dilute the antigen solution in the consecutive rows with H_2O to give 5–0.04 µg total antigen in each well, and incubate the plates at 37°C overnight to allow the antigen to dry onto the surface of the plates (*see* Notes 5–7).
3. Block the remaining binding sites on the plates with 10% GG free horse serum in PBS (100 µL/well) (*see* Note 8).
4. Incubate the plates at 37°C for a minimum of 30 min and a maximum of 2 h.
5. Remove the blocking solution by applying vacuum.
6. Add primary antibody solution (50 µL/well) and incubate the plates at 37°C for 30–60 min (*see* Note 9).
7. Remove the primary antibody by applying vacuum, wash the plates 4 times in PBS, and vacuum dry.
8. Add secondary antibody solution (100 µL/well); then incubate the plates at 37°C for 30–60 min (*see* Note 10).
9. Remove the secondary antibody by applying vacuum, wash the plates 5 times in PBS, and vacuum dry.

10. Prepare the developer. For each plate you will need: 20 mL of 0.1*M* citrate buffer, pH 4.5, 20 mg of *o*-phenylenediamine, 8 µL of hydrogen peroxide (*see* Note 3).
11. Add 200 µL/well of developer. Positive (antigen–antibody reaction) wells will turn yellow.
12. Terminate the reaction after 5–10 min by the addition of 25 µL/well of 0.1*M* sodium fluoride.
13. Read the absorbance at 450 nm on the microplate reader.

4. Notes

1. It is fair to say that every laboratory using ELISA sooner or later develops its own protocol. The protocol provided in this chapter works very well in our hands, and should work well for beginners. Because the readout of the assay is highly sensitive to the actual conditions, we encourage the readers to find an optimal procedure for their lab conditions, and to repeat its use in strictly identical conditions.
2. In this chapter, we recommend the purchase of HRP already coupled to the secondary (reporter) antibody to reduce the risk of failure from inappropriate conjugation by inexperienced users. Researchers performing ELISA on a daily basis may want to consider in-house conjugation of enzyme and secondary antibody *(1)*.
3. The citrate buffer should not be used more than a month after it is opened. Use clean glassware and sterile reservoirs. Rinse both with distilled H_2O before using. If the developer solution turns yellow, it cannot be used any longer. The *o*-phenylenediamine is carcinogenic; always wear gloves when handling it.
4. Because individual experiments vary, only experiments done on the same day can be compared. Each row of the plate must have its own background and negative control wells to form background and negative control lanes on the plate as a whole. As negative control, we use unrelated peptides of approximately the same size.
5. Generally, increasing amounts of peptide antigens bind increasing amounts of antibody and result in a linear ELISA curve. This is indicative, however, of sub-optimal antigen–antibody binding. In an ideal case, a maximum curve like the one in Fig. 3, 100% trifluoroethanol concentration is detected. This curve reflects the often-observed "pro-zone" binding behavior of antibodies at higher antigen concentrations.
6. When applying the peptide antigens, cover the wells with 50 µL distilled water except for the first row. Add 100 µL/well of the antigen solution to the first row. Take out 50 µL/well (5 µg peptide remains in the first row wells) and add it to the second row (total volume in the second row is now 100 µL). Take out 50 µL/well from the second row (2.5 µg peptide remains in the second row wells) and add it to the third row. Repeat this procedure until at least the fifth row. Each row represents a 50% dilution of the previous row.

7. The rows always represent different antigen amounts. In the first step of the analysis, the lanes represent different dilutions of the antibody preparation. We usually use 1:10, 1:100, 1:1000, 1:5000, and 1:10,000 primary antibody dilutions. Select the best antigen dilution, and use this dilution in the next step. In the second step of the analysis, the lanes represent different antigens, like peptide analogs, overlapping sequences, different regions of the same protein, and so on.
8. Store 10-mL aliquots of the GG free horse serum at temperatures below 0°C. The 10% GG free horse serum in PBS should be filtered through a 0.22-μm filter before use. This solution should not be used more than a week after the solution has been diluted.
9. Dilute the primary antibody preparation with 10% GG free horse serum in PBS (as described in Note 6) and spin for 5 min at 7600*g* in a microcentrifuge before adding to the wells.
10. Dissolve the lyophilized powder of the secondary antibody in 2 mL of distilled H_2O (the protein concentration is 10.8 mg/mL). Store this solution at temperatures below 0°C in 15-μL aliquots to avoid repetitive freeze-thawing. Dilute the HRP conjugated secondary antibody with 10% GG free horse serum in PBS (1:1000) in two steps. First mix 10 μL secondary antibody solution with 990 μL of blocking solution in a 1-mL microcentrifuge tube. Spin the mixture for 1 min at 7600*g*, remove the supernatant, and add an additional 9 mL of the blocking solution. Select the secondary antibody carefully according to animal source. While polyclonal primary antibodies are usually developed in rabbits (use the antirabbit HRP-conjugated IgG), monoclonal primary antibodies are developed in mice or rats (use antimouse or antirat HRP-conjugated IgG).

Acknowledgment

The authors thank Hildegund C. J. Ertl for critical reading of the manuscript.

References

1. Goding, J. W. (1986) *Monoclonal Antibodies: Principles and Practice.* Academic, Orlando, FL.
2. Campbell, A. M. (1991) *Monoclonal Antibody and Immunosensor Technology.* Elsevier, New York, NY.
3. Porstmann, T. and Kiessig, S. T. (1992) Enzyme immunoassay techniques. *J. Immunol. Meth.* **150,** 5–21.
4. Kemeny, D. M. (1992) Titration of antibodies. *J. Immunol. Meth.* **150,** 57–76.
5. Otvos, L., Jr., Feiner, L., Lang, E., Szendrei, G. I., Goedert, M., and Lee, V. M.-Y. (1994) Monoclonal antibody PHF-1 recognizes tau protein phosphorylated at serine residues 396 and 404. *J. Neurosci. Res.* **39,** 669–673.
6. Arnon, R. (1973) Immunochemistry of enzymes, in *The Antigens* (Sela, M., ed.), vol. 1, Academic, New York, pp. 87–159.

7. Atassi, M. Z. (1975) Antigenic structure of myoglobin: the complete immunochemical anatomy of a protein and conclusions relating to antigenic structures of proteins. *Immunochemistry* **12,** 423–438.
8. Reim, D. F. and Speicher, D. W. (1992) Microsequence analysis of electroblotted proteins II. Comparison of sequence performance on different types of PVDF membranes. *Anal. Biochem.* **207,** 19–23.
9. Lang, E., Szendrei, G. I., Lee, V. M.-Y., and Otvos, L., Jr. (1994) Spectroscopic evidence that monoclonal antibodies recognize the dominant conformation of medium-sized synthetic peptides. *J. Immunol. Meth.* **170,** 103–115.
10. Szendrei, G. I., Lee, V. M.-Y., and Otvos, L., Jr. (1993) Recognition of the minimal epitope of monoclonal antibody Tau-1 depends upon the presence of a phosphate group but not its location. *J. Neurosci. Res.* **34,** 243–249.

24

Sample Preparation for Peptide Immunocytochemistry

Kathy M. Pogue and Colin F. Johnston

1. Introduction

Immunocytochemistry is the localization of a tissue constituent *in situ* by means of a specific antigen–antibody reaction tagged by a visible label *(1)*. For many years, the technique was not considered sufficiently reliable for the *de novo* identification of substances, but was used to assess histopathological alterations in substances previously identified and characterized by other means. Recent improvements in technique, particularly of specificity control and of sample preparation, have allowed immunocytochemistry to be used for the identification of substances before their biochemical identification. Immunocytochemistry is now used in almost all areas of biomedical research. Whereas subsequent chapters deal with specific immunocytochemical techniques, this chapter describes the methods of sample preparation that facilitate the immunocytochemistry of peptides at the light microscopical level.

Appropriate sample preparation is of paramount importance if consistent successful immunostaining results are to be achieved *(2)*. The initial stage in specimen preparation is fixation. This preserves morphology by minimizing postmortem deterioration and reduces dispersal and loss of the peptides of interest. Choice of fixative is dependent on the peptides for which immunocytochemistry is to be attempted. Paraformaldehyde is the unicomponent fixative of choice for the immunocytochemistry of neuropeptides such as calcitonin gene-related peptide, neuropeptide Y, and vasoactive intestinal polypeptide. This agent lightly crosslinks peptides via the N-terminal α-amino function and by the ε-amino functions of lysyl residues *(3)*. Problems with loss of peptide antigenicity may be encountered if either of these groups resides within the epitope to which the antiserum is raised, and it may be necessary to resort to other fixation regimes. Alternative crosslinking agents have been suggested,

From: *Methods in Molecular Biology, Neuropeptide Protocols*
Edited by: G. B. Irvine and C. H. Williams Humana Press Inc., Totowa, NJ

including carbodiimide and parabenzoquinone *(4)*. Multicomponent fixatives that allow immunocytochemistry, particularly for robust gastrointestinal or pancreatic regulatory peptides such as gastrin, insulin, and somatostatin, include Bouin or modified Susa *(5)*.

Subsequent to fixation, specimens may be embedded in a supporting medium such as epoxy and methacrylate resins or paraffin, or prior to sectioning, frozen for cryosectioning or, provided they are of a suitable size, treated as whole mounted preparations (Table 1). Resin embedding provides the best subcellular detail. Unfortunately, good light microscope immunocytochemical staining is difficult to achieve owing to the physical barrier to antibody penetration posed by the resin's matrix. Wax embedding, which involves dehydration of the specimen in alcohols and infiltration with hot wax, preserves tissue morphology well but is not suitable for all peptides. Neuropeptides in particular are difficult to immunostain following this procedure. Several regimes for unmasking antigens concealed during the preparative process have been suggested. These include the partial proteolytic digestion of tissue sections with pronase or trypsin. However, these regimes are difficult to control and rarely produce consistent results *(6)*. The technique of microwave antigen retrieval *(7)* can, in some circumstances, prove more useful when the only samples available are from archival libraries *(8)*. Frozen sections of paraformaldehyde-fixed tissues, although yielding poorer subcellular detail, are appropriate for the immuno-cytochemical demonstration of most peptides. Whole mounted immunofluorescent specimens, which are not amenable to conventional microscopy, can prove the most useful, both in terms of antigen retention and resolution, if examined by confocal laser scanning microscopy as described in Chapter 25.

2. Materials

1. Apparatus required is standard histology laboratory equipment including hot wax dispenser, heated water bath, 60°C oven, microtome, and cryostat. An H2500 microwave oven (Energy Beam Sciences, Agawam, MA) is suitable for antigen retrieval, although a cheaper model may be substituted.
2. Modified Susa: Mix 50 g trichloroacetic acid, 10 g sodium chloride, 100 mL glacial acetic acid, 400 mL formaldehyde solution, and 1.5 L distilled H_2O; store at 4°C.
3. Paraformaldehyde: Add 16 g paraformaldehyde (Agar Scientific, Essex, UK) to 200 mL distilled H_2O, cover, and heat to 55–60°C for 1 h. Add 1*M* sodium hydroxide dropwise until precipitate clears. Cool, add 200 mL 2X phosphate-buffered saline (PBS) (*see* item 5), pH to 7.4 with sodium hydroxide, filter, and store at 4°C. Use within 1 wk.
4. PBS: Dissolve 85 g sodium chloride, 34.5 g sodium dihydrogen orthophosphate 1-hydrate, and 107 g anhydrous disodium hydrogen orthophosphate in 2 L distilled H_2O; pH to 7.2 with ammonia. Make up to 10 L with distilled H_2O.

Table 1
Tissue Processing—Postfixation Alternatives

Preparation	Section thickness	Detection system	Microscopy	Morphology	Retention of antigenicity
Smears/cultures	N/A	Fluorescence	Light/CLSM	+	++
Whole mounts	N/A	Fluorescence	CLSM	+++	+++
Frozen section	10–30 μm	Fluorescence, enzyme methods	Light/CLSM	+	+++
Wax section	3–7 μm	Fluorescence, enzyme methods	Light	++	++
Resin section (semithin)	1–3 μm	Enzyme methods, immunogold	Light	+++	+
Resin section (thin)	90 nm	Enzyme methods, immunogold	EM	+++	+

The common processing strategies for the immunocytochemical localization of peptides. With the exception of whole mounted specimens amenable to examination by confocal laser scanning microscopy (CLSM), regimes that preserve morphology adversely affect antigen availability. Increasingly sensitive detection systems are required as antigenicity becomes increasingly compromised. +, poor; ++, moderate; +++, good. EM, electron microscopy.

5. 2X PBS for paraformaldehyde: Dissolve 3.4 g sodium chloride, 1.38 g sodium dihydrogen orthophosphate 1-hydrate, and 4.28 g anhydrous disodium hydrogen orthophosphate in 200 mL distilled H_2O.
6. 0.01*M* citrate buffer pH 6.0: Dissolve 2.1 g citric acid in 1 L distilled H_2O and pH to 6.0 with 1*M* sodium hydroxide. Store at 4°C.
7. Embedding and support media: Cryo-M-Bed and Ralwax 1, pastillated (BDH Chemicals, Poole, UK).
8. Subbing solution: Dissolve 2.5 g gelatin and 0.25 g chromic potassium sulfate in distilled H_2O at 80°C. Store at 4°C.
9. APES solution: Mix 2% 3-aminopropyltriethoxysilane (Sigma, Poole, UK) in acetone.

All chemicals should be analar quality.

3. Methods

3.1. Tissue Processing for Wax Sections

1. Fix tissue in modified Susa at room temperature overnight (*see* Note 1).
2. Wash in 70% industrial methylated spirits for 30 min, then repeat (*see* Note 2).
3. Trim the tissue if necessary.
4. Wash in 95% industrial methylated spirits for 30 min, then repeat.
5. Wash in 100% industrial methylated spirits for 30 min, then repeat for 1 h.
6. Wash in 100% ethanol for 30 min, then repeat for 1 h.
7. Wash in a clearing agent, e.g., xylene, for 30 min, then repeat for 1 h (*see* Note 3).
8. Infiltrate with Ralwax at 60°C overnight (*see* Note 3).
9. Block out (*see* Note 4).
10. Cut 5-µm sections and lift onto coverslips (*see* Notes 5 and 6).
11. Dry sections at 50°C for a minimum of 1 h (*see* Note 5).
12. Rack coverslips and incubate at 60°C overnight.

3.2. Tissue Processing for Frozen Sections

1. Fix tissue in paraformaldehyde at 4°C overnight (*see* Notes 1 and 2).
2. Wash in 5% sucrose in PBS at 4°C overnight (*see* Note 2).
3. Cryoprotect in 30% sucrose in PBS at 4°C overnight.
4. Freeze onto stubs using Cryo-M-Bed (BDH) as tissue support (*see* Note 7).
5. Equilibrate in the cryostat for at least 20 min (*see* Note 7).
6. Cut 7–20-µm sections onto subbed coverslips (*see* Method 3.3.).
7. Air dry at room temperature for 30 min.

3.3. Subbing Coverslips

1. Rack coverslips and dip into subbing solution for 30 s.
2. Separate and blot dry.
3. Oven dry at 60°C for approx 30 min.
4. Repeat steps 1–3 twice more.

3.4. APES Coating Slides

1. Rack slides and dip in APES solution for 30 s.
2. Wash in acetone for approx 10 s and repeat using fresh acetone.
3. Air dry.

3.5. Microwave Antigen Retrieval

1. Cut 5-μm wax sections and lift onto APES-coated slides (*see* Section 3.4. and Note 8).
2. Dry sections at 50°C for a minimum of 1 h (*see* Note 5).
3. Rack slides and incubate at 60°C overnight.
4. Cool to room temperature.
5. Dewax in xylene (2 × 3 min).
6. Wash in 100% ethanol (2 × 3 min).
7. Wash in 95% ethanol (3 min).
8. Wash in 70% ethanol (3 min).
9. Dip in tap water (10 s).
10. Transfer slides to coplin jars containing citrate buffer.
11. Microwave at 100°C for 25 min (*see* Note 8).
12. Replace hot buffer with distilled H_2O until cool.
13. Transfer slides to PBS at room temperature for 5 min.

4. Notes

1. Fixatives: For routine histology, a variety of agents are employed including crosslinking agents such as aldehydes, organic precipitants, and protein-precipitant metals such as mercury. These agents may be used either singly or in combination. However, in immunocytochemistry, fixation is always a compromise between the preservation of morphology and retention of peptide antigenicity. For this reason, fixatives containing glutaraldehyde and protein precipitant metals should be avoided.
2. Tissue processing: Write labels for all pieces of tissue using pencil on small pieces of card. These are processed along with the tissue. Trim larger pieces of tissue before fixation and do not overfix as this can reduce peptide antigenicity. Whereas modified Susa has a long shelf life, paraformaldehyde should be <1 wk old. Use the minimum quantities of fixatives, solvents, and so on, use beakers for all solution transfers, and never work directly over a sink. (It is always the irreplaceable samples that disappear down the drain!) Infiltrate for no longer than 18 h in hot wax, as this too can reduce antigenicity.
3. As xylene carries the "Irritating Substance" warning label, appropriate precautions should be taken. Work inside a fume cupboard or a solvent extractor box and always wear gloves. Replace with hot Ralwax using the wax dispenser and collect xylene for safe disposal.
4. Blocking out means transferring the wax-impregnated tissues into a mold containing molten wax and allowing the block to set. A small meth burner is used to

warm the forceps employed in the transfer. Setting can be speeded up by placing the completed mold on ice. The solid block is prized out of the mold, trimmed, and then melted onto a metal stub by means of a hot spatula, before positioning in the microtome for sectioning.

5. Microtome sectioning: Keep blocks on ice prior to sectioning. Float off sections onto a water bath at 50°C, and lift onto coverslips with a vertical action. Dry sections flat initially (around the rim of the water bath is ideal). These measures should ensure that the sections are not wrinkled and will remain attached to the coverslips during the prolonged incubation steps used in some immunocytochemical methods. Work logically and methodically as labeling coverslips is not practical.
6. Sections are lifted onto coverslips rather than onto slides, as with most conventional histochemical processes, because during the immunostaining procedure, antisera (50–100 µL) are placed as drops onto the tissue sections. The size of the coverslip limits the spreading of the drop, thus ensuring that the tissue section remains coated with antiserum.
7. Cryostat sectioning: When mounting tissue onto cryostat stubs using Cryo-M-Bed, avoid trapping air bubbles as these will make sectioning difficult. Allow adequate equilibration time before sectioning—the larger the tissue, the longer the time.
8. For antigen retrieval, position sections slightly to one end of the slide when lifting so that tissues remain immersed in buffer throughout the microwaving period. Should the buffer volume decrease significantly during boiling, top up with distilled H_2O.

References

1. Coons, A. H., Leduc, E. H., and Connolly, J. (1955) Studies on antibody production. I. A method for the histochemical demonstration of specific antibody and its application to the hyperimmune rabbit. *J. Exp. Med.* **102,** 49–60.
2. Beesley, J. E. (1993) Immunocytochemical avenues, in *Immunocytochemistry: A Practical Approach* (Beesley, J. E., ed.), Oxford University Press, Oxford, pp. 7–13.
3. Pearse, A. G. E. (1980) *Histochemistry, Theoretical and Applied, 4th ed., vol. 1. Preparative and Optical Technology,* Churchill Livingstone, London.
4. Pearse, A. G. E. and Polak, J. M. (1975) Bifunctional reagents as vapour- and liquid-phase fixatives for immunohistochemistry. *Histochem. J.* **7,** 179–186.
5. Johnston, C. F., O'Neill, A. B., O'Hare, M. M. T., and Buchanan, K. D. (1986) Neuroendocrine cells within colorectal tumours induced by dimethylhydrazine. An immunocytochemical study. *Cell Tissue Res.* **246,** 205–210.
6. Brandtzaeg, P. (1982) Tissue preparation methods for immunohistochemistry, in *Techniques in Immunocytochemistry,* vol. 1 (Bullock, G. R. and Petrusz, P., eds.), Academic, London, pp. 2–75.
7. Kok, L. P. and Boon, M. E. (1992) *Microwave Cookbook for Microscopists. Art and Science of Visualization,* 3rd ed., Coulomb Press Leyden, Leiden.
8. Wang, D.-G., Johnston, C. F., Anderson, N., Sloan, J. M., and Buchanan, K. D. (1995) Overexpression of the tumour suppressor gene p53 is not implicated in neuroendocrine tumour carcinogenesis. *J. Pathol.* **175,** 397–401.

25

Immunocytochemical Methods for Regulatory Peptides

Kathy M. Pogue and Colin F. Johnston

1. Introduction

Immunocytochemistry employs antibodies to detect peptides or proteins in tissue preparations *(1)*. However, this in itself is not a visible reaction. Several methods, employing secondary antibodies labeled in some way to render them visible, generally with fluorescent compounds or enzymes fed with suitable chromogenic substrates, are in current use *(2)*. The method chosen depends on a number of factors. The greater sensitivity of enzyme methods is rarely an advantage in the immunocytochemical demonstration of regulatory peptides, as these substances tend to occur at relatively high concentrations in tissues. Advantages of enzyme methods are that the preparations are permanent and the underlying tissue morphology is available for study along with the labeled antigen. This is of particular importance in diagnostic histopathology.

Indirect immunofluorescence, although less sensitive than enzyme methods, is suitable for most regulatory peptide immunocytochemical research programs and is the method described here (Fig. 1). If required, additional amplification of the fluorescent signal may be achieved by using biotinylated secondary antibodies and fluoresceinated streptavidin *(3)* (Fig. 2). Simultaneous localization of two antigens in single specimens is possible provided that primary antisera from different donor species, e.g., rabbit and guinea pig, are available along with appropriate secondary antibodies labeled with either fluorescein or rhodamine (Fig. 3). If primary antisera are only available from a single species, immunostaining and colocalization of multiple antigens in single specimens must be performed sequentially. The specimen is immunostained for one antigen, photographed, and the bound antibodies eluted oxidatively prior to immunostaining for the second antigen *(4)*.

From: *Methods in Molecular Biology, Neuropeptide Protocols*
Edited by: G. B. Irvine and C. H. Williams Humana Press Inc., Totowa, NJ

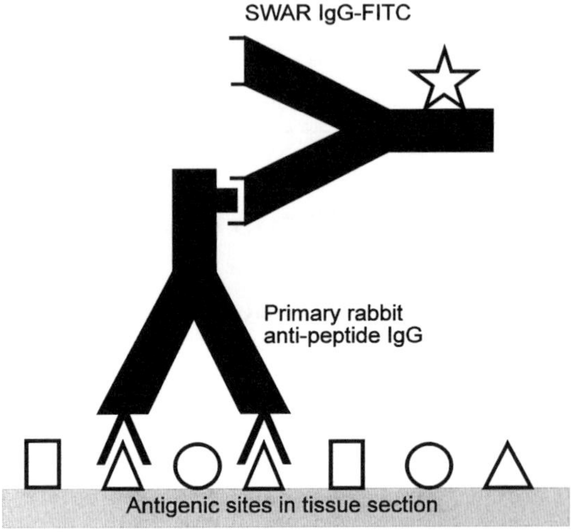

Fig. 1. Diagrammatic representation of the indirect immunofluorescence technique. The primary rabbit antibody bound to peptide antigens in the tissue section is detected by the secondary porcine antibody, which is specific for rabbit IgG and has been previously labeled with fluorescein isothiocyanate, swine antirabbit (SWAR) FITC.

Immunofluorescent images are degraded by the out-of-focus fluorescent flare from antigens both above and below the focal plane of the objective lens. This is a particular problem at higher magnifications or when thicker sections or whole-mounted specimens are employed. The confocal laser scanning microscope has eliminated this problem and revolutionized immunofluorescence microscopy in the process *(5)*. This microscope, by rejecting information from outside the focal plane of the objective lens, provides much clearer images of higher resolution and contrast than has previously been possible. The ability of the confocal laser scanning microscope to optically section (i.e., to collect images from different levels in the z-dimension) relatively large whole-mounted specimens, and to digitally store and process such images on a computer, makes possible the three-dimensional reconstruction of composite images from many levels within the specimen *(6)*. Because the optical sectioning capability is nondestructive, the specimen may be re-examined repeatedly during the life of the fluorescent label.

Suitable controls for staining procedure and specificity of staining must be used for the critical interpretation of immunocytochemistry results *(7)*. These include the omission of primary or secondary antisera or replacement of the primary antiserum with preimmune serum from the donor animal, and controls for nonspecific ionic binding of antibodies to highly charged molecules

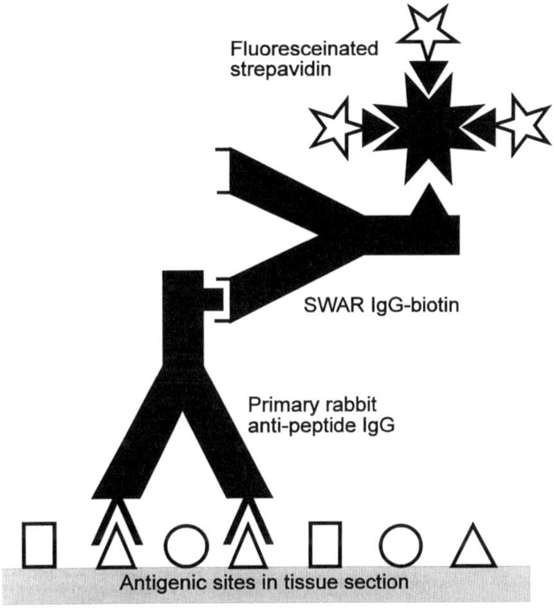

Fig. 2. Indirect immunofluorescence using a biotinylated secondary antibody that is detected using fluoresceinated streptavidin. Streptavidin, which is prepared from the culture supernatant of *Streptomyces avidinii,* possesses four high-affinity binding sites for biotin, one of which is free. Thus, as each molecule of streptavidin delivers several moieties of fluorescein to each unit of biotin on the secondary antibody, the signal is amplified and the sensitivity of detection is increased.

in the test tissue *(8,9)*. Background staining is minimized by using the primary antiserum at its highest usable working dilution. Controls for staining specificity include liquid-preabsorption or solid-phase preadsorption of the antiserum with a range of concentrations of the appropriate peptide and other related and unrelated peptides. Preabsorbed antibodies have no free antigen binding sites and so fail to interact with tissue antigens (Fig. 4).

2. Materials

1. Apparatus required is standard histology laboratory equipment including staining troughs for dewaxing and rehydrating sections, a flat-bottomed tray or chamber with lid, in which a humid atmosphere can be maintained for antiserum incubations, and microscopes. A transmitted light microscope is used for viewing sections immunostained using enzyme methods, and an incident light microscope with a mercury vapor lamp is used to view immunofluorescent samples. The latter should be fitted with exciter filters producing blue light (492 nm) and green light (520 nm) for the excitation of fluorescein and rhodamine or Texas

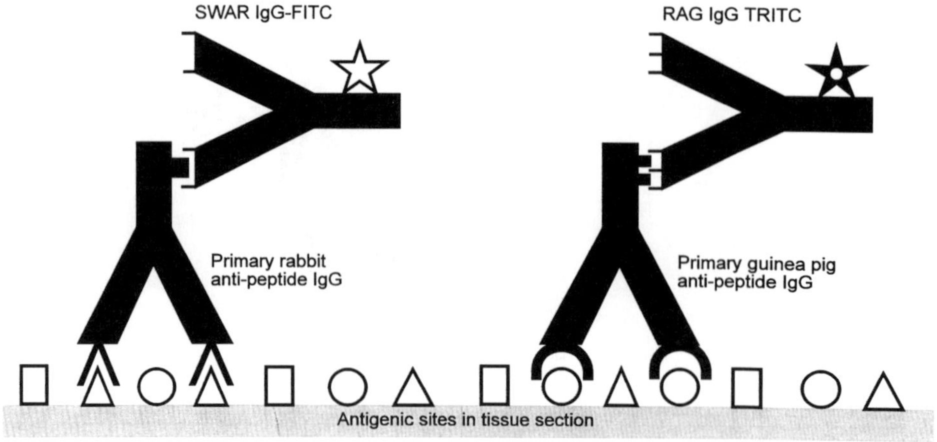

Fig. 3. Double labeling of two peptide antigens is achieved using primary antibodies raised in different donor species. These are detected using differently labeled secondary antisera specific for the IgGs of the primary antisera donor species. In the example illustrated, the rabbit primary is detected using SWAR-FITC, which fluoresces green, whereas the guinea pig primary is detected using rabbit anti-guinea pig IgG (RAG) labeled with tetramethylrhodamine isomer R (TRITC), which fluoresces red.

red, respectively. Similarly, the laser of a confocal laser scanning microscope must produce bright lines and be filtered to these wavelengths.

2. Paraformaldehyde: Add 16 g paraformaldehyde (Agar Scientific, Essex, UK) to 200 mL distilled H_2O, cover, and heat to 55–60°C for 1 h. Add 1*M* sodium hydroxide dropwise until precipitate clears. Cool, add 200 mL 2X phosphate-buffered saline (PBS) (*see* item 10), pH to 7.4 with sodium hydroxide, filter, and store at 4°C. Use within 1 wk.
3. PBS: Dissolve 85 g sodium chloride, 34.5 g sodium dihydrogen orthophosphate 1-hydrate, and 107 g anhydrous disodium hydrogen orthophosphate in 2 L distilled H_2O and pH to 7.2 with ammonia. Make up to 10 L with distilled H_2O.
4. Antibody diluent: Dissolve 0.5 g sodium azide, 0.5 g bovine serum albumin Fraction V (Sigma, Poole, UK) in 500 mL PBS, add 2.5 mL Triton X-100, and store at 4°C.
5. PBS/glycerol: Dissolve 1.25 g 1,4-diazabicyclo [2,2,2] octane (as free radical scavenger antifade agent) in 5 mL PBS, and 45 mL glycerol, and store at 4°C in the dark.
6. High molarity salt wash: Dissolve 5.18 g sodium chloride in 250 mL PBS.
7. Poly-L-lysine: Dissolve 200 mg poly-L-lysine (MW 3800) (Sigma) in 100 mL antibody diluent. Store as 500-µL aliquots at –20°C.
8. Secondary antisera: Swine antirabbit-fluorescein, swine antirabbit-rhodamine, rabbit anti-guinea pig-fluorescein, and rabbit antimouse-fluorescein (all Dakopatts, Glostrup, Denmark) are aliquoted (25-µL), stored at –20°C, and used

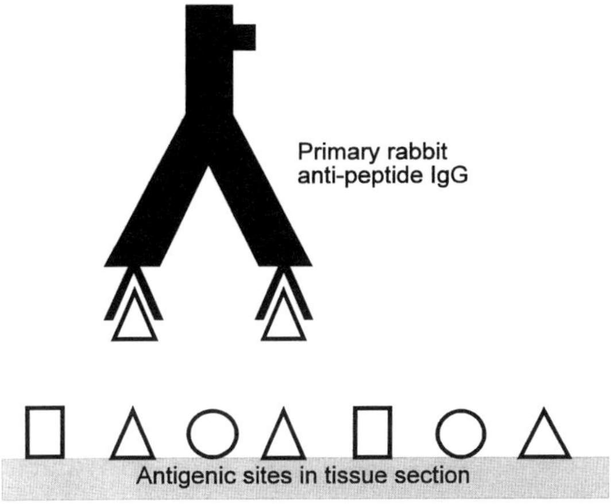

Fig. 4. Liquid-phase preabsorption test of antiserum specificity is carried out by preincubating the antiserum with a range of concentrations of antigen. Preabsorbed antibodies have no free antigen binding sites and so fail to interact with tissue antigens.

at dilutions from 1/40 to 1/200. Biotinylated donkey antirabbit, biotinylated sheep antimouse, streptavidin-fluorescein, and streptavidin-Texas red (Amersham, Buckinghamshire, UK) are stored at 4°C and used at 1/200 dilution.

9. Acidified $KMnO_4$ solution: Solution A: Dissolve 0.5 g $KMnO_4$ in 100 mL distilled H_2O. Solution B: Mix 0.5 mL H_2SO_4 in 100 mL distilled H_2O. Mix A and B equally and use fresh.
10. 2X PBS for paraformaldehyde: Dissolve 3.4 g sodium chloride, 1.38 g sodium dihydrogen orthophosphate 1-hydrate, and 4.28 g anhydrous disodium hydrogen orthophosphate in 200 mL distilled H_2O.
    All chemicals should be analar quality.

3. Methods

3.1. Immunostaining for Frozen Sections

1. Air dry sections at room temperature for 30 min.
2. Rehydrate in PBS for 20 min.
3. Arrange coverslips supported on embryo dishes in immunostaining chamber (*see* Notes 1 and 2).
4. Apply primary antiserum (50–100 µL) and incubate at 4°C overnight (*see* Notes 2 and 3).
5. Wash in PBS for 5 min.
6. Apply secondary antiserum and incubate in the dark, at room temperature for 30 min.
7. Wash in PBS for 5 min in the dark (*see* Note 4).
8. Mount using PBS/glycerol (*see* Note 5).

3.2. Immunostaining for Wax Sections

1. Dewax sections in xylene (2 × 3 min).
2. Wash in 100% ethanol (2 × 3 min).
3. Wash in 95% ethanol (3 min).
4. Wash in 70% ethanol (3 min).
5. Dip in distilled water (10 s).
6. Wash in PBS (5-min minimum).
7. Arrange coverslips supported on embryo dishes in immunostaining chamber (*see* Notes 1 and 2).
8. Apply primary antiserum and incubate at 4°C overnight (*see* Notes 2 and 3).
9. Repeat step 6.
10. Apply secondary antiserum and incubate in the dark, at room temperature for 30 min.
11. Wash in PBS for 5 min in the dark (*see* Note 4).
12. Mount using PBS/glycerol (*see* Note 5).

3.3. Immunostaining for Whole Mounted Specimens

1. Fix fresh tissue of a suitable size (*see* Note 6) in paraformaldehyde overnight at 4°C.
2. Wash in antibody diluent 3 or 4 times and incubate overnight at 4°C.
3. Incubate with primary antiserum for a minimum of 2 d at 4°C.
4. Repeat step 2.
5. Incubate with secondary antiserum in the dark, overnight at 4°C.
6. Repeat step 2. Keep dark (*see* Note 4).
7. Mount using PBS/glycerol (*see* Note 5).
8. View using a confocal laser scanning microscope (*see* Notes 7 and 8).

3.4. Immunostaining Using the Biotin/Streptavidin Detection System

1. For frozen sections, follow immunostaining procedures from Section 3.1. up to step 5; for wax sections, follow immunostaining procedure from Section 3.2. to step 9.
2. Apply biotinylated secondary antiserum and incubate for 30 min at room temperature (*see* Note 3).
3. Wash in PBS (5-min minimum).
4. Apply streptavidin-FITC and incubate for 30 min in the dark at room temperature.
5. Wash in PBS for 5 min in the dark (*see* Note 4).
6. Mount using PBS/glycerol (*see* Note 5).

3.5. Oxidative Elution of Antibodies for Sequential Immunostaining for Two or More Antigens

1. Immunostain for the first antigen as described in Sections 3.1. and 3.2.
2. Mount using PBS/glycerol.
3. Examine and photograph regions of interest (*see* Note 9).
4. Immerse slide in trough of PBS and gently remove coverslip.
5. Apply acidified $KMnO_4$ for 1 min.

6. Wash in double-distilled water for 10 s.
7. Wash in PBS (5-min minimum).
8. Immunostain for second antigen.
9. Mount using PBS/glycerol.

3.6. Primary Antiserum Titer

1. Divide the neat antiserum into 100-μL aliquots and store at –20°C.
2. Add 900 μL antibody diluent to 100 μL neat serum, divide this 1/10 dilution into 100-μL aliquots, and store at –20°C.
3. Add 400 μL of antibody diluent to 100 μL 1/10 antiserum to give 500 μL 1/50 dilution.
4. Pipet 200 μL antibody diluent to each of 6 tubes.
5. Add 200 μL 1/50 diluted antiserum to tube 1 and mix to give 400 μL 1/100 dilution.
6. Transfer 200 μL 1/100 diluted antiserum to tube 2 and mix to give 400 μL 1/200 dilution.
7. Continue double diluting to produce antisera at 1/400, 1/800, 1/1600, 1/3200. On rare occasions with antisera of especially high titer, it may be necessary to produce further dilutions.
8. Use these dilutions in the chosen immunocytochemical staining method.
9. Examine the stained sections and determine the highest usable dilution.

3.7. Elimination of Nonspecific Ionic Binding Using a High Molarity Salt Wash

1. For frozen sections, follow immunostaining procedures from Section 3.1. up to step 5; for wax sections, follow immunostaining procedures from Section 3.2. to step 9.
2. Wash in 0.5*M* NaCl/PBS for 30 min.
3. Wash in PBS for 5 min.
4. For frozen sections, continue immunostaining from step 6 of Section 3.1. For wax sections, continue immunostaining from step 10 of Section 3.2.

   Alternatively, poly-L-lysine 3800 mol wt at a concentration of 2 μg/mL can be included in the primary antiserum.

3.8. Preabsorption Test of Primary Antiserum Specificity (see Note 10)

1. Ascertain primary antiserum working titer as described in Section 3.6.
2. Prepare sections as described above following steps 1–3 of Section 3.1. for frozen sections or steps 1–7 of Section 3.2. for wax sections.
3. Make up antigen 100 μg/mL in primary antiserum at its working dilution.
4. Dilute with primary antiserum at its working dilution to give a series of 10-fold dilutions of antigen, i.e., 100 μg, 10 μg, 1 μg, 100 ng, 10 ng, 1 ng/mL.
5. Incubate at room temperature for 1 h.
6. Use these antiserum/antigen dilutions in the chosen immunocytochemical staining method.

7. Examine the stained sections and determine the lowest concentration of antigen that abolishes antibody binding, i.e., no visible immunostaining.
8. Repeat the above with related and nonrelated antigens to determine crossreactions.

4. Notes

1. A flat-bottomed cat litter tray makes a cheap immunostaining chamber; a second tray serves as a lid. Wet, folded tissue paper in the base of the tray ensures the humid atmosphere required to prevent the evaporation of antisera during prolonged incubations.
2. Draw a plan beforehand if several different primary or secondary antisera are to be used or when several different tissues are to be immunostained. Arrange coverslips on the corners of embryo dishes in the immunostaining tray according to the plan. Group together sections that will subsequently require the same secondary antiserum; i.e., group all sections receiving rabbit primary antisera away from those receiving guinea pig primary antisera, or those that will require biotinylated secondary antisera. Label all slides according to the tray plan before mounting the sections. Countless immunostaining runs have been ruined by not following these simple procedures!
3. Store primary antisera in convenient aliquots at suitable dilutions so that the addition of 500 μL or 1 mL of antibody diluent brings them to their working dilutions. Diluted antibodies store well at –20°C provided the overall protein concentration is maintained and they are not subjected to repeated thawing and refreezing. Make up biotinylated secondary antibodies weekly and store, ready to use, at 4°C.
4. Fluorochromes and fluorescein in particular are light-sensitive. When possible, sections should remain in the dark, with washing troughs and slide trays covered with tin foil. Immunofluorescent specimens will retain their fluorescence for several weeks if stored at 4°C in the dark.
5. PBS/glycerol: 1,4-diazabicyclo [2,2,2] octane is slow to dissolve, so leave stirring overnight when making up, otherwise the undissolved crystals will not allow coverslips to lie flat on the slides. Use only the minimum of mountant to prevent coverslips from sliding on the slides and wrinkling the tissue sections.
6. Flat fixing is best for whole-mounted preparations. Specimens up to 1 mm in thickness are placed between two glass slides in a tray and fixative introduced at the sides. Immunostaining is carried out in small plastic tubes with successive solutions added and removed by pipet. The tubes are kept dark where necessary by wrapping in tin foil.

   Larger specimens require longer incubation periods. Inclusion of a mild detergent, such as Triton X-100, in the antibody diluent, removes membrane lipids, which would otherwise hinder antibody penetration.
7. Routinely, confocal laser scanning microscopy is used to retrieve high-resolution, high-contrast internal images from depths of almost 100 μm within whole-mounted biological specimens. This varies from specimen to specimen and is largely dependent on the opacity of the preparation. Some specimens

will have to be sliced to 50 μm prior to immunostaining. The thickness of the optical section is dependent on, among other things, the quality of the objective lenses employed. These should be of the highest numerical aperture available. With objectives of numerical aperture >0.6, the optical section thickness is <1 μm.

8. Photobleaching of fluorochromes is a particular problem when specimens are examined by confocal laser scanning microscopy. Neutral density filters will reduce the intensity of the laser light impingent on the specimen, thus prolonging the useful life of the label. Specimens should be examined and recorded at low magnifications before higher magnifications.
9. Whereas 100 ASA film is suitable for recording images produced by confocal laser scanning microscopy, faster film, such as 400 or even 1000 ASA, is necessary when using standard fluorescence microscopy if exposure times are to be within reasonable limits.
10. Preabsorption studies can be extended to assess the regional specificity of antibodies by employing peptide fragments. The regional specificity of a given polyclonal antiserum as assessed by other methods, such as radioimmunoassay, should not be assumed to apply when that antiserum is employed in immunocytochemistry.

References

1. Sternberger, L. A. (1979) *Immunocytochemistry,* 2nd ed., Wiley, New York.
2. Jackson, P. and Blythe, B. (1993) Immunolabelling techniques for light microscopy, in *Immunocytochemistry: A Practical Approach* (Beesley, J. E., ed.), Oxford University Press, Oxford, UK, pp. 15–41.
3. Bonnard, C., Papermaster, D. S., and Kraehenbuhl, J.-P. (1984) The streptavidin-biotin bridge technique: application in light and electron microscope immunocytochemistry, in *Immunolabelling for Electron Microscopy* (Polak, J. M. and Varndell, I. M., eds.), Elsevier, Amsterdam, pp. 95–111.
4. Tramu, G., Pillez, A., and Leonardelli, J. (1978) An efficient method of antibody elution for the successive or simultaneous localization of two antigens by immunocytochemistry. *J. Histochem. Cytochem.* **26,** 322–324.
5. Shotton, D. and White, N. (1989) Confocal scanning microscopy: three dimensional biological imaging. *TIBS* **14,** 435–439.
6. Johnston, C. F., Shaw, C., Halton, D. W., and Fairweather, I. (1990) Confocal scanning laser microscopy and helminth neuroanatomy. *Parasitol. Today* **6,** 305–308.
7. van Leeuwen, F. (1982) Specific immunocytochemical localization of neuropeptides: a utopian goal?, in *Techniques in Immunocytochemistry,* vol. 1 (Bullock, G. R. and Petrusz, P., eds.), Academic, London, pp. 283–299.
8. Grube, D. (1980) Immunoreactivities of gastrin (G-)cells. II. Non-specific binding of immunoglobulins to G-cells by ionic interactions. *Histochemistry* **66,** 149–167.
9. Scopsi, L., Wang, B.-L., and Larson, L.-I. (1986) Non-specific immunocytochemical reactions with certain neurohormonal peptides and basic peptide sequences. *J. Histochem. Cytochem.* **34,** 1469–1476.

26

Ultrastructural Localization of Peptides Using Immunogold Labeling

David W. Halton and Gerard P. Brennan

1. Introduction

The subcellular localization of bioactive peptides, including hormones and neurotransmitters, has immense value, not only in understanding how cells function, but in correlating biochemical and clinical data from tissues in both normal and diseased states. The pioneering immunocytochemical work of Coons et al. *(1)* exploited the specificity of antigen–antibody interactions by using fluorescent-labeled antibodies as probes to determine the spatial distribution of antigens. It established the principles of immunocytochemistry, and thereby revolutionized the means of identifying peptide-containing cells by light microscopy. The application of immunocytochemistry at the ultrastructural level combines the specificity and sensitivity it offers with the fine spatial definition of electron microscopy, providing an exceedingly precise means of localizing the position of specific peptide antigens to a resolution of 5 nm or better. As a technique, electron microscopic immunocytochemistry was first accomplished some 35 yr ago *(2)*, with the development of ferritin–antibody conjugates. Since then, the introduction of colloidal gold *(3)* as a tracer in electron microscopy, and its use as a marker for antisera *(4)*, has established immunogold labeling as by far the most popular choice for ultrastructural localization of bioactive peptides and most other immunogens. This is not surprising since gold spheres are easily prepared in the laboratory, and are readily adsorbed by a variety of proteins, including immunoglobulins, with which they form simple but stable antibody–gold complexes without involving chemical conjugation. Their shape and high electron density make gold particles clearly identifiable in tissue sections in the electron microscope, and, moreover, their labeling can be quantified simply by particle counting. Finally, since gold par-

From: *Methods in Molecular Biology, Neuropeptide Protocols*
Edited by: G. B. Irvine and C. H. Williams Humana Press Inc., Totowa, NJ

ticles can be prepared in various sizes of 1–150 nm, they make techniques of multiple immunolabeling possible, thereby enabling the simultaneous ultrastructural demonstration of two or more peptides in the same cell or tissue. Through successive modifications and refinements, colloidal gold technology *(5)* has made relatively cheap but exceptionally versatile probes available for localizing specific peptides in cells. It has also been an important factor in helping to establish electron microscopic immunocytochemistry as a reliable and fairly routine technical procedure, with numerous applications in the life sciences *(6)*.

The method of choice for immunogold labeling is largely determined by the distribution and lability of the antigen in the sample, and by the nature of the primary antibody to be used. For surface membrane antigens, labeling can be done prior to fixation and embedment, a method referred to as pre-embedding. However, where the antigen is inaccessible without first sectioning, as is generally the case with bioactive peptides, labeling is done on fixed and embedded samples, and is known as postembedding labeling. Successful localization requires the all-important compromise between preserving acceptable ultrastructure and retaining antigens *in situ* with sufficient antigenicity for reactivity with the antibodies. Fortunately, peptides are usually quite robust and many are well resistant to the fixatives, solvents, and resins used in tissue preparation techniques; moreover, they are strongly immunogenic, allowing the generation of potent antisera. Postembedding labeling gives high resolution of fine structure and is straightforward and adaptable, enabling a single tissue sample to be sectioned and tested with a variety of antisera; multiple labeling techniques can also be applied and are relatively simple to perform.

Of the many immunogold procedures, the indirect methods for antigen localization, where a secondary antibody or sequence of antibodies is used to visualize the primary antibody (Fig. 1A), are considered to be the most sensitive, both in single- and double-labeling experiments. This is because of the signal amplification, since anti-IgG sera are usually of very high titer and avidity, and several labeled anti-IgG molecules can bind to each primary antibody. One disadvantage of the indirect method is the risk of nonspecific labeling, although with suitable controls and the use of highly specific, well-characterized antisera whose labeling can be totally abolished by the antigen (Fig. 1B), this can be kept to a minimum (*see* Section 3.3.3.).

The protocols described in this chapter are for the indirect immunogold labeling of resin-embedded material, and have been applied successfully to peptide antigens in our laboratory for a number of years (Fig. 2). However, to save time and effort, it is advisable that, when evaluating particular procedures and antisera for electron microscopic immunocytochemistry, the various parameters (e.g., fixation, dehydration, antibody binding characteristics) are established first at the light microscope level.

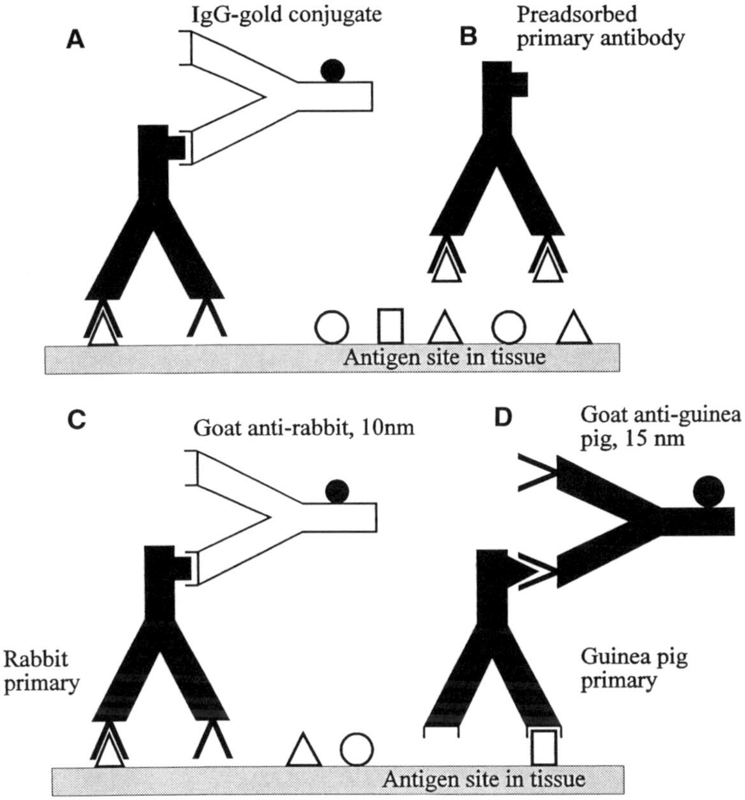

Fig. 1. (A) In the indirect single labeling of a tissue antigen (Δ), the primary antibody (black) is not labeled but is visualized using a second labeled reagent, in this case a gold (●)-conjugated immunoglobulin (IgG). **(B)** Preadsorption of the primary antibody (black) by specific antigen (Δ) is one essential control. **(C,D)** Simultaneous double-labeling requires a mixture of two primary antisera raised in different donor species, e.g., rabbit anti-Δ (C) and guinea pig anti-□ (D), which are then revealed using species-specific secondary antisera, e.g., goat antirabbit and goat antiguinea pig, each conjugated with different sized gold probes (10 and 15 nm, respectively).

2. Materials

2.1. Fixation and Dehydration

The most widely used fixatives for adequately preserving cell ultrastructure in immunoelectron microscopy are glutaraldehyde and formaldehyde (freshly prepared from paraformaldehyde). The former crosslinks tissue proteins more effectively and produces superior preservation; however, its rate of penetration into tissue is slower than that of formaldehyde. For this reason, a buffered

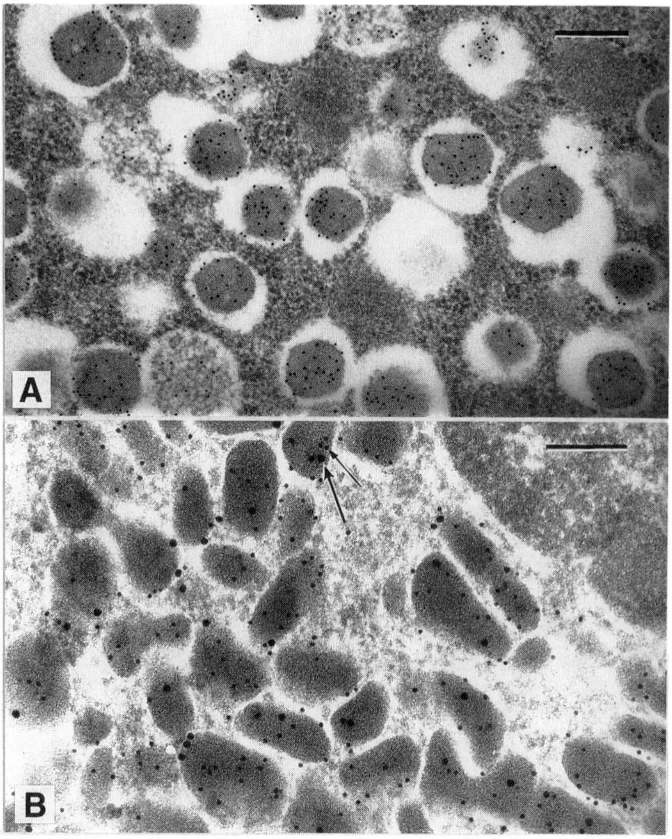

Fig. 2. Sections of frog *(Rana temporaria)* pancreas, **(A)** showing immunoreactivity for insulin in the β-cells, using a 10-nm gold probe (scale bar, 1 μm); and **(B)** colocalization, by simultaneous double-labeling, of immunoreactivities for glucagon (10-nm gold probe, small arrow) and pancreatic polypeptide (15-nm gold probe, large arrow) in the α-cells (scale bar, 0.5 μm). Note that in both cells, the gold particles are concentrated over secretory granules and that nuclear and cytoplasmic labeling is negligible. Tissue was fixed with a mixture of glutaraldehyde/paraformaldehyde and embedded in Epon resin; sections were finally stained with uranyl acetate and lead citrate.

mixture of the two is preferred. Whereas most tissues fixed this way give adequate ultrastructural detail, some, such as nerve tissue, can benefit from secondary fixation with osmium tetroxide to preserve the more delicate membrane structures. However, relatively few tissue antigens can be labeled successfully following osmium postfixation, without some pretreatment of the sections with an oxidizer to restore antigenicity (*see* Note 1). Note that osmic-acid postfixation is omitted when embedding in acrylic resins.

1. Glutaraldehyde (GTA): 70% double-distilled monomeric GTA (Agar Scientific, Stansted, UK).
2. Paraformaldehyde (PFA; BDH Chemicals, UK): 4% in 0.1*M* cacodylate buffer, pH 7.2–7.4. Under a ventilated hood, add 16 g solid PFA to 200 mL deionized water, heat to 55–60°C until completely dissolved. Add 5*M* NaOH solution dropwise until clear. When cool, make up to 400 mL with 0.2*M* cacodylate buffer. Filter before use. Can be stored at 4°C for up to 2 wk.
3. 2% paraformaldehyde and 2% double-distilled glutaraldehyde in 0.1*M* sodium cacodylate pH 7.2 buffer, containing 100 m*M* sucrose.
4. 0.1*M* sodium cacodylate, pH 7.2 buffer, containing 100 m*M* sucrose.
5. Ethanol.
6. Propylene oxide.

2.2. Resin Infiltration and Polymerization

As already mentioned, peptide antigens are perhaps more robust than most, so that the conventional electron microscopic procedures of fixation, dehydration, and embedding in epoxy resins, such as Araldite or Epon, can often be applied quite successfully (*see* refs. *7* and *8* for details), but require the sections to be etched with an oxidizing agent prior to processing (*see* Note 1). As an alternative procedure, immunolabeling of peptide antigens, particularly the more sensitive ones, can be successfully achieved using acrylic resins, such as LR White or Lowicryl K4M. These water-miscible resins can be polymerized at near ambient temperature (e.g., LR White), or at very low temperatures (e.g., –20 to –40°C) (e.g., Lowicryl resins) by UV light, causing minimal denaturation of tissue and loss of antigenicity. This is because their hydrophilic nature allows them to be cured without complete dehydration of the protein/peptide molecules, enabling a closer interaction of antibodies with the resin surface, thus enhancing the efficiency of the labeling *(9)*.

2.2.1. Epoxy Resins

1. Araldite or Epon resin (TAAB, Watford, UK).
2. Propylene oxide.
3. Rotator.
4. Embedding molds (Agar).
5. Oven for thermal polymerization.

2.2.2. Acrylic Resins

1. Lowicryl K4M (TAAB, Watford, UK).
2. Ethanol.
3. BEEM capsules, size 00 (Agar).
4. UV light source (360 nm wavelength).

2.3. Sectioning and Immunolabeling

Ultrathin sections of materials to be immunolabeled are cut on an ultramicrotome, following standard procedures described in electron microscopy texts *(7,8)*. Sections of the epoxy plastics (Araldite or Epon), if cut rather thick (70–90 nm; interference color: gold) to maximize labeling, can be collected directly on to clean, bare nickel grids and allowed to dry prior to processing. However, thinner sections (60–80 nm; interference color: silver) allow better definition, and those cut from acrylic blocks are best transferred on to nickel grids that have been coated with a plastic support film, such as Formvar. This is because the hydrophilic acrylic sections are somewhat less stable under the electron beam than conventional epoxy plastics, and using a support film means that the sections are less likely to be damaged; it also promotes their adherence to the grid. However, it must be noted that support films can contribute to nonspecific background labeling.

The immunolabeling is best carried out by incubating the grids in multiwell microtest plates or simply on droplets of the immunoreagents on parafilm sheets inside a sealable container, such as a Petri dish, ensuring a dust-free environment.

1. Nickel grids (200 mesh) (Agar).
2. Ultramicrotome.
3. Diamond or glass knives.
4. Multiwell culture plates/parafilm and Petri dishes.
5. Watchmaker's forceps (size 7) to handle the grids.
6. 10% hydrogen peroxide (for epoxy plastic sections only).
7. 20 m*M* Tris-HCl, pH 8.2 buffer.
8. 0.1% bovine serum albumin (BSA) in 20 m*M* Tris-HCl, pH 8.2 buffer.
9. Tween-20 (1:20 dilution).
10. Normal goat serum (1:20 dilution in 20 m*M* Tris-HCl, pH 8.2).
11. 20 m*M* glycine in 20 m*M* Tris-HCl, pH 8.2 buffer.
12. Primary antisera (Peninsula Laboratories, St. Helens, UK). The dilution of antisera used will have to be determined by a dilution series (*see* Note 6).
13. Gold-labeled secondary antisera (British BioCell, Cardiff, UK).
14. 2% double-distilled glutaraldehyde (GTA) in 0.1*M* cacodylate, pH 7.2 buffer.
15. Double-distilled water.
16. Uranyl acetate and lead citrate solutions (*see* Note 2 for composition).

3. Methods

3.1. Fixation and Dehydration

1. Wash tissue in an appropriate saline (e.g., cacodylate or phosphate-buffered saline) to remove any serum proteins or other contaminants.
2. Slice tissue (approx 0.5–1 mm in thickness) in fixative at ambient temperature, using a mixture of 2% paraformaldehyde and 2% double-distilled glutaralde-

hyde in 0.1*M* cacodylate buffer (pH 7.2), containing 100 m*M* sucrose. Leave for 1 h at 4°C.

3. Wash slices in 3 × 10-min changes of cold 0.1*M* cacodylate, pH 7.2 buffer containing 100 m*M* sucrose.
4. For embedment in epoxy resins, dehydrate as follows at ambient temperature:
   a. 70% ethanol, 10 min.
   b. 90% ethanol, 10 min.
   c. 100% ethanol, 10 min.
   d. Propylene oxide, 5 min.
5. For embedment in acrylic resins, dehydrate as follows:
   a. 35% ethanol, 15 min at 4°C.
   b. 50% ethanol, 30 min at –20°C.
   c. 75% ethanol, 30 min at –20°C.
   d. 95% ethanol, 30 min at –20°C.
   e. 100% ethanol, 2 × 30-min changes at –20°C.

3.2. Resin Infiltration and Polymerization

3.2.1. Epoxy Resins

1. Infiltrate specimens in a 1:1 mixture of propylene oxide and Araldite or Epon for 2–3 h at ambient temperature, using a rotator.
2. Infiltrate specimens in 100% pure resin for 15–20 h, using a rotator.
3. Transfer specimens to fresh resin in embedding molds and polymerize for 48 h at 60°C.
4. Allow resin to cure for 48 h at room temperature.
5. Remove plastic block from mold and section.

3.2.2. Acrylic Resins

1. Infiltrate specimens in a 1:1 mixture of 100% ethanol and Lowicryl K4M resin for 12 h at –20°C.
2. Infiltrate in 100% pure resin for 12–18 h at –20°C.
3. Transfer to BEEM capsules containing fresh K4M resin, capped to exclude air, and polymerize under UV light (360 nm wavelength) for 30 h at –20°C.
4. Allow resin to cure for a further 72 h under UV light at room temperature.
5. Remove plastic block from capsule and section.

3.3. Sectioning and Immunolabeling

3.3.1. Single Indirect Immunogold Labeling

1. Trim blocks and cut sections on to nickel grids and allow to dry. Araldite or Epon sections (but not acrylic) should be first etched with 10% hydrogen peroxide for 10 min, then dip-washed in 20 m*M* Tris-HCl, pH 8.2 buffer, 5 × 1 min.
2. Using forceps, float grids section-side down (*see* Note 3) on the following, as indicated in steps 3–14, at ambient temperature (unless stated otherwise). Ensure grids are kept wet throughout the entire labeling procedure (*see* Note 4).

3. 50-μL droplets of 20 m*M* Tris-HCl buffer, pH 8.2, containing 0.1% bovine serum albumin (BSA) and Tween-20 (1:20 dilution), 5 × 1-min washes.
4. 30-μL droplet of normal goat serum (1:20 dilution in 20 m*M* Tris-HCl, pH 8.2 buffer) to block nonspecific proteins, 30 min (*see* Note 5).
5. 50-μL droplets of 20 m*M* Tris-HCl, pH 8.2, 5 × 1 min.
6. 30-μL droplet of 20 m*M* Tris-HCl, pH 8.2, containing 0.02*M* glycine, to block free aldehyde groups in the tissue, 5 min (*see* Note 5).
7. 50-μL droplets of 20 m*M* Tris-HCl, pH 8.2, 2 × 1 min.
8. 30-μL droplet of primary antibody suitably diluted (*see* Note 6) with 0.1% BSA in 20 m*M* Tris-HCl, pH 8.2, at 4°C overnight.
9. 50-μL droplets of 0.1% BSA in 20 m*M* Tris-HCl, pH 8.2, 5 × 1 min.
10. 30-μL droplet of secondary antibody-gold conjugate, 1 h (*see* Note 7).
11. 50-μL droplets of 0.1% BSA in 20 m*M* Tris-HCl, pH 8.2, 5 × 1 min.
12. Postfix on 20-μL droplet of 2% double-distilled glutaraldehyde in 0.1*M* cacodylate buffer, 2 min.
13. Wash on 50-μL droplets of 20 m*M* Tris-HCl, pH 8.2, 5 × 1 min.
14. Thoroughly rinse on 50-μL droplets of double-distilled water, 7 × 1 min.
15. Stain grids by immersion in watch glass of uranyl acetate (10 min); dip-wash in distilled water (10 times), and dry on filter paper; immerse in lead citrate (8 min); dip-wash, and dry as before. Examine in the electron microscope (*see* Notes 8 and 9).

3.3.2. Double Indirect Immunogold Labeling

The availability of different-sized gold probes makes immunogold methods particularly suitable to multiple-labeling experiments on the same tissue section, although in practice crossreactivity and steric hindrance generally limit the procedure to using only two different sized probes (e.g., 10 and 15 nm). A fairly simple and quick method of double-labeling is to simultaneously apply to the tissue section a mixture of two specific primary antisera from different donor species (e.g., rabbit and guinea pig), followed, after appropriate washes, by two species-specific secondary antisera (e.g., a mixture of goat antirabbit and goat antiguinea pig), each conjugated to different sizes of gold probes *(10)* (Fig. 1C,D; Fig. 2B). Care must be taken to ensure that there is no cross-reaction between the labeled antibodies The method is known as simultaneous double-labeling (*see* Note 10). Alternative procedures of double-labeling are necessary when the primary antisera to be used have been raised in the same donor species. One such method is to label the two antigens in question sequentially, but before applying the second antibody and probe, saturate any free Fc sites on the tissue with unconjugated IgG. In other words, block any possible crossreactions by the secondary IgG-gold probe. To further minimize crossreactions, it is recommended that the first antigen be localized with the smaller (e.g., 10-nm size) of the two gold probes.

3.3.2.1. Simultaneous Double-Labeling

1. Follow steps 1–7, as described in Section 3.3.1.
2. Float grids (section side down) on a 30-µL mixture of two 15-µL droplets of the two primary antisera (e.g., from rabbit and guinea pig). Incubate overnight at ambient temperature.
3. Wash on 50-µL droplets of 0.1% BSA in 20 m*M* Tris-HCl buffer, pH 8.2, 5 × 1 min.
4. Float grids on a 30-µL mixture of the two appropriate secondary antisera (e.g., goat antirabbit and goat antiguinea pig), each conjugated with a different-sized gold probe (e.g., 10 and 15 nm), 1 h.
5. Wash, postfix, and stain, as described in steps 11–15 of Section 3.3.1.

3.3.2.2. Sequential Double-Labeling

1. Follow steps 1–7, as described in Section 3.3.1.
2. Float grids on a 30-µL droplet of one of the primary (e.g., rabbit) antisera. Incubate overnight at ambient temperature.
3. Wash on 50-µL droplets of 0.1% BSA in 20 m*M* Tris-HCl, pH 8.2 buffer, 5 × 1 min.
4. Float grids on a 30-µL droplet of secondary (e.g., goat antirabbit) antibody-gold conjugate (10-nm size), 1 h.
5. Wash on 50-µL droplets of 0.1% BSA in 20 m*M* Tris-HCl, pH 8.2 buffer, 5 × 1 min.
6. Float grids on a 30-µL droplet of normal (e.g., rabbit) serum (1:10 dilution) for 30 min.
7. Float grids on a 30-µL droplet of the second of the primary antisera, overnight.
8. Wash on 50-µL droplets of 0.1% BSA in 20 m*M* Tris-HCl, pH 8.2 buffer, 5 × 1 min.
9. Float grids on a 30-µL droplet of secondary (goat antirabbit) antibody–gold conjugate (15-nm size), 1 h.
10. Wash, postfix, and stain, as described in steps 11–15 of Section 3.3.1.

3.3.3. Controls

Since the localization of the peptides is dependent on the specificity of the binding of the immunoreagents, it is essential to assess the immunostaining by a series of controls; these should be performed on sections cut from the same blocks that gave positive reactions and preferably at the same time as the test labeling procedure. The following controls are essential procedures in all immunogold labeling experiments and should give negative results.

1. Omit the primary antiserum to determine if there is nonspecific binding of secondary antibodies to Fc receptors or to other nonspecific binding sites.
2. Substitute nonimmune serum from the donor species in which the secondary antiserum was raised, in place of the primary antiserum.

3. Preadsorb the primary antiserum with different concentrations of its homologous antigen (*see* Note 11), so as to confirm the specificity of antibodies used.
4. Controls for double-labeling must include omission of each of the primary antibodies in turn, and include both secondaries.

4. Notes

1. Epoxy plastic sections of tissue require treatment with an oxidizing agent to allow penetration of the antibody. This is done by etching the sections with a 10% aqueous solution of H_2O_2 for 10 min or, if tissue has been osmicated, by exposure to a saturated solution of sodium metaperiodate *(6)*. Sections are then thoroughly rinsed in an appropriate buffer (e.g., Tris-HCl), before being processed with the immunoreagents.
2. Uranyl acetate is made up as a saturated solution in 50% ethanol. Protect from light. Lead citrate solution is made by dissolving 1.33 g lead nitrate and 1.76 g sodium citrate in 30 mL boiling water; add 8 mL of 1*M* NaOH and dilute to 50 mL with distilled water. Protect from CO_2.
3. Be careful when transferring grids that the droplets of antiserum are not unduly diluted by liquid carried over from the previous step. Drain grids by carefully touching their edges with filter paper. Also, be sure the grids are floated section-side-down on the droplets of immunoreagents and that there are no air bubbles trapped under the grid.
4. Minimize evaporation of the immunoreagents by placing a strip of water-soaked filter paper within the Petri dish.
5. In trials, treatment with normal goat serum and glycine steps can be omitted until labeling has been achieved.
6. As in all immunocytochemical procedures, it is essential to use the highest possible dilution of the primary antiserum, without sacrificing any specific staining, so as to achieve a good tissue-to-background labeling (i.e., a favorable signal-to-noise ratio).
7. It is recommended to keep the dilution (1:100) and gold probe size (e.g., 10 nm) constant throughout different experiments, and vary the dilution of the antiserum, so as to obtain optimum labeling efficiency with minimum background labeling. The size of the gold probe selected is generally determined by the size of the specimen to be examined and by the magnification to be used. However, it must be remembered that the larger the diameter of the gold probe used, the lower the observed binding intensity, owing to steric hindrance.
8. Where background labeling is unacceptably high, as judged by examining regions of the tissue in which the peptide antigen would not be expected to be present, e.g., the nuclei or mitochondria, increase dilution of the antibodies and consider an extra blocking step (step 4) between the two antibody incubations (steps 8 and 10).
9. Where there is little or no labeling, ensure that epitopes have not been destroyed by fixation: try shorter periods of fixation at lower fixative concentrations. Check the quality of the primary antibody and that of the gold conjugate. Excellent sources of gold probes are commercially available and, if stored correctly and

used within their shelf-life at the correct concentration, are extremely reliable. Remember, it is always worth doing an immunocytochemical screen for the peptides of interest at the light microscope level, so as to establish suitable conditions of fixation and antibody binding for successful labeling of electron microscope sections.

10. Double-labeling may be less than satisfactory if one or other of the two antigens has a low distribution count and is dominated by the other, or there is a generally poor level of labeling. As a rule, to succeed in double-labeling, it is important to first establish good single-labeling for each antigen.
11. For liquid-phase preadsorption of the primary antiserum, determine the optimal dilution of the primary antibody for immunolabeling a given peptide and divide into two aliquots. Incubate one of the aliquots with excess (usually 1 n*M*) purified peptide antigen. Use each aliquot for immunolabeling, and note that preadsorption should quench or markedly reduce the labeling reaction.

References

1. Coons, A. H., Creech, H. J., Jones, R. N., and Berliner, E. (1942) The demonstration of pneumococcal antigen in tissue by the use of fluorescent antibody. *J. Immunol.* **45,** 159.
2. Singer, S. J. (1959) Preparation of an electron-dense antibody conjugate. *Nature* **183,** 1523,1524.
3. Feldherr, C. M. and Marshall, J. M. (1962) The use of colloidal gold for studies of intracellular exchange in amoeba *Chaos chaos*. *J. Cell Biol.* **12,** 640–645.
4. Faulk, W. P. and Taylor, G. M. (1971) An immunocolloid method for the electron microscope. *Immunocytochemistry* **8,** 1081–1083.
5. Roth, J. (1983) The colloidal gold marker system for light and electron microscopic cytochemistry, in *Immunocytochemistry,* vol. 2 (Bullock, G. R. and Petrusz, P., eds.), Academic, London, pp. 217–284.
6. Polak, J. M. and Priestley, J. V. (1992) *Electron Microscopic Immunocytochemistry.* Oxford University Press, Oxford.
7. Dykstra, M. J. (1993) *A Manual of Applied Techniques for Biological Electron Microscopy*. Plenum, New York.
8. Hayat, M. A. (1989) *Principles and Techniques of Electron Microscopy: Biological Applications*. 3rd ed. CRC, Boca Raton, FL.
9. Newman, G. R. and Hobot, J. A. (1987) Modern acrylics for post-embedding immunostaining techniques. *J. Histochem. Cytochem.* **35,** 971–981.
10. Tapia, F. J., Varndell, I. M., Probert, L., De Mey, J., and Polak, J. M. (1983) Double immunogold staining method for the simultaneous ultrastructural localization of regulatory peptides. *J. Histochem. Cytochem.* **31,** 977–981.

27

Preparation of a Membrane Fraction for Receptor Studies and Solubilization of Receptor Proteins with Retention of Biological Activity

Mark Wheatley, John Howl, Nicola J. Yarwood, Andrew R. L. Davies, and Rosemary A. Parslow

1. Introduction

1.1. Why Solubilize Receptors?

Hormones, neurotransmitters, and growth factors generate their multifarious effects by binding to specific receptor proteins located on the plasma membranes of target tissues (the soluble DNA binding proteins that are receptors for steroid hormones and thyroid hormones are outside the scope of this chapter). The receptors under consideration are integral membrane proteins and are usually glycosylated. Receptors for different hormones and growth factors are pharmacologically distinct. Indeed, multiple subtypes of receptor for a single agonist are often expressed. However, this heterogeneous group of proteins can be conveniently classified into three basic categories: receptors that possess only one transmembrane domain (1TMRs), receptors with an integral ion channel (RICs), and receptors whose effects are mediated by coupling to guanine nucleotide-binding proteins (G-protein-coupled receptors; GPRs). It is fundamentally important that we understand these receptors in molecular detail, study the architecture of the protein, and define how they recognize hormones and generate their effects. In order to do this, it is necessary to remove the receptor from the complex native membrane environment, to study it in solution, to purify the protein, and to reconstitute it into defined systems. A prerequisite to these approaches is the effective solubilization of receptor with retention of its biological functions. After more than 20 yr of research addressing solubilization of a wide range of receptors, it

From: *Methods in Molecular Biology, Neuropeptide Protocols*
Edited by: G. B. Irvine and C. H. Williams Humana Press Inc., Totowa, NJ

is apparent that there is no one protocol, or detergent, that is applicable in all cases. Methodologies employed for any given receptor have to be developed empirically; however, certain common features important to successful receptor solubilization have evolved and these are presented in this chapter.

1.2. Preparation of a Membrane Fraction as a Source of Receptors

Prior to initiating studies on any receptor, it is important that a good source of that receptor be available. This usually involves preparing a plasma membrane fraction from a target tissue that is known to express the receptor in high abundance. Alternatively, the receptor may be expressed by a cultured cell line. This is becoming increasingly common as more and more receptors are cloned and subsequently expressed in transfected cultured cells lines, either transiently (e.g., in COS-7 cells) or stably (e.g., in CHO cells). As the nature of the receptor source can vary so much, three protocols are provided. These have been developed for preparing a crude plasma membrane preparation, which is suitable for receptor studies, from a relatively soft tissue (liver), a fibrous tissue (heart), and a cultured cell line (COS-7 cells), respectively.

1.3. Solubilization of Receptors by Detergents

As the receptor proteins under consideration are integral proteins, they exhibit a high degree of interaction with the membrane lipids. Consequently, solubilization requires the use of detergent to molecularly disperse the membrane components. Other treatments such as sonication, exposure to high salt concentrations (e.g., 1*M* NaCl), or chelating agents (e.g., 10 m*M* EDTA) are inappropriate for receptor solubilization. However, if receptor purification is the ultimate aim, then these protocols can have a role in enriching the receptor preparation as they remove peripheral proteins.

Detergents are analogous to phospholipids in that they are amphipathic molecules, possessing both a hydrophilic head and a hydrophobic tail (or a hydrophilic face and a hydrophobic face) *(1–3)*. The nature of the head group varies and can be anionic, cationic, zwitterionic (composed of both positive and negative moieties but possessing a net null charge), or nonionic *(3)*. Prior to developing a successful solubilization protocol for a receptor, it is useful to know what is actually occurring at the molecular level when a detergent is added to a suspension of plasma membranes. The following four points are in reference to Fig. 1:

1. When present at low concentrations, the detergent will exist in the aqueous solution as monomers. These will partition into the membranes where they will compete with the membrane lipids for the hydrophobic domains of integral membrane proteins including receptors (Fig. 1B).

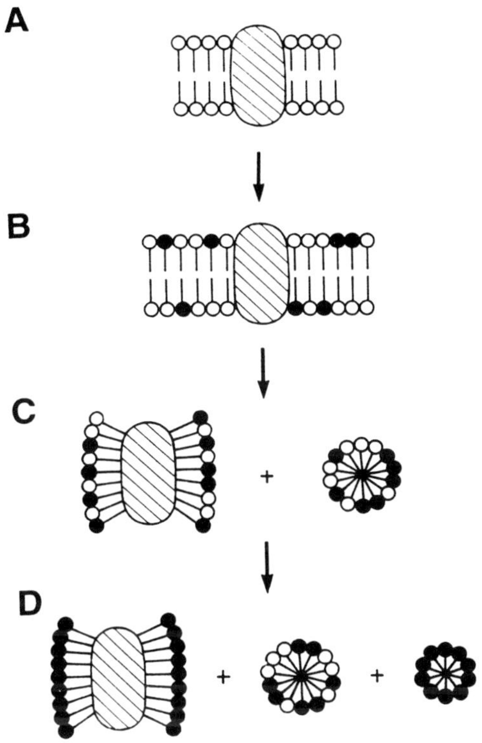

Fig. 1. Solubilization of membranes by detergent. (◍) integral membrane protein; (○——) membrane lipid; (●——) detergent monomer. This is a schematic representation of the molecular events that occur when membrane components are dispersed by the action of a detergent. Panels **A–D** represent the key stages of this process as discussed in Section 1.3. Detergent is absent in panel A and is present at increasing concentration in panels B, C, and D.

2. As the detergent concentration increases, the concentration of detergent monomers in both the solution and in the membranes will increase and some disruption of the membranes may occur.
3. Above a certain concentration (or concentration range) termed the critical micellar concentration (CMC), the detergent molecules aggregate to form micelles. The number of monomers forming the micelle is the aggregation number. The CMC and micelle size are established characteristics for any detergent, but they are affected by temperature, ionic strength, and by the presence of impurities. The micelles are in equilibrium with detergent monomer, and the monomer concentration is equal to the CMC. Extensive disruption of membranes will occur above the CMC, with integral proteins being solubilized in the form of mixed detergent:phospholipid:protein micelles (Fig. 1C).

4. Further increases in detergent concentration increase the concentration of micelles. In addition to detergent micelles, mixed micelles composed of detergent:protein and detergent:lipid will occur. This removal of endogenous lipid from the protein can be deleterious to the functioning of some receptors *(4,5)*. A ramification of this is that when solubilizing receptors, there is usually a balance to be struck between extracting the maximum amount of receptor from the membranes and preserving receptor function (*see* Note 1).

In addition to the stripping of lipid from the receptor, another source of receptor damage is proteolytic degradation. This is often exacerbated in the presence of detergent owing to disruption of subcellular compartmentalization. Consequently, it is important to take precautions to prevent the degradation of receptors by proteolytic enzymes throughout the solubilization process.

1.4. Selecting a Suitable Detergent

By far the easiest strategy is to adopt a published protocol that has been shown to be useful for solubilizing an analogous receptor. Whether you are using a previously published method or developing a new protocol, it is useful to give some thought to the ultimate use of the receptor preparation. Certain types of study or assay are incompatible with some detergents. So, regardless of how good a detergent is at solubilizing receptors, it may be precluded from your list of candidate detergents. For example, the Triton series contain aromatic groups that absorb strongly in the UV region. Consequently, some fluorometric analyses and assaying protein content by absorbance at 280 nm will be compromised in the presence of Triton (however, *see* Note 2). If detergent removal is required postsolubilization, in reconstitution experiments for example, then a detergent that is dialyzable and has a high CMC is preferable. Ionic detergents will bind to proteins during solubilization and can swamp the protein's native charge. This may restrict the usefulness of any charge-based protein purification methods such as isoelectric focusing. The use of nonionic or zwitterionic detergent would be more appropriate in this case. Some detergents are very expensive, which may preclude the use of a detergent purely on the grounds of cost. A range of potentially useful detergents is presented in Table 1 together with some of their characteristics. These have been selected because they have already proved successful in published receptor solubilization protocols. Their structures are given in Fig. 2.

2. Materials

2.1. Solutions and Media

2.1.1. Solutions Required for a Membrane Preparation from Rat Liver

1. Leupeptin/antipain stock.
2. Dissolve 5 mg leupeptin plus 5 mg antipain together in 10 mL distilled water to give a final concentration of 0.5 mg/mL of each. Aliquot 0.5-mL samples into plastic microcentrifuge tubes and store at –20°C.

Table 1
Characteristics of Some Detergents that Have Been Successfully Employed for Receptor Solubilization

Detergent	CMC, m*M*	Aggregation number	Micellar weight	Monomer mol wt	Dialyzable	Class of receptor solubilized	Ref.
CHAPS[a]	8	4–14	2500–8600	615	+	1TM,RIC,GPR	*7,8*
CHAPSO	8	10–16	10,000	631	+	1TM,RIC,GPR	*9*
Cholate[b]	14	2–4	900–1800	431	+	1TM,RIC,GPR	*6,10*
Deoxycholate[b]	5	3–12	1700–5000	415	+	1TM,RIC	*11*
Digitonin[c]	—	60	75,000	1229	–	GPR	*12*
Dodecylmaltoside	0.1–0.2	98	50,000	511	–	GPR	*13*
Lysophosphatidyl choline (C_{16})	0.02–0.2	approx 190	92,000	495	–	GPR	*12,14*
Octylglucoside	23	approx 27	8000	292	+	1TM,RIC,GPR	*15,16*
Triton X-100[d]	0.3	100–150	90,000	625	–	1TM,RIC	*17*
Triton X-405	0.3	approx 50	90,000	2142	–	GPR	*18*

Properties of the detergents that are pertinent to their use for solubilizing receptors are presented. The data for the CMC, aggregation number and micellar weight are approximate because these parameters are influenced by various factors as discussed in Section 1.3. The "receptor class solubilized" column gives a guide to the classes of receptor that have been solubilized in an active form by the detergents cited. 1TM, receptors with 1 transmembrane domain; RIC, receptors with an integral ion channel; GPR, G-protein-coupled receptors. CHAPS, 3-[(3-cholamidopropyl)dimethylammonio]-1-propanesulfonate; CHAPSO, 3-[(3-cholamidopropyl)dimethylammonio]-2-hydroxypropane-1-sulfonate. Triton is a trade name of Union Carbide Chemicals and Plastics Co. The Triton series of detergents are polyethylene glycol octylphenyl ethers.

[a]*See* Note 3; [b]*see* Notes 4 and 5; [c]*see* Note 6; [d]*see* Note 7.

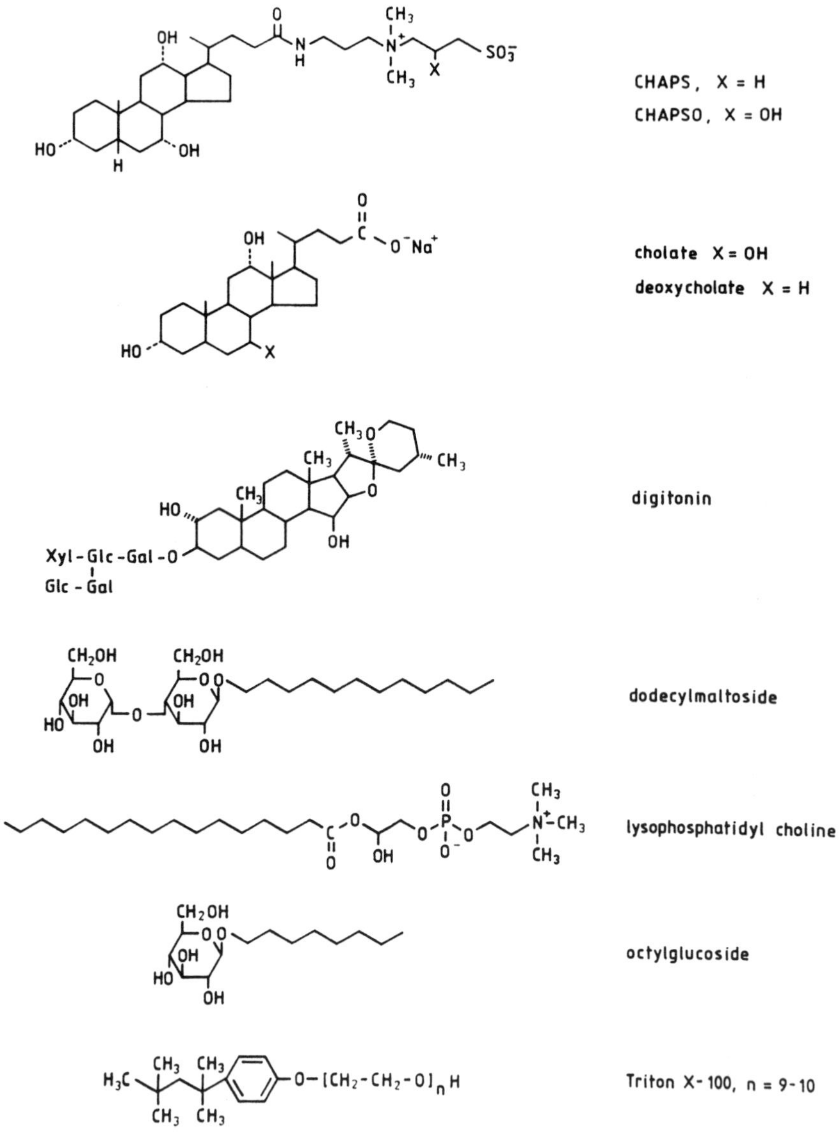

Fig. 2. Detergent structures.

3. Percoll: This is supplied by Pharmacia (St. Albans, Herts, UK) and by Sigma (Poole, Dorset, UK).
4. All media should be ice cold when used.
5. Medium A: 5 m*M* HEPES, 250 m*M* sucrose, 1 m*M* EGTA, pH 7.4.
6. Medium B: 50 m*M* Tris, 250 m*M* sucrose, pH 8.0.

7. Medium C: 20 mM HEPES, 250 mM sucrose, 10 mM $Mg(CH_3COO)_2$, 1 mM EGTA, pH 7.4.

2.1.2. Solutions for a Membrane Preparation from Rat Heart

1. Medium D: 20 mM HEPES, 0.1 mM EDTA, pH 7.5.
2. Medium E: 3M KCl, 200 mM disodium pyrophosphate.
3. Medium F: 20 mM HEPES, 10 mM EDTA, pH 7.5.
4. Medium G: 20 mM HEPES, 1 mM $Mg(CH_3COO)_2$, pH 7.5.

2.1.3. Solutions for a Membrane Preparation from COS-7 Cells

1. Medium H (phosphate-buffered saline; PBS): 2.30 g of Na_2HPO_4 (anhydrous), 0.593 g of $NaH_2PO_4.2H_2O$, 1.6 g NaCl, pH 7.4 made up to 200 mL in water.
2. Medium I: 20 mM HEPES, 10 mM $Mg(CH_3COO)_2$, 1 mM EGTA, pH 7.4.
3. Medium J: Medium I supplemented with 250 mM sucrose plus bacitracin (0.1 mg/mL).
4. Medium K: Medium I supplemented with bacitracin (0.1 mg/mL).
5. Medium L: Medium I supplemented with 250 mM sucrose.

2.1.4. Solutions for Receptor Solubilization

1. Medium I (*see* Section 2.1.3.).
2. Protease inhibitors cocktail: Dissolve 0.5 mg Pepstatin A, 5 mg leupeptin, 5 mg chymostatin, 5 mg antipain, and 5 mg aprotinin in distilled water and make up to 10 mL. The Pepstatin A and chymostatin are insoluble and are employed as a suspension. All of these inhibitors are marketed by Sigma. The cocktail can be stored as 100-µL aliquots at –20°C for several months and is used at a 50-fold final dilution. In addition, EDTA or EGTA (1 mM) should be present in the buffers to inhibit metalloproteases and calcium-activated proteases (*see* Medium I composition). EGTA is preferable to EDTA if Mg^{2+} ions are required for receptors coupling to G-proteins (*see* Note 8).
3. Phenylmethylsulfonyl fluoride (PMSF) solution: Dissolve 34.8 mg PMSF in 1 mL of ethanol. This is better prepared fresh on the day of use and is employed at 1 µL/mL of solution (*see* Note 9). **Care:** PMSF is toxic and should be handled accordingly.

2.1.5. Solutions for Determining the Protein Content of the Solubilized Preparations

1. Trichloroacetic acid (TCA; trichloroethanoic acid) solution: 25% (w/v) TCA in water.
2. Sodium dodecylsulfate (SDS) solution: 10% (w/v) in water.
3. Lowry solution A: anhydrous sodium carbonate (2% w/v) and NaOH (0.1M) in water.
4. Lowry solution B: 0.5% (w/v) $CuSO_4.5H_2O$, 1% (w/v) sodium potassium tartrate, pH 7.4.

5. Lowry solution C: Folin-Ciocalteau's phenol reagent (Fisons, Loughborough, UK) diluted 1:1 with distilled water on the day of use. **Care:** This reagent contains phenol and should be handled appropriately.
6. Standard protein solution: 4 mg/mL bovine serum albumin (BSA) in water. Aliquots (e.g., 1 mL) can be stored at –20°C for several months.

2.2. Apparatus

A glass tube/teflon pestle homogenizer is used for all the membrane preparation protocols given. The size of this will obviously be dictated by the amount of tissue being processed. A Brinkman Polytron homogenizer is required for making a preparation of plasma membranes from a fibrous tissue.

A standard Minisart filter unit with a pore size of 200 nm (Sartorius, Göttingen, Germany) is required to establish the molecularly dispersed nature of a truly solubilized preparation. These units are designed to fit onto the end of a syringe.

3. Methods

3.1. The Preparation of a Membrane Fraction Suitable for Studying Receptors

3.1.1. A Membrane Preparation from Rat Liver

1. Supplement Medium A with protease inhibitors. For example, to 50 mL of Medium A add 0.5 mL of leupeptin/antipain stock (this is subsequently referred to as Medium A^+).
2. Store 50-mL centrifuge tubes and the glass/teflon homogenizer on ice until required. Sacrifice the rat, rapidly remove the liver, and transfer it to a beaker (on ice) containing approx 20 mL/liver of Medium A^+. Roughly chop the tissue with scissors (*see* Note 10).
3. Gently homogenize (using only medium power) with a glass/teflon homogenizer. Five up and down strokes is usually sufficient (*see* Note 11).
4. Rinse the pestle with Medium A^+ and add this to the homogenate. Adjust the homogenate volume to 30–40 mL/liver by adding Medium A^+.
5. Centrifuge this homogenate at 2500*g* for 10 min at 4°C.
6. Discard the supernatant. Resuspend the loose pellet to 24 mL/liver with Medium A^+ and rehomogenize the suspension with five strokes of the pestle rotating at slow/medium speed.
7. Add 3.2 mL Percoll/24 mL of liver suspension, mix thoroughly, and centrifuge at 35,000*g* for 30 min at 4°C (*see* Note 12).
8. Using a Pasteur pipet, remove and discard the shallow surface band as this is predominantly fat. Transfer the middle "fluffy" layer, which constitutes the bulk of the supernatant, into a clean 50-mL centrifuge tube (remove to just above the pellet). The discarded pellet is composed of a firm membranous layer on top of a clear viscous glycogen pellet.

9. Add Medium B until the tube is full and mix. Centrifuge at 70,000*g* for 1 h at 4°C.
10. Resuspend the pellet in Medium C by gentle homogenization using 6–8 mL buffer/liver.
11. Transfer 1-mL aliquots of this membrane preparation to microcentrifuge tubes and store at –20°C. These are stable for up to 1 yr in many cases (*see* Note 13).

3.1.2. A Membrane Preparation from Rat Heart

1. Fill a 10-mL syringe fitted with a hypodermic needle with ice-cold Medium D.
2. Sacrifice the rat, expose the heart, then use the syringe to perfuse the heart *in situ* with buffer to flush out any blood. Transfer the hearts into a preweighed beaker in an ice bucket.
3. Weigh the hearts and measure 9 vol of Medium D (e.g., if the hearts weigh 10 g, measure out 90 mL of Medium D).
4. Chop the tissue coarsely with scissors and add sufficient Medium D (from the 9 vol) to cover the pieces. Homogenize the heart with a Polytron using setting 5 for 3 × 15-s bursts. Cool on ice between bursts.
5. Add Medium D from the 9 vol if required to stop the preparation from becoming too viscous.
6. Homogenize the tissue further using 10 up and down strokes of a glass/teflon pestle homogenizer.
7. Filter the homogenate through 1 layer of cheese-cloth to remove poorly homogenized and fibrous material and measure the volume collected.
8. Add 0.8 vol (that is 0.8 vol of what is collected in step 7) of Medium D plus 0.2 vol of Medium E, mix, and centrifuge at 100,000*g* for 30 min at 4°C (*see* Note 14).
9. Resuspend the pellet in the original 9 vol (from step 3) of Medium F, incubate on ice for 30 min, and centrifuge at 100,000*g* for 30 min at 4°C (*see* Note 15).
10. Discard the supernatant, resuspend the pellet in the original 9 vol of Medium D, then centrifuge at 100,000*g* for 30 min at 4°C.
11. Repeat step 10.
12. Resuspend the pellet in the original 9 vol of Medium G. Aliquot as 1-mL samples and store at –20°C.

3.1.3. A Membrane Preparation from COS-7 Cells

1. Remove the culture medium from the cells and wash once with cold Medium H.
2. Add approx 1 mL of Medium J, then scrape the cells off the dish into the medium using a rubber scraper.
3. Transfer the cell suspension into a 50-mL centrifuge tube on ice and centrifuge at 20,000*g* for 30 min at 4°C.
4. Discard the supernatant, then add approx 40 mL of Medium K to the tube to resuspend the pellet. Leave on ice for 20 min for the cells to lyse.
5. Homogenize the preparation using 5 up-and-down strokes of a small teflon/glass homogenizer. Centrifuge at 20,000*g* for 30 min at 4°C.

6. Resuspend the pellet in a small volume (approx 3–4 mL) of Medium L and store as 0.5-mL aliquots at –20°C.

3.2. Solubilizing Receptors

3.2.1. Determination of the Optimum Detergent Concentration for Receptor Solubilization

1. Suspend membranes at a known concentration of 5–10 mg/mL in ice-cold Medium I (*see* Note 16).
2. Using Medium I, make a range of dilutions of the detergent of your choice (refer to Table 1 and *see* Notes 17 and 18).
3. Add protease inhibitor cocktail to the membranes, while stirring the suspension, using 20 μL cocktail/mL membrane suspension.
4. Aliquot 4 mL of membranes into 6 × 10-mL centrifuge tubes on ice.
5. Add PMSF (1 μL/mL) to each aliquot, then add an equal volume of detergent solution to each tube to produce the final concentration range required. This should be done rapidly as PMSF is unstable in aqueous solutions and its presence is required when the detergent is added.
6. Incubate at 4°C for 30 min, with mixing throughout this period. A rotating wheel that produces an end-over-end action is useful for this, although a magnetic stir-bar can be used.
7. Centrifuge the tubes at 100,000*g* for 60 min at 4°C. The supernatant from this spin is used as the solubilized preparation and should be clear and possibly straw-colored.
8. Freeze a small sample (approx 0.5 mL) of the solubilized preparation produced by each of the detergent concentrations for a protein determination (*see* Section 3.3.). Use the remainder in binding assays to establish if active receptors are present in the solubilized preparations and at what abundance (Chapters 28 and 29; *see* Note 19).
9. From the binding data and the protein assays, the optimum detergent concentration for solubilizing the receptors can be determined (refer to Fig. 3; *see* Note 20). By comparing the receptor abundance of the original membrane preparation and the solubilized preparation, the yield of solubilized receptors can be calculated (however, *see* Note 21).

3.2.2. Routine Protocol for Receptor Solubilization

The procedure is as described in Section 3.2.1., except that only the optimum detergent concentration is utilized. Consequently, only one solubilizing incubation is performed, but this will probably be on a larger scale than that described in Section 3.2.1.

3.3. Determination of the Protein Content of Solubilized Preparations (see Note 22)

1. Aliquot 0, 10, 20, 30, and 40 μL of the BSA standard solution into tubes (in quadruplicate). This will be used to construct a standard curve. Aliquot 20 and 40 μL of each solubilized preparation into quadruplicate tubes.

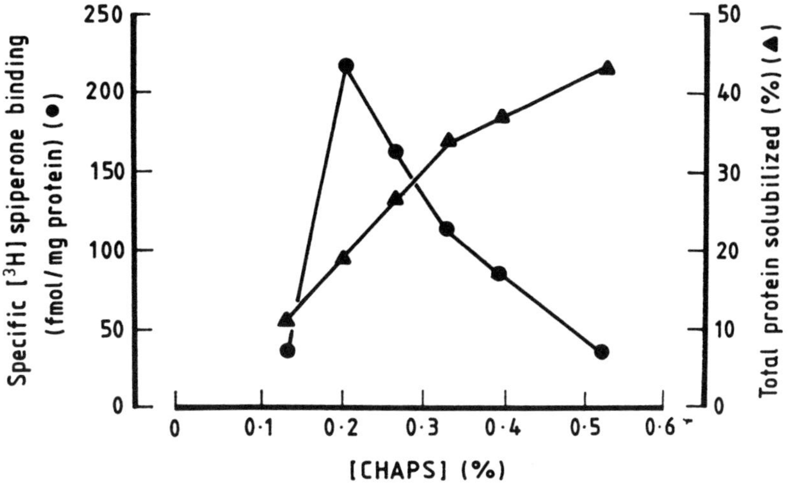

Fig. 3. Solubilization of [^{3}H]spiperone binding sites from bovine caudate nucleus membranes by CHAPS. The active binding sites extracted, together with the total protein solubilized, by varying concentrations of CHAPS is presented. Adapted from ref. *19* with permission.

2. Add 1 mL of 25% (w/v) TCA solution to each tube and mix. After 15 min at room temperature, centrifuge at 2000*g* for 10 min.
3. Discard the supernatant, invert the tubes, and leave them to drain for 15 min.
4. Add 1-mL of Lowry solution B to 50 mL of Lowry solution A and mix. Add 1 mL of Lowry solution A/B to each tube and mix.
5. Add 1 mL of 10% (w/v) SDS solution to each tube, mix, and leave overnight.
6. Add 100 μL of Lowry solution C to each tube, mix, and incubate for 30 min.
7. Add 1 mL of water to each tube and determine the absorbance at 750 nm using a spectrophotometer. Using the mean value of the quadruplicate determinations, the BSA samples are used to construct a standard curve from which the protein content of the unknowns is obtained (*see* Note 23).

3.4. Establishing that Receptors Are Solubilized

It is important to establish that molecular dispersal of the membrane components has been achieved by the solubilization protocol rather than mere fragmentation of the membrane. This can be demonstrated in several ways.

3.4.1. Centrifugation

1. Determine the amount of receptor in the putative solubilized preparation by radioligand binding (Chapters 28 and 29).
2. Centrifuge the preparation at 100,000*g* for 60 min at 4°C.

3. Determine the amount of receptor in the supernatant by radioligand binding. There should be no significant difference in the receptor abundance pre-, and postcentrifugation (*see* Note 24).

3.4.2. Ultrafiltration

1. Determine the amount of receptor (using radioligand binding, Chapters 28 and 29) and the protein content (*see* Section 3.3.) of the putative solubilized preparation.
2. Transfer approx 8 mL of the preparation into a syringe and, using gentle pressure, pass the solubilized preparation through a Minisart 200-nm filter into a collecting tube on ice.
3. Determine the amount of receptor radioligand binding and protein in the filtered sample. For a solubilized preparation, these postfiltration values should not be significantly different from the prefiltration values (however, *see* Notes 25 and 26).

4. Notes

1. The detrimental effect of the detergent stripping closely associated lipid from receptors can sometimes be ameliorated by adding exogenous lipids to the solubilizing incubation described in Section 3.2. For example, the $GABA_A$ receptor has been solubilized with 1.5% (w/v) CHAPS + 0.15% (w/v) soya bean asolectin *(8)*; the addition of glycerol (10–30% v/v) and cholesterol have also been employed *(20)*.
2. For monitoring protein in the presence of Triton X-100 (TX-100), it may be beneficial to measure the absorbance at 254 nm rather than at the conventional 280 nm. The absorbance of the detergent is markedly lower at this wavelength and the protein signal is still good. Alternatively, hydrogenated Triton X-100 is marketed by Pierce (Madison, IL), by Calbiochem (Nottingham, UK), and by Sigma (Poole, Dorset, UK). This has very similar detergent properties to normal TX-100 but it does not absorb at 280 nm *(21)*.
3. CHAPS was specifically designed to solubilize membrane proteins and to preserve their biological activity *(22)*. It is a chimeric molecule of two distinct detergent classes. It incorporates the bile salt cholic acid as the hydrophobic tail joined to a sulfobetaine head group *(22,23),* and possesses the beneficial properties of both of these detergents *(22)*. Being zwitterionic, it has no net charge between pH 2 and pH 12, so it does not migrate in an electric field, and is fully compatible with ion-exchange chromatography and isoelectric focusing. Its use is not detrimental to protein monitoring, as it exhibits very little absorbance at 280 nm and does not interfere with the Lowry assay.
4. The CMC of anionic detergents like cholate and deoxycholate is lowered by increasing the ionic strength of the medium *(1,3)*. This phenomenon has been utilized to reduce the amount of detergent required to solubilize receptors. For example, D_2 dopamine receptors can be solubilized by 0.2% (w/v) cholate + 1*M* NaCl *(10)*. If solubilization is performed in the presence of high salt, it is

important to remember that the detergent properties of the cholate will alter if the salt concentration is changed in subsequent manipulations.

5. Handling problems may be encountered with bile acid detergents. Cholate and deoxycholate have a pK_a value of approx 6.5. At pH <6.5 they become insoluble owing to protonation. They can also precipitate by forming complexes with metal ions and deoxycholate precipitates at 4°C.
6. Digitonin is a cardiac glycoside and so it must be handled with care. It binds specifically to cholesterol and consequently solubilizes cholesterol-containing membranes. Unlike most of the detergents in Table 1, commercial preparations of digitonin are not pure but contain digitonin and gitonin *(24)*. Furthermore, it is the ratio of these compounds that is important for generating an effective solubilizing medium. This probably is the basis of the notorious batch variation in water solubility and solubilizing characteristics experienced with digitonin.

   To prepare a stock solution of digitonin, it may be necessary to boil an aqueous suspension for 5–10 min. If the vessel containing the hot liquid is then plunged into ice, the digitonin should remain in solution. Conversely, digitonin can be used in conjunction with a second detergent, e.g., digitonin (1% w/v) plus cholate (0.1% w/v). The cholate is added to the digitonin solution just before cooling to 4°C after boiling. In general, the CMC of a detergent mix will approximate to that of the lower of the two individual CMCs. Alternatively, the WAKO Pure Chemical Company (Osaka, Japan) markets water-soluble (up to 10% w/v) digitonin, but it is very expensive.
7. Light-induced auto-oxidation of TX-100 can lead to an accumulation of peroxide contaminants. These can modify proteins solubilized by the detergent, for example, by oxidizing sulfhydryl groups. Although these peroxides can be removed *(25)*, Calbiochem, Pierce, and Sigma market peroxide-free TX-100. Nevertheless, it is still recommended that butylated hydroxytoluene (BHT) be added to the detergent (1 mol BHT/500 mol TX-100) as a free radical scavenger.
8. The cocktail of inhibitors given includes inhibitors of all the main classes of protease (i.e., serine, thiol, aspartic, and metalloproteases) and has proved useful in our studies. Obviously, components can be added or removed as appropriate. A more comprehensive list of available inhibitors is presented as Table 2, together with the protease class that is inhibited. It should be noted that although sulfhydryl reagents will inhibit thiol proteases, they can also modify receptor proteins and so are best avoided.
9. It should be established that the final concentration of ethanol after the addition of PMSF (0.1% v/v ethanol) is not detrimental to receptor binding.
10. Tissue should be fresh if possible to avoid a freeze/thaw cycle that can be detrimental to the integrity of some receptors. If the tissue is transported from an abattoir it should be surrounded by ice. Dissection and homogenization steps should be done rapidly and, if possible, performed at 4°C so as to reduce receptor damage.
11. Do not homogenize the tissue too vigorously, otherwise the plasma membranes form a pellet, rather than a suspension, in the subsequent centrifugation with Percoll.

Table 2
Protease Inhibitors

Inhibitor	Protease class inhibited	Recommended final concentration
Amastatin	Aminoexopeptidases	1–10 μg/mL
Antipain	Trypsin, papain, cathepsin A,B	1–10 μg/mL
APMSF (4-amidinophenyl) methanesulfonyl fluoride	Trypsin-like serine proteases	10–20 μ*M*
Aprotinin	Serine proteases	1–10 μg/mL
Bacitracin	General inhibitor	100 μg/mL
Benzamidine	Serine proteases	1–10 m*M*
Benzethonium chloride	Thiol proteases, arylaminopeptidases	0.1 m*M*
Benzylmalic acid	Carboxypeptidases	1–10 μg/mL
Bestatin	Aminopeptidases	1–10 μg/mL
Chymostatin	Chymotrypsin, papain, cathepsin A,B,D	1–10 μg/mL
3,4,-dichloroisocoumarin	Many serine proteases	5–200 μ*M*
Di-isopropylfluorophoshate	Serine proteases	0.1 m*M*
E-64	Thiol proteases	1 μg/mL
EDTA	Metalloproteases, Ca^{2+}-dependent thiol proteases	1–10 m*M*
EGTA	Metalloproteases, Ca^{2+}-dependent thiol proteases	1–10 m*M*
Elastinal	Elastases	10 μg/mL
Iodoacetate	Thiol proteases	1–10 m*M*
Iodoacetamide	Thiol proteases	1–10 m*M*
Leupeptin	Serine and thiol proteases	1–10 μg/mL
Pepstatin A	Aspartic proteases	1–10 μg/mL
N-ethylmaleimide	Thiol proteases	1 m*M*
PMSF (phenlymethane-sulfonylfluoride)	Serine proteases	0.1–1 m*M*
Phosphoramidon	Thermolysin, metalloproteases	1–10 μg/mL
Sodium tetrathionate	Thiol proteases	5 m*M*
TLCK (L-1-chloro-3-[4-tosylamido]-7-amino-2-heptanone	Trypsin, some serine, and thiol proteases	10 m*M*
TPCK (L-1-chloro-3-[4-tosylamido]-4-phenyl-2-butanone	Chymotrypsin, some thiol proteases	10 m*M*
Trypsin inhibitor	Trypsin	100–500 μg/mL
2-phenanthroline	Metalloproteases, Ca^{2+}-dependent thiol proteases	1–10 m*M*

A list of protease inhibitors is presented that have been used to restrict proteolytic damage to receptors during solubilization. A cocktail of several should be used as discussed in Section 2.1.4.

12. Percoll consists of small silica particles coated with nondialyzable polyvinylpyrrolidone. It is nontoxic, chemically inert, and does not adhere to membranes. In this protocol, it forms a density gradient during the centrifugation step that enables a fraction enriched in plasma membranes to be obtained.
13. The suspension should be constantly stirred during the aliquoting process to prevent the membranes from sedimenting. If this is not done, the samples will contain varying amounts of protein that will introduce inconsistencies into subsequent binding data.
14. This treatment is thought to remove some cytoskeletal components and has been found to reduce the nonspecific binding of some radioligands in subsequent binding studies.
15. This step removes divalent cations from the medium and, when Mg^{2+} ions are reintroduced, it has been found to increase the observed effects of GTP on the binding of agonists to G-protein-coupled receptors. This step will probably be superfluous for the study of receptors not of the GPR type.
16. To prevent interexperimental variation in the character of the solubilized preparation, it is important that subsequent solubilizations use the same protein:detergent ratio (really lipid:detergent ratio) as this pilot experiment. Hence, although a protein range is given, the precise protein concentration should be known.
17. Calbiochem markets some detergents as "ULTROL grade" or "PROTEIN grade." These are purified preparations designed for use with membrane proteins.
18. The appropriate range will vary from detergent to detergent owing to the differences in properties discussed in Section 1.3. and presented in Table 1. However, a good starting point will be 0, 0.2, 0.4, 1.0, 2.0, and 4.0% (w/v), which will give final detergent concentrations of 0, 0.1, 0.2, 0.5, 1.0, and 2.0% (w/v). This initial range could then be amended as appropriate in subsequent experiments in order to optimize the protocol for a specific detergent and receptor.

    A formula for determining the minimum concentration of detergent required to solubilize membrane proteins has been published *(26)*:

    $$\rho \text{ (rho)} = \text{(detergent)-CMC}_{\text{eff}}/\text{(phospholipid)}$$

    where CMC_{eff} is the effective CMC under solubilizing conditions. This is usually slightly lower than the CMC owing to the presence of membrane lipids. However, as a guide, the published CMCs presented in Table 1 can be used. Rho values of 1.5–10 seem to be most successful. It has also been suggested that detergent:protein ratios over the range 10:1 to 0.1:1 (w/w) be used in the initial solubilization trial *(27)*.
19. It is probable that the solubilized receptors will have to be freshly prepared for each experiment as it is generally found (although not always) that freezing/thawing solubilized receptors is deleterious to the receptor binding.
20. The concentration of a detergent that proves optimal for extracting receptor from the membrane may not be optimal for maintaining the protein confor-

mation postsolubilization. Some detergent must always be present to prevent solubilized receptors from aggregating, but detergent concentration may be reduced postsolubilization to increase receptor stability and to save cost. For example, muscarinic acetylcholine receptors were solubilized in 1% (w/v) digitonin but were stable on columns equilibrated with 0.1% (w/v) digitonin *(28)*.

21. This calculation obviously only accounts for active receptors. This will probably be an underestimate of the percentage of receptor protein solubilized, as upon solubilization some of the receptors may loose their ability to bind ligands for the reasons outlined in Section 1.3.
22. This method is based on that of Lowry et al. *(29)*. As many detergents interfere with this assay, the proteins are first precipitated from the solubilized preparation and then sodium dodecylsulfate (SDS) is subsequently added. This ionic detergent has been reported to prevent other detergents from interfering with the assay *(30)*.
23. Protein assay kits are available commercially that are not as sensitive to detergents as the Lowry method. For example, Quantify Protein Assay Systems (Promega, Madison, WI) is not affected by CHAPS (0.1% w/v) or octylglucoside (0.1% w/v). However, the background intensity is raised (and therefore sensitivity reduced) by Triton X-100 (0.01% w/v), and the assay is inapplicable in the presence of 0.05% (w/v) of either Triton X-100 or deoxycholate. The BCA Protein Assay (BCA, bicinchoninic acid; Pierce, Rockford, IL) is more compatible with detergents. For example, Triton X-100 (1% w/v), CHAPS (1% w/v), and octylglucoside (1% w/v) do not interfere. In addition, the assay is rapid and a useful instruction/information booklet is supplied when the kit is purchased. If a detergent is suspected of interfering with a protein assay, then it is good technique to construct the BSA standard curve in water containing the same detergent concentration as is present in the solubilized preparation. Alternatively, the protein can be TCA precipitated as described in Section 3.3.
24. This is a simple but nevertheless useful test, as cell membranes (including fragments) will be sedimented under the conditions described. However, it is important to remember that buffer components that markedly increase the density of the medium such as high concentrations of NaCl, sucrose, glycerol, and so on may prevent effective membrane sedimentation *(31)*.
25. The filter unit cited exhibits low, nonspecific adsorption of proteins, but some protein loss may still occur. If protein is depleted during filtration, then a proportional decrease in binding would be expected (assuming that different proteins within the preparation are adsorbed equally).
26. Examination of a solubilized preparation for membrane fragments by an electron microscope is a stringent test of true solubilization. This requires specialist knowledge and is outside the scope of this chapter. However, it should be remembered that sample preparation protocols that reduce the detergent concentration by dilution, can result in artifactual reconstitution of membranes.

References

1. Helenius, A. and Simons, K. (1975) Solubilization of membranes by detergents. *Biochim. Biophys. Acta* **415,** 29–79.
2. Tanford, C. and Reynolds, J. A. (1976) Characterization of membrane proteins in detergent solutions. *Biochim. Biophys. Acta* **457,** 133–170.
3. Helenius, A., McCaslin, D. R., Friess, E., and Tanford, C. S. (1979) Properties of detergents. *Methods Enzymol.* **56,** 734–749.
4. Hjelmeland, L. M. and Chrambach, A. (1981) Electrophoresis and electrofocusing in detergent-containing media: a discussion of concepts. *Electrophoresis* **2,** 1–11.
5. Kirilovsky, J. and Schramm, M. (1983) Delipidation of a β-adrenergic receptor preparation and reconstitution by specific lipids. *J. Biol. Chem.* **258,** 6841–6849.
6. Anholt, R., Fredkin, D. R., Deerinck, T., Ellisman, M., Montal, M., and Lindstrom, J. (1982) Incorporation of acetylcholine receptors into liposomes. *J. Biol. Chem.* **257,** 7122–7134.
7. Simonds, W. F., Koski, G., Streaty, R. A., Hjelmland, L. M., and Klee, W. A. (1980) Solubilization of active opiate receptors. *Proc. Natl. Acad. Sci. USA* **77,** 4623–4627.
8. Sigel, E. and Barnard, E. A. (1984) A γ-aminobutyric acid/benzodiazepine receptor complex from bovine cerebral cortex. *J. Biol. Chem.* **259,** 7219–7223.
9. Poyner, D. R., Birdsall, N. J. M., Curtis, C. A. M., Eveleigh, P., Hulme, E. C., Pedder, E. K., and Wheatley, M. (1989) Binding and hydrodynamic properties of muscarinic receptor subtypes solubilized in 3-(3-cholamidopropyl) dimethylammonio-2-hydroxy-1-propane sulfonate. *Mol. Pharmacol.* **36,** 420–429.
10. Wheatley, M., Hall, J. M., Frankham, P. A., and Strange, P. G. (1984) Improvement in conditions for solubilization and characterisation of brain D_2 dopamine receptors using various detergents. *J. Neurochem.* **43,** 926–934.
11. Sigel, E., Stephenson, F. A., Mamalaki, C., and Barnard, E. A. (1983) A γ-aminobutyric acid/benzodiazepine receptor complex of bovine cerebral cortex. *J. Biol. Chem.* **258,** 6965–6971.
12. Berrie, C. P., Birdsall, N. J. M., Haga, K., Haga, T., and Hulme, E. C. (1984) Hydrodynamic properties of muscarinic acetylcholine receptors solubilized from rat forebrain. *Brit. J. Pharmacol.* **82,** 839–851.
13. Peterson, G. L., Rosenbaum, L. C., and Schimerlik, M. I. (1988) Solubilization and hydrodynamic properties of pig atrial muscarinic acetylcholine receptor in dodecyl β-D-maltoside. *Biochem. J.* **255,** 553–560.
14. Wheatley, M. and Strange, P. G. (1983) Characterisation of brain D2 dopamine receptors solubilized by lysophosphatidylcholine. *FEBS Letts.* **151,** 97–101.
15. Panayotou, G. N., Magee, A. I., and Geisow, M. J. (1985) Reconstitution of epidermal growth factor receptor in artificial lipid bilayers. *FEBS Letts.* **183,** 321–325.
16. Gonzalez-Ros, J. M., Paraschos, A., and Martinez-Carrion, M. (1980) Reconstitution of functional membrane-bound acetylcholine receptor from isolated torpedo californica receptor protein and electroplax lipid. *Proc. Natl. Acad. Sci. USA* **77,** 1796–1800.

17. Franklin, G. I. and Potter, L. T. (1972) Studies of the binding of α-bungarotoxin to membrane-bound and detergent-dispersed acetylcholine receptors from torpedo electric tissue. *FEBS Letts.* **28,** 101–106.
18. Peterson, G. L., Rosenbaum, L. C., Broderick, D. J., and Schimerlik, M. I. (1986) Physical properties of the purified cardiac muscarinic acetylcholine receptor. *Biochemistry* **25,** 3189–3202.
19. Wheatley, M. (1983) The Isolation, characterisation and reconstitution of dopamine receptors from bovine caudate nucleus. Ph.D. thesis, University of Nottingham.
20. Klein, U. and Fahrenholz, F. (1994) Reconstitution of the myometrial oxytocin receptor into proteoliposomes. *Eur. J. Biochem.* **220,** 559–567.
21. Tiller, G. E., Mueller, T. J., Dockter, M. E., and Struve, W. G. (1984) Hydrogenation of Triton X-100 eliminates its fluorescent and ultraviolet light absorption while preserving its detergent properties. *Anal. Biochem.* **141,** 262–266.
22. Hjelmeland, L. M. (1980) A non-denaturing zwitterionic detergent for membrane biochemistry: design and synthesis. *Proc. Natl. Acad. Sci. USA* **77,** 6368–6370.
23. Hjelmeland, L. M., Nebert, D. W., and Osborne, J. C. (1983) Sulphobetaine of bile acids: nondenaturing surfactants for membrane proteins. *Anal. Biochem.* **130,** 72–82.
24. Repke, H. (1987) Muscarinic receptor-detergent complexes with different biochemical properties—selective solubilization, lectin affinity chromatography and ligand binding studies. *Biochim. Biophys. Acta* **929,** 47–61.
25. Ashani, Y. and Catravas, G. N. (1980) Highly reactive impurities in Triton X-100 and Brij 35: partial characterization and removal. *Anal. Biochem.* **109,** 55–62.
26. Rivnay, B. and Metzger, H. (1982) Reconstitution of the receptor for immunoglobulin E into liposomes. *J. Biol. Chem.* **257,** 12,800–12,808.
27. Hjelmeland, L. M. and Chrambach, A. (1984) Solubilization of functional membrane proteins. *Methods Enzymol.* **104,** 305–318.
28. Berrie, C. P., Birdsall, N. J. M., Dadi, H. K., Hulme, E. C., Morris, R. J., Stockton, J. M., and Wheatley, M. (1985) Purification of the muscarinic acetylcholine receptor from rat forebrain. *Biochem. Soc. Trans.* **13,** 1101–1103.
29. Lowry, O. H., Rosebrough, N. J., Farr, A. L., and Randall, R. J. (1951) Protein measurement with a folin phenol reagent. *J. Biol. Chem.* **193,** 265–275.
30. Dulley, J. R. and Grieve, P. A. (1975) A simple technique for eliminating interference by detergents in the Lowry method of protein determination. *Anal. Biochem.* **64,** 136–141.
31. Gorissen, H., Aerts, G., Ilien, B., and Laduron, P. (1981) Solubilization of muscarinic acetylcholine receptor from bovine brain: An analytical approach. *Anal. Biochem.* **111,** 33–41.

28

Radioligand Binding Using ^{125}I-Labeled Peptides

Peter W. Abel, David Waugh, and William B. Jeffries

1. Introduction

The radioligand binding assay is a useful tool to characterize peptide receptors and to study the interaction of endogenous and synthetic peptides with those receptors. The most commonly used method is the membrane filtration technique that involves binding of the radioligand to membranes that contain the receptor of interest. The bound and free radioligand are then separated by rapid filtration through glass fiber filters. This technique can be used for many different peptides as long as a high-affinity radioligand is available.

The most frequently used assay type is the competition binding assay. In the competition assay, one concentration of the peptide radioligand and various concentrations of unlabeled peptide ligand compete for binding to their receptors. Usually, the unlabeled peptide ligand is a synthetic peptide that has a chemical structure different from the radiolabeled peptide ligand. This type of binding assay is called a heterologous competition binding assay. The competition assay is used to calculate the affinity (the equilibrium dissociation constant or K_d) of the unlabeled peptide for the receptors. Since the K_d value is a constant for the interaction of a specific peptide with a specific type of receptor, that value can be used to characterize peptide receptor types. It is also possible to determine whether or not a novel synthetic peptide binds to a certain peptide receptor type and the affinity with which it binds.

If the unlabeled peptide ligand is the nonradioactive form of the radiolabeled ligand, a homologous competition binding assay, also called a cold saturation assay, is performed. This assay is conducted in the same general way as the competition assay described above. If certain criteria are met, this assay can be used to calculate the density of receptors on the membranes (B_{max}) and the dissociation constant (K_d) of the radioligand for the receptors (*see* Chapter 29).

From: *Methods in Molecular Biology, Neuropeptide Protocols*
Edited by: G. B. Irvine and C. H. Williams Humana Press Inc., Totowa, NJ

This chapter will outline the methods for performing a homologous or cold saturation competition binding assay using the membrane filtration technique. Competition between human ^{125}I-calcitonin gene related peptide (^{125}I-CGRP) and unlabeled human calcitonin gene related peptide (CGRP) for binding to porcine endothelial cell membranes will be described.

2. Materials

2.1. Apparatus

The basic equipment necessary for radioligand binding assays using the membrane filtration method will be described. The degree of automation of the assay system can vary greatly depending on the specific equipment available.

1. To prepare cell membranes, a Teckmar Tissuemizer (Cincinnati, OH; SDT-1810 motor, DZM-5 tachometer, PT-10 probe) is used.
2. For incubation of membranes and ligands, a standard shaking water bath (Dubnoff type) with temperature and shaking speed adjustments is used.
3. To separate the bound from the free radioligand, a system capable of rapid vacuum filtration is required. A manually operated Millipore filtration manifold is sufficient for this technique. However, radioligand binding assays usually involve duplicate or triplicate measurements using numerous ligands, each of which may require 10–15 separate filtrations. It is more efficient to purchase an automated system (Brandel Cell Harvester, Gaithersburg, MD) that provides rapid vacuum filtration of up to 48 samples simultaneously.
4. The filters containing radioligand bound to membrane receptors can be counted in a γ-counter or in a liquid scintillation counter.

2.2. Reagents and Solutions

1. Radioligand: Most commonly used radioiodinated peptides can be purchased from commercial sources (DuPont, Wilmington, DE; Amersham, Arlington Heights, IL). Peptides can also be radioiodinated in the laboratory using commercially available Na ^{125}I (*see* Note 1). Dilutions of the radioligand are made using a 5 m*M* HCl solution (*see* Note 2). The HCl solution for diluting the peptide radioligand is prepared by adding 5 mL of 1*M* HCl to 955 mL of distilled water.
2. Sample collection buffer: Prepare a 50 m*M* Tris, 100 m*M* NaCl solution. Dissolve 6.61 g of Trizma HCl (Sigma, St. Louis, MO) and 0.97 g of Trizma base (Sigma) in 1 L of distilled water and then add 5.84 g NaCl. This will give a solution of pH 7.4 at room temperature (25°C).
3. Homogenization buffer: This solution contains 50 m*M* Tris, prepared with Trizma HCl, as in item 2, and 5 m*M* EDTA. Add 1.86 g of Na_2 EDTA $2H_2O$ to 1 L of a 50 m*M* Tris solution. The pH is 7.4 at room temperature.
4. Incubation buffer: The incubation buffer contains 50 m*M* Tris, prepared with Trizma HCl, as in item 2, 100 m*M* NaCl, 0.2% bovine serum albumin (BSA), and 0.1% bacitracin (*see* Notes 3 and 4). Prepare 50 m*M* Tris as described in item 2. Then add 5.84 g NaCl, 2 g BSA, and 1 g bacitracin (50,000 U/g). This solution is pH 7.4 at room temperature.

5. Wash buffer: The wash buffer is 50 m*M* Tris-HCl, pH 7.4, prepared as described in item 3.
6. Coating agents: To inhibit the binding of peptides to plastic or glass surfaces, these surfaces are treated with an organosilane compound such as Prosil-28 (PCR Incorporated, Gainesville, FL) and/or polyethylenimine (*see* Note 5). Take 1 mL of the Prosil-28 concentrate and dissolve in 100 mL of distilled water to prepare a 1% (v/v) working solution. Take 1 mL of polyethylenimine solution (50% w/v aqueous solution; Sigma) and dissolve in 100 mL of distilled water to prepare a 1% (v/v) working solution (*see* Note 5 for coating procedure).

3. Methods

3.1. Membrane Preparation

A variety of methods can be used to prepare membranes for radioligand binding assays (*see* Chapter 27). The method described here is used to prepare membranes from porcine endothelium.

1. Cut open porcine aorta and gently scrape the intimal surface with a number 10 scalpel blade. The luminal scrapings from 5 aortae are placed in a 25-mL centrifuge tube containing approx 10 mL of ice-cold sample collection buffer. Wash the scrapings by centrifugation at 3000*g* at 4°C for 15 min.
2. Resuspend the pellet in 10 mL of ice-cold homogenization buffer and homogenize three times using a Tekmar Tissuemizer at a speed setting of 50 for 10 s. Centrifuge the homogenate at 35,000*g* at 4°C for 30 min. This sequence is then repeated a second time.
3. Resuspend the pellet and homogenize in 5 mL of incubation buffer. 100 µL of the membrane suspension is used for determination of the protein concentration using the method of Lowry *(1)*. Dilute the remaining membrane suspension with incubation buffer to contain a final protein concentration of approx 30 µg protein in 150 µL of membrane suspension (*see* Note 6).

3.2. Incubation

1. Prepare a single concentration of radioligand by diluting the stock concentration supplied by the manufacturer with incubation buffer. The final concentration of radioligand used should be less than the K_d value of that ligand for the receptor of interest. Commercial ^{125}I-CGRP is supplied as specific activity of 2000 Ci/mmol. Prepare a 200 p*M* solution by adding 8 µL of stock ^{125}I-CGRP to 2 mL of incubation buffer. This solution will be diluted fivefold in the final incubation.
2. Dilute the unlabeled competing ligand, CGRP, with incubation buffer. As a rough estimate, 5 concentrations below and 5 concentrations above the estimated K_d value of the competing drug are used. These concentrations should be prepared such that they span a range of approx 1000-fold. Ten dilutions of CGRP (0.15, 0.3, 0.625, 1.25, 2.5, 5.0, 10.0, 20.0, 40.0, and 500 n*M*) in incubation buffer are made for the competition assay. An additional dilution of 5 µ*M* CGRP is also made to define nonspecific binding (*see* Note 7). These solutions will be diluted by fivefold in the final incubation.

3. To perform the assay in duplicate, 24 polypropylene plastic incubation tubes (12 × 75 mm) coated with 1% (v/v) Prosil-28 are used. Pipet 150 µL of the membrane suspension into all 12 pairs of tubes. Into 10 pairs of tubes, also pipet 50 µL of one of the CGRP competition assay dilutions. Into one remaining pair of tubes, pipet 50 µL of the 5 µ*M* CGRP solution (used to determine nonspecific binding). To the last pair of tubes, add 50 µL of incubation buffer. This set of tubes will be used to measure the total amount of ^{125}I-CGRP bound.
4. To start the incubation, add 50 µL (40 p*M*) of the diluted ^{125}I-CGRP solution to each tube. The total volume of incubation is 250 µL. Incubate the tubes for 50 min in a shaking water bath maintained at 25°C (*see* Note 8).
5. To stop the reaction, dilute the contents of the tubes with 5 mL of ice-cold wash buffer and rapidly filter through glass fiber filters (Whatman GFB) using a Brandell Cell Harvester (*see* Note 9). Wash the tubes twice with 5 mL of ice-cold wash buffer.
6. Dry the filters overnight. Place each filter in a separate 12 × 75 mm polystyrene tube. Count the radioactivity on the filters in a γ-counter.

4. Notes

1. Radioiodination of peptides: The large majority of peptide radioligands available are labeled with ^{125}I. Radioiodinated peptide ligands usually have a high specific activity of 2000 Ci/mmol and, therefore, are quite useful in labeling receptors in tissues with relatively low receptor densities.

   Radioiodinated peptides can be less costly if they can be radioiodinated in the laboratory. A number of common methods are available for this purpose. The choice of a radioiodination method depends on the structure of the peptide. Peptides that have an unsubstituted tyrosyl or histidyl residue available for oxidative iodination, such as CGRP, can be iodinated using the chloramine T method *(3)*. This is an oxidative method that involves reacting chloramine T (*p*-toluene sulfonochloramide) with Na^{125}I (DuPont NEN) in the presence of CGRP *(4)*. Many peptides that contain a free amino group can react with the Bolton-Hunter reagent (DuPont) to produce a high specific activity mono- or di-iodinated peptide ligand. This is a nonoxidative procedure that can be used for peptides that contain methionines or other residues that must remain in a reduced state to bind to their receptors. The Iodogen reagent is another oxidative method that is less likely to oxidize methionines or other residues than chloramine T and therefore, may be a better method for iodination of some peptides (*see* Chapter 20). The radioiodinated peptide is purified by reverse-phase high performance liquid chromatography as described in Chapters 20 and 22.

   One disadvantage to radioiodinated ligands is that they have a relatively short half-life of 60 d and, therefore, should be used within 5–6 wk after they are prepared.
2. Radioligand solutions: Solutions of radioiodinated ligands keep best when stored at low temperatures. Unless specifically recommended by the manufacturer, radioligands should not be frozen since this can form aggregates that hasten

decomposition. For many peptide radioligands, a solution of 5 m*M* HCl in water is recommended for dissolving the peptide and making dilutions. The acid is present to reduce the loss of radioligand by binding to the glass container or tubes containing the radioligand. To reduce decomposition, store the radioligand at the lowest specific activity possible.

3. Effects of ions: Certain divalent cations can increase or decrease the binding of some peptide radioligands. Cations can also either enhance or inhibit the binding of unlabeled competing peptides at their receptors. The effect of ions is dependent on the concentration of the ion, as well as the specific ligand and peptide receptor being studied. For instance, low concentrations of Mg^{2+} (1–10 m*M*) are known to increase the affinity of radioligands for their receptors. Mg^{2+} does this by promoting the formation of the high-affinity form of the receptor to which agonists preferentially bind. In some systems, Mg^{2+} and Mn^{2+} have been reported to increase the amount of radiolabeled peptide binding *(7)*. High concentrations of Mg^{2+} and other cations can reduce the binding of some radiolabeled peptide ligands *(8)*. Unless the effect of ions is known and the effect is wanted in the binding assay, it is best to limit the ions present in the binding buffer.
4. Peptide degradation: For some peptides, tissue proteases can significantly degrade the labeled or unlabeled peptide ligand. Radioactivity associated with peptide products that are unable to bind to the receptor being studied will cause an overestimation of the free radioligand concentration. In this situation, calculation of the percent of radioligand bound from the measured values will be less than the actual percent bound and the calculated K_d value will be in error. To test whether or not the radiolabeled peptide is degraded, the incubation is terminated by addition of trichloroacetic acid (5% v/v final concentration) and the TCA precipitate is counted in a γ-counter. The radioactivity associated with the radiolabeled peptide is precipitated by TCA, whereas radioactivity released by proteolysis is not.

   If significant degradation occurs, then the incubation buffer should contain one or a number of protease inhibitors, which are used to inhibit the breakdown of the labeled or unlabeled peptides (*see* Chapter 33). Commonly used peptidase inhibitors include bacitracin (0.001–1 mg/mL; nonspecific protease inhibitor), leupeptin (1–50 μg/mL; inhibits trypsin-like serine and cysteine proteases), aprotonin (0.01–0.5 mg/mL; inhibits serine proteases), phosphoramidon (0.5–50 μg/mL; inhibits neutral endopeptidase 24.11), and pepstatin A (0.5–10 μg/mL; inhibits aspartic proteases). Whereas peptidase inhibitors may be necessary to prevent proteolysis, indiscriminate use of these compounds must be avoided since they may inhibit binding of radiolabeled peptides in some cases *(6)*. It is thus important for the investigator to determine the optimal concentration of protease inhibitor necessary to inhibit peptide degradation without affecting peptide binding.
5. Binding of peptides to glass and plastic surfaces: Peptides are highly prone to bind nonspecifically to glass and plastic test tubes used in the assay, plastic pipet tips, and to glass fiber filters. An organosilane compound such as Prosil-28 (1% aqueous solution) is often used to coat test tubes to inhibit this nonspecific

binding. To use Prosil-28, submerge tubes in a 1% Prosil-28 solution for 20 s, remove, and then rinse tubes once with distilled water. Allow tubes to air dry for 24 h before use. For some peptides, it is useful to treat the binding tubes with a mixture of 1% Prosil-28 and 1% polyethylenimine *(6)*. The negatively charged substance polyethylenimine has also been used to inhibit nonspecific binding of peptides to plastic and glass surfaces. A 1% solution of polyethylenimine is prepared and the glass fiber binding filters are submerged in this solution for 2 h. The filters are then used immediately in the binding assay.

6. Protein concentration: In general, the higher the protein concentration in the assay, the more receptors available for binding by the peptide radioligand. More receptors leads to higher specific binding to the receptors relative to nonspecific binding and a larger signal-to-noise ratio. However, a protein concentration that is too high can cause the affinity of the competing ligand to be underestimated if the amount of bound radioligand significantly reduces the free concentration of ligand available to bind to the receptors *(5)*. In general, if >10% of the added radioligand is bound, then the protein concentration in the assay is too high. In this case, the amount of added protein should be reduced or the incubation volume of the assay can be increased.
7. Defining specific binding: The purpose of radioligand binding assays is to make quantitative measurements of binding of a ligand to a specific receptor. This is called specific binding and is saturable as the concentration of radioligand is increased. Radioligands may also bind to other sites that can include the glass fiber filter, the incubation tube, and components of the tissue. Binding other than to the receptor of interest is called nonspecific binding and is nonsaturable. In the binding assay, nonspecific binding is defined as binding of the radioligand in the presence of an excess of an unlabeled drug that will occupy all of the receptors. Generally, 100–1000 times the K_d concentration of the unlabeled drug at the receptors of interest is used. If the concentration of unlabeled drug is too high, it may begin to inhibit nonspecific binding as well. If possible, the unlabeled drug should also be structurally different from the radioligand. This will reduce the chance that the unlabeled drug will inhibit nonspecific binding of the radioligand.

   The radioligand should have a high affinity, preferably in the p*M* range. A high affinity means that a lower concentration of radioligand can be used in the binding assay and, therefore, nonspecific binding will be reduced. Nonspecific binding of the peptide radioligand should be <50%.
8. Incubation time and temperature: One assumption in radioligand binding assays is that the binding reaction reaches equilibrium. Practically, binding assays are designed to be performed at steady-state. Two important factors that determine the time to reach steady-state are the concentration of radioligand and the temperature. The lower the concentration of radioligand and the higher the temperature, the sooner steady-state will be reached. At room temperature (25°C) many radioligands reach steady-state within 30–60 min. These are commonly used conditions for radioligand binding assays using radiolabeled peptides. If the peptide

radioligand is degraded, it may be useful to perform the incubation on ice (4°C). In some cases, this improves the quality of the binding assay by decreasing degradation of the radioligand and/or competing peptide. If a low temperature is used, it is best to increase the incubation time to assure that steady state will be reached.

9. Filtration: Radioligand binding assays depend on a method for the efficient separation of membrane-bound radioligand from free radioligand. Vacuum filtration through glass fiber filters provides a rapid and convenient method for this separation. In this technique, membranes, and radioligand bound to them, are trapped on the filter, whereas the free radioligand passes through. During this separation, it is essential that the bound radioligand not be allowed to dissociate from the receptors. To minimize dissociation, a high-affinity ligand (K_d in the pM range) should be used. Peptide radioligands with high affinity for their receptors are advantageous because the rate of dissociation of the ligand from the receptor is relatively slow. This means that less bound peptide ligand will dissociate during the time necessary for separation of bound from free radioligand in the filtration assay *(2)*. If the K_d is above 10 nM, the filtration method cannot be used. The filtration should be as rapid as possible and should be done using ice-cold wash buffer, since low temperature will also slow the rate of dissociation of the radioligand from the receptor.

References

1. Lowry, O. H., Rosebrough, N. J., Farr, A. L., and Randall, R. J. (1951) Protein measurement with the folin phenol reagent. *J. Biol. Chem.* **193,** 265–275.
2. Bennett, J. P., Jr. (1978) Methods in binding studies, in *Neurotransmitter Receptor Binding* (Yamamura, H. I., Enna, S. J., and Kuhar, M., eds.), Raven, New York, pp. 57–90.
3. Hunter, W. M. and Greenwood, F. C. (1962) Preparation of iodine-131 labelled human growth hormone. *Nature* **194,** 495,496.
4. McGillis, J. P., Humphreys, S., and Reid, S. (1991) Characterization of functional calcitonin gene-related peptide receptors on rat lymphocytes. *J. Immunol.* **147,** 3482–3489.
5. Kenakin, T. (1993) Radioligand Binding Experiments, in *Pharmacological Analysis of Drug-Receptor Interaction* (Kenakin, T., ed.), Raven, New York, pp. 385–410.
6. Turner, J. T. and Bylund, D. B. (1987) Characterization of the vasoactive intestinal peptide receptor in rat submandibular gland: radioligand binding assay in membrane preparations. *J. Pharmacol. Exptl. Ther.* **242,** 873–881.
7. Buck, S. H. and Shatzer, S. A. (1988) Agonists and antagonist binding to tachykinin peptide NK-2 receptors. *Life Sci.* **42,** 2701–2708.
8. Stallone, J. N., Nishimura, H., and Khosla, M. C. (1989) Angiotensin II vascular receptors in fowl aorta: binding specificity and modulation by divalent cations and guanine nucleotides. *J. Pharmacol. Exptl. Ther.* **251,** 1076–1082.

29

Analysis of Data from "Cold Saturation" Radioligand Binding Experiments

William B. Jeffries, David Waugh, and Peter W. Abel

1. Introduction

In the previous chapter, we described the laboratory methods needed for the generation of data used to calculate receptor density and ligand affinity. The purpose of the present chapter is to outline the methods for the careful analysis of data obtained from competition radioligand binding experiments using actual data from our laboratory. The detailed theoretical basis for our approach is not discussed here but is the subject of several excellent reviews *(1–3)*. There are two ways to determine the density of a receptor population and to measure the affinity of the radioligand in binding assays: hot saturation binding and homologous competition. In hot saturation studies, aliquots of receptor-containing membrane are incubated to equilibrium with a series of increasing radioligand concentrations (usually from 5- to 10-fold lower up to 5- to 10-fold higher than the K_d) in the presence and absence of a large concentration of an unlabeled, structurally dissimilar competing ligand. The incubation is terminated and binding of the radioligand to the receptor population is then measured. B_{max} and K_d are then calculated using a Rosenthal (Scatchard) analysis or nonlinear curve fitting. The hot saturation method is the most accurate way of determining B_{max} and K_d of the radioligand. However, the hot saturation assay consumes relatively large amounts of radioligand, which can render it impractical and/or prohibitively expensive in peptide research. Thus, most peptide researchers prefer to use homologous competition assays, also known as "cold saturation" assays, to investigate peptide receptors. With an optimal radioligand (*see* Note 1), these assays will yield data comparable to those obtained from hot saturation studies.

From: *Methods in Molecular Biology, Neuropeptide Protocols*
Edited by: G. B. Irvine and C. H. Williams Humana Press Inc., Totowa, NJ

In a cold saturation assay, each membrane aliquot contains the same amount of added radioligand and a varied amount of cold competing ligand. Unlike the hot saturation binding assay, the competing ligand is simply the unlabeled form of the radioligand. In simple systems, the B_{max} and K_d can be estimated using the Cheng-Prussoff equation *(4)*:

$$IC_{50} = K_{dc}\ (1+[L]/K_{dh}) \tag{1}$$

where IC_{50} is the concentration of competing ligand that gives a half maximal inhibition of specific binding, K_{dc} is the dissociation constant for the competing (cold) ligand, [L] is the radiolabeled ligand concentration, and K_{dh} is the dissociation constant for the radioligand. The calculations in this chapter are based on the assumption that the labeled and unlabeled form of the ligand possess *equal affinity* for the receptor. This assumption is stated mathematically as $K_{dc} = K_{dh} = K_d$, which when substituted into equation 1 yields Eq. (2):

$$K_d = IC_{50} - [L] \tag{2}$$

B_{max} can be calculated by the law of mass action:

$$B_{max} = B_0(K_d+[L])/[L] \tag{3}$$

where B_0 is the specific binding of the radioligand (difference between binding in the absence of competitor and the presence of a saturating concentration of competitor; *see* Fig. 1). Since $IC_{50} = K_d + L$ (*see* Eq. [2]), Eq. (3) becomes:

$$B_{max} = B_0(IC_{50})/[L] \tag{4}$$

Through the use of these equations, the number of binding sites present and the affinity of the radioligand can be calculated *(3)*. In addition to the assumption that the cold and hot form of the ligand have identical affinities, several other assumptions are made for this type of analysis. First, we assume that the total concentration of ligand is unaffected by the receptor binding event (i.e., the amount of bound ligand is negligible compared to the total present). Second, we assume that only one class of receptors is present. Finally, we assume no binding cooperativity. If more than one class of sites is present, it is necessary to use computer analysis to dissect individual binding values from the raw data. In practice, it is best to use a nonlinear curve fitting program for the analysis of all radioligand binding data. The mathematical basis for computer-assisted nonlinear curve fitting is described in ref. *2*. In this chapter, we will outline the method for both manual and computer-assisted data analysis.

2. Materials

To calculate the B_{max} and K_d of uncomplicated binding data taken from a single receptor population, the primary materials required are a pencil and paper. However, as described in Section 3.2., it is more accurate and conve-

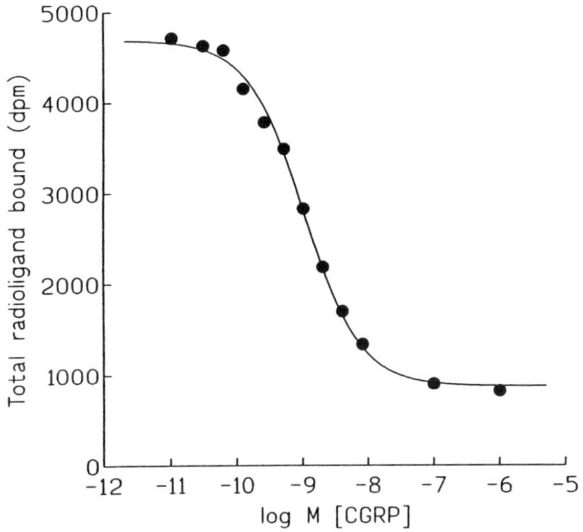

Fig. 1. Competition curve for hCGRP at porcine aorta endothelial cell CGRP receptors. Various concentrations of unlabeled hCGRP were added in the presence of a single concentration of ^{125}I-[His^{10}]-hCGRP as described in Chapter 28. Total ligand binding is taken as the value for the upper plateau, and nonspecific binding is the value for the lower plateau. The B_0 is calculated as the difference between total and nonspecific binding (Eq. [4]). The IC_{50} value is the concentration of cold hCGRP that inhibits 50% of the specific binding.

nient to analyze binding data with a nonlinear curve fitting program. There are several available; here we will describe the use of an updated version of Ligand *(3,5),* which was the first program specifically developed for analysis of radioligand binding data. The updated version we use is commercially available as a package called "Kinetic, EBDA, Ligand & Lowry, version 4.0" (KELL). KELL can be obtained from Biosoft (Ferguson, MO). The minimum requirements for this program are an IBM compatible personal computer with DOS 3.0 or later and 640 K RAM. An Apple Macintosh version is available as well.

3. Methods

3.1. Manual Calculation

1. Perform a cold saturation assay as outlined in Chapter 28. Examine the data obtained from the cold saturation experiment to ensure that saturation is achieved with low specific binding and a small fractional total binding (*see* Note 2). (The radioligand [^{125}I-CGRP] concentration in our assay is 40 p*M*, with a specific activity of 3.865×10^6 DPM/pmol.) Convert radioligand data from CPM to DPM

and arrange the data as shown in Table 1. Calculate B_0 by subtracting the total binding found in the absence of competitor from the total binding found in the presence of a saturating concentration (1 μ*M*) of competitor (nonspecific binding). Thus, from Table 1:

$$B_0 = 4709 - 822 = 3887 \text{ DPM}$$

2. Plot the total radioligand binding to the filters as a function of the cold ligand concentration. Determine the IC_{50} value by interpolation of the curve at 50% inhibition of specific binding (i.e., concentration corresponding to [0.5 × specific binding] + nonspecific binding; *see* Fig. 1). From Fig. 1, IC_{50} is estimated at 910 p*M*.
3. Eq. (2) and (4) are used to calculate B_{max} and K_d. For K_d calculation:

$$K_d = IC_{50} - [L]$$
$$K_d = 9.1 \times 10^{-10}M - 4.0 \times 10^{-11}M \quad (2)$$
$$K_d = 8.7 \times 10^{-10}M$$

For B_{max}:

$$B_{max} = B_0(IC_{50})/[L]$$
$$B_{max} = 3887 \text{ DPM} \times 9.1 \times 10^{-10}M/4.0 \times 10^{-11}M \quad (4)$$
$$B_{max} = 88{,}429 \text{ DPM}$$

For this calculation, it is necessary to convert the B_{max} value from DPM to mass units by dividing by the specific activity of the radioligand:

$$B_{max} = 88{,}429 \text{ dpm}/3{,}865{,}245 \text{ dpm}/10^{-12}\text{mol} = 2.29 \times 10^{-14}\text{mol}$$

Normalize the calculated B_{max} for the amount of added protein obtained from the Lowry assay (47.75×10^{-6}g/tube):

$$B_{max} = 22.9 \text{ fmol}/47.75 \times 10^{-6} \text{ g}$$
$$B_{max} = 479 \text{ fmol/mg protein}$$

3.2. Nonlinear Regression Analysis Using KELL

1. Using the EBDA (equilibrium binding data analysis) software (*see* Note 3), enter the following data sequentially for each point:
   a. Amount of unlabeled ligand in pmol.
   b. Replicate values of total radioligand bound (DPM).
   c. Replicate values of nonspecific bound (DPM, these values are the same for each point).
   d. Replicate values of radioligand added to each tube (DPM, these values are the same for each point).

   These raw data are found in Table 1. Once entered, the EBDA program calculates the additional data shown in Tables 2 and 3.
2. Select a model (number of sites, nonspecific binding fixed or fitted, and so on) for analysis. We usually model (in separate analyses) for one and two sites. Both

Table 1
Raw Experimental Data

Tubes	Unlabeled ligand, *M*	Unlabeled ligand, pmol	CPM (duplicate values)		DPM (mean)[a]
1	0 (total radioligand)	0	53341.4	54635.7	70108.55
2	0 (total bound)	0	3556.8	3696.0	4709.6
3	1.0×10^{-6} (nonspecific)	250	635.7	630.4	822.15
4	3.0×10^{-11}	0.0075	3539.9	3731.3	4624.2
5	6.0×10^{-11}	0.015	3492.1	3554.75	4575.0
6	1.2×10^{-10}	0.03	3267.2	3124.8	4150.55
7	2.5×10^{-10}	0.0625	2834.5	2978.4	3774.65
8	5.0×10^{-10}	0.125	2814.3	2549.3	3482.9
9	1.0×10^{-9}	0.25	2147.7	2203.8	2825.65
10	2.0×10^{-9}	0.5	1648.6	1710.3	2181.1
11	4.0×10^{-9}	1.0	1240.3	1372.6	1696.7
12	8.0×10^{-9}	2.0	1104.4	953.0	1336.0
13	1.0×10^{-7}	25	713.8	673.6	901.2

[a]Counter efficiency 77.2%.

Table 2
Initial Data Processing by the EBDA Program

Point	Unlabeled, pmol	Bound (T), DPM	N-Specific, DPM	Total, DPM	% Specific	% Total
1	0	4709.6	822.15	70108.55	82.54	5.54
2	0.0075	4624.2	822.15	70108.55	82.22	5.42
3	0.15	4575.0	822.15	70108.55	82.03	5.35
4	0.03	4150.55	822.15	70108.55	80.19	4.75
5	0.0625	3774.65	822.15	70108.55	78.22	4.21
6	0.125	3482.9	822.15	70108.55	76.39	3.8
7	0.25	2825.65	822.15	70108.55	70.9	2.86
8	0.5	2181.1	822.15	70108.55	62.31	1.94
9	1.0	1696.7	822.15	70108.55	51.54	1.25
10	2.0	1336.0	822.15	70108.55	38.46	0.73
11	25.0	901.0	822.15	70108.55	8.75	0.11

Data file, B109REV.EBD; data title, B109 CGRP pig endothelial cells; experiment type, saturation study (cold); data type, DPM; specific activity (DPM/pmol), 3865245; volume of incubation (mL), 0.25; calculate free using, specific bound.

Bound (T), total radioligand bound in absence or presence of unlabeled ligand; N-Specific, defined as the amount of radioligand (^{125}I-CGRP) bound in the presence of 1 μ*M* CGRP; Total, the amount of radioactive ligand added to each tube.

Table 3
Ligand Binding Parameters Calculated by EBDA Program

Point	Bound (T), *M*[a]	Nsp., *M*[b]	Total, *M*[c]	Bound (S), *M*[d]	Free, *M*[e]	B/F (Sp.)[f]
1	4.8738E-12	8.5081E-13	7.2553E-11	4.0230E-12	6.8530E-11	0.058704
2	6.7641E-12	1.2026E-12	1.0255E-10	5.5615E-12	9.6991E-11	0.057341
3	8.6499E-12	1.5544E-12	1.3255E-10	7.0954E-12	1.2546E-10	0.056557
4	1.1399E-11	2.2580E-12	1.9255E-10	9.1414E-12	1.8341E-10	0.049841
5	1.7366E-11	3.7825E-12	3.2255E-10	1.3584E-11	3.0897E-10	0.043965
6	2.8444E-11	6.7142E-12	5.7255E-10	2.1729E-11	5.5082E-10	0.039449
7	4.3228E-11	1.2578E-11	1.0726E-09	3.0650E-11	1.0419E-09	0.029418
8	6.4478E-11	2.4304E-11	2.0726E-09	4.0173E-11	2.0324E-09	0.019767
9	9.8560E-11	4.7758E-11	4.0726E-09	5.0802E-11	4.0218E-09	0.012632
10	1.5383E-10	9.4665E-11	8.0726E-09	5.9167E-11	8.0134E-09	0.007383
11	1.2861E-09	1.1735E-09	1.0007E-07	1.1255E-10	9.9960E-08	0.001126

[a]Concentration of bound radioligand. Calculated by multiplying Total *M* by the percentage of total binding for each point. *M*, concentration in molar units.

[b]Concentration of radioligand bound nonspecifically. Calculated by multiplying Total *(M)* by the percentage of nonspecific binding for each point.

[c]Concentration of ligand (radiolabeled + unlabeled) in tube. Calculated by adding constant concentration of radioligand (pt. 1) and the increasing concentration of cold ligand (*see* Table 1). Total concentration of hot label calculated using specific activity value.

[d]Concentration of hot ligand bound specifically to receptor. Calculated by multiplying total binding value by specific binding percentage (*see* Table 2).

[e]Concentration of free ligand that is capable of binding to receptor. Calculated by subtracting specific binding from total concentration.

[f]Ratio of bound/free radioligand concentrations.

of these models are analyzed by either fixing nonspecific binding to our measured value or "floating" the nonspecific binding as a fitted parameter (*see* Note 4). Thus for each experiment, we usually analyze the data using four models. The results often show a good fit for more than one model. The data in this experiment do not conform to a two-site fit and are not shown (*see* Notes 5 and 6). However, good fits were obtained for the two models in which nonspecific binding was either fixed or floated (Table 4). In Fig. 2, the data for the model with fitted nonspecific binding are calculated by the EBDA program as a standard Rosenthal (Scatchard) transformation, where the negative reciprocal of the slope is the K_d and the X intercept is the B_{max}. When the B_{max} for a nonspecific fixed fit is calculated (Table 4) and normalized for protein (as shown in Section 3.1.), a value of 578 fmol/mg protein is obtained. The comparable B_{max} value for a nonspecific floated analysis is 322 fmol/mg protein (*see* Note 7).

3. The best model for the data is determined statistically by using the F-test, which is performed on the combined data for all of the models. In this example, the best fit of the data is accomplished when the nonspecific binding is treated as a fitted, rather than a fixed parameter. We know this because the fitted analysis is significantly better than the fixed analysis (Table 5).

4. Notes

1. Criteria for determining the suitability of the radioligand: Prior to the routine use of a radioligand in cold saturation studies, one should demonstrate that the radioligand is actually binding to the receptor in question (*see* ref. *1* for review). The analysis described here assumes that the ligands used fulfill certain criteria. First, the binding of the radioligand must be saturable and of high affinity. The best way to determine these two parameters is to perform a hot saturation experiment using membranes from a native tissue or cell line of known receptor density. Good radioiodinated peptide ligands usually have K_d values of 1–100 p*M*, whereas the K_d values of nonpeptide ligands are usually 0.1–10 n*M*. A suitable concentration of radioligand is selected based on the results of the hot saturation assay. Second, the peptides used in the receptor binding assay should demonstrate a specificity in the binding assay that is similar to that seen in physiologic systems. Thus, the rank order of affinity of agonists in the binding assay should be similar to the rank order of potency in a bioassay of receptor function. Likewise, the rank order of affinity of antagonists should correspond to the rank order of PA_2 or IC_{50} values in bioassays. The affinity of competing agents can be determined in competition binding experiments. Third, the kinetics of binding should be consistent with the time-course of the biologic action of the radioligand. The kinetics of the receptor can be determined in association and dissociation binding experiments *(1)* and compared to temporal measurements made in bioassays.
2. Initial data analysis: criteria for a good experiment. Upon completion of the radioligand binding experiment, the data should be initially evaluated to determine if the criteria of a valid saturation study have been met. The best way to do this is to plot the cold saturation curve. The resultant curve should resemble

Table 4
Results of Curve Fitting Analysis by EBDA[a]

	Nonspecific binding floated		Nonspecific binding fixed	
Parameter	Estimate	Error	Estimate	Error
K_d (Site 1), M	1.091 E-9	4%	4.1169 E-9	6%
$B_{\max}$ (Site 1), M	6.189 E-11	1%	1.1042 E-10	2%
Point distribution[b]	Random		Not random	

[a]Data file, B109REV; data title, B109 CGRP pig endothelial cells; experiment type, saturation study (cold); data type, DPM; specific activity (DPM/pmol), 3865245; volume of incubation (mL), 0.25; calculate free using, specific bound; data used, specific binding; model type, saturation—site number:1; model weighting, none; iterations, 4.

[b]This parameter indicates the results of a runs test, wherein the randomness of data points at each fit is determined. A random fit is acceptable; a nonrandom fit is not.

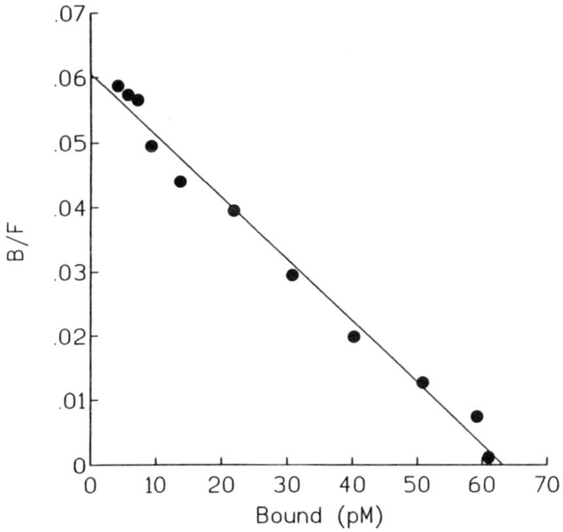

Fig. 2. Rosenthal (Scatchard) analysis of hCGRP binding to porcine aorta endothelial cell membranes, plotted from data generated by the EBDA program. The X intercept value corresponds to $B_{\max}$. When normalized for protein content, CGRP receptor density is calculated at 322 fmol/mg protein. This figure is plotted in the EBDA program with the "view results/fit" option.

Fig. 1. When these data are calculated and plotted, several characteristics should be seen:

a. Saturation should be obtained. If the total binding curve does not plateau at the bottom, this indicates that an insufficient amount of competing ligand may have been used or that nonspecific binding has not been defined properly (*see also* Note 1).

Table 5
F-Test of Data Analysis from EBDA Program[a]

Fit	Model	SSQ	DF	F	P
NSB fixed	1 Site	4.77E-22	9	1345.48	<0.001[b]
NSB floated	1 Site	2.81E-24	8	Reference	

[a]Statistical comparison results: B109REV.
[b]The reference fit is significantly better.
SSQ, sum of squares; DF, degrees of freedom; P, probability value.

b. Nonspecific binding should be <50% of total binding. If nonspecific binding is too high, it will be difficult to accurately determine specific binding and whether saturation is attained. High nonspecific binding may indicate that the agent used to define it is not suitable. Alternatively (or in addition), further steps can be taken to limit nonspecific binding, as described in the previous chapter. It is very important that nonspecific binding be properly defined, since overestimation or underestimation can lead to large errors in calculation of B_{max} and K_d.
c. Total binding should be <10% of the total ligand concentration at each data point. An assumption of this procedure is that the concentration of bound ligand is negligible. If total binding is >10% of total radioligand concentration, this will cause an underestimation of the affinity of the radioligand. Thus the membrane protein concentration in the assay should be sufficiently dilute as to render bound ligand concentration negligible (*see* Chapter 28).

3. Data analysis software: Originally, EBDA was a program designed to process raw binding data (e.g., DPM) and prepare initial parameter estimates for iterative nonlinear regression by the computer program Ligand. The newest version of EBDA contained in the KELL software package has been improved to perform iterative nonlinear regression for most binding experiments without the need for the more cumbersome Ligand program. Ligand must still be used for complex fits and for fitting more than one curve simultaneously. This latter feature is quite useful when determining the final mean binding parameters for a similar group of experiments. When Ligand is needed, EBDA is run first and the data file thus generated is used as the data input for the Ligand program.
4. Fitted parameters: The EBDA and Ligand programs allow for each of the binding parameters to be either fitted or fixed. The parameters in the EBDA analysis are K_d, B_{max}, and nonspecific binding. In Ligand, the data parameters are K1 ($1/K_d$), R1 (B_{max} in molar concentration), N1 ("background," which is equivalent to nonspecific binding), and C, which is fixed to the number of curves in the analysis. We have found that the calculated values of the fitted parameters for most analyses are similar in EBDA and Ligand, making it unnecessary to use Ligand in most cases. In the present experiment, we treat K_d and B_{max} as fitted parameters, since these are the values we seek to determine from the analysis. In our experi-

ence, it is usually best to let N1 "float," as well (*see* Tables 4 and 5). Most curve fitting programs require initial estimates of parameters to be fitted. EBDA suggests initial estimates for uncomplicated analyses. It is wise to use these same estimates as a starting point for the iterative analysis of the data by Ligand as well.

5. Multiple sites: If the cold saturation curve looks multiphasic or data analysis indicates that multiple sites are present, several additional experiments should be performed. If the radioligand is an agonist and the receptor is coupled to a G-protein, then it is likely that the receptor will be present in two interconvertable forms: a high-affinity (G-protein-coupled) state and a low-affinity (uncoupled) state. Agonist ligands often produce biphasic saturation curves. Aside from changing to an antagonist radioligand, there are two experimental options when this occurs. First, if you wish to carefully study the agonist-receptor interaction with both forms of the receptor, you must increase the number of data points in your experiment (we recommend at least 16 competing ligand concentrations). This will allow sufficient data to accurately resolve the density of each form of the receptor and affinity of the agonist ligand for each site. The total density is the sum of the B_{max} values obtained for each site. If you are interested only in receptor density, the best strategy is to eliminate the formation of the high affinity form of the receptor. This can be accomplished by using a buffer solution that does not include magnesium, since Mg^{2+} promotes formation of the high-affinity form of the receptor. The buffer should also include GTP (100 μ*M*) or (preferably) a nonhydrolyzable form of GTP such as GppNHp. The GTP will promote dissociation of the high-affinity form of the receptor to the low-affinity form. Thus the pool of receptors will all be present in the low-affinity state, which will produce a monophasic saturation curve and simplify the data analysis.
6. Hot and cold ligand affinity: As stated in Section 1., Eq. (2) assumes that the affinities of the radiolabeled and unlabeled form of the ligand are the same. If there is a significant difference in affinity between the two species, the cold saturation analysis will yield an inaccurate K_d value. Furthermore, the analyses contained herein assume radiochemical purity of the hot ligand. Thus, if the hot ligand exists as a mixture of variably iodinated species, the cold saturation analysis may generate apparent multiple sites where there is in fact only one. The affinities of the hot and cold species can be evaluated by comparing the K_d values generated in hot vs cold saturation experiments, or by comparing the affinities of each species in a bioassay.
7. A considerable variation is obtained when calculating B_{max} by hand (479 fmol/mg protein) and by EBDA analysis with nonspecific fixed (578 fmol/mg protein) and nonspecific floated (322 fmol/mg protein). Based on the data analysis, we conclude that the correct value is 322 fmol/mg protein. This is owing to the fact that direct statistical comparison (Table 5) of the data reveals that, when the nonspecific binding is treated as a fitted parameter, a significantly better fit is obtained vs when nonspecific binding is fixed at the experimentally observed value. These disparate results demonstrate one of the pitfalls of cold saturation techniques, i.e., properly defining nonspecific binding. As pointed out in

Chapter 28, specific binding (and thus B_{max}) may be overestimated when the ligand used to define nonspecific binding is simply the unlabeled form of the radioligand. One can generally avoid this problem in binding assays by using a ligand that is structurally dissimilar to the radioligand to define nonspecific binding. However, in a cold saturation assay, it is more appropriate to treat nonspecific binding as a fitted parameter and eliminate the guesswork associated with defining nonspecific binding.

References

1. Limbird, L. E. (1996) Identification of receptors using direct radioligand binding techniques, in *Cell Surface Receptors: A Short Course on Theory and Methods,* Kluwer Academic, Boston, MA, pp. 61–119.
2. McPherson, G. A. (1986) A mathematical approach to receptor characterization, in *Receptor Pharmacology and Function* (Williams, M., Gleenon, R. A., and Timmermans, P. B. M. W. M., eds.), Marcel Dekker, New York, pp. 47–84.
3. DeBlasi, A., O'Reilly, K. O., and Motulsky, H. J (1989) Calculating receptor number from binding experiments using same compound as radioligand and competitor. *Trends Pharmacol. Sci.* **10,** 227–229.
4. Cheng, Y. and Prusoff, W. H. (1973) Relationship between inhibition constant (K_i) and the concentration of an inhibitor that causes 50% inhibition (I_{50}) of an enzymatic reaction. *Biochem. Pharmacol.* **22,** 3099–3108.
5. Munson, P. J. and Rodbard, D. (1980) LIGAND, a versatile computerized approach for characterization of ligand-binding systems. *Anal. Biochem.* **107,** 220–239.

30

Organ/Tissue Preparations for the Assessment of Agonist/Antagonist Activity

Mark Murnin, Sándor Lovas, James M. Allen, and Richard F. Murphy

1. Introduction

Pharmacologists have two major tools to evaluate the biological properties of peptides, receptor binding assays (*see* Chapter 28) and functional assays. Binding assays determine if a peptide binds to a receptor, but in functional assays, which measure the response of a tissue to a peptide, both binding and resultant biological actions can be examined. When the response to a peptide is plotted against the logarithm of the concentration administered, the sigmoidal response curve can be of therapeutic and mechanistic importance (*see* Note 1). Values obtained from the curve, such as the concentrations of peptide that produce maximal (E_{max}, or the efficacy) and half maximal tissue response (EC_{50}, or the potency) are used to compare peptides (*see* Note 2).

In practice, biochemical or physiological changes may be measured in an intact organism or any of its isolated components such as an organ, tissue, cell, or cell component. Peptides that bind to a receptor and cause a response are referred to as agonists, whereas those that block the action of the agonist are antagonists. The two properties of a peptide that determine its biological activity are affinity and intrinsic efficacy. Affinity is the tenacity with which a peptide binds to the receptor. Receptor binding is a dynamic process in which the peptide and receptor continually associate and dissociate. The affinity is calculated as the reciprocal of the dissociation constant that should be determined experimentally *(1)*. The intrinsic efficacy of a peptide is the innate ability of the peptide to stimulate the receptor, resulting in production of a biological response *(1,2)*. Agonists, thus, have both affinity for and efficacy at the receptor, whereas antagonists have affinity but no efficacy.

From: *Methods in Molecular Biology, Neuropeptide Protocols*
Edited by: G. B. Irvine and C. H. Williams Humana Press Inc., Totowa, NJ

Competitive antagonists reversibly bind and compete with agonists for the same sites in the receptor, altering the EC_{50} but not the maximal response to the agonist. Thus, a greater concentration of agonist is required to elicit the same response in the presence of the antagonist; the concentration–response curve is shifted to the right (*see* Note 3). Noncompetitive antagonists act on the same receptor as do the agonists, but not always at the same site. Noncompetitive antagonists irreversibly bind to the receptor and permanently reduce the population of available receptors. They do not significantly reduce the EC_{50} but do decrease the maximal tissue response induced by the agonist (*see* Note 4).

Many of the phenomena associated with the activation and contractile behavior of smooth muscle are most conveniently examined under isometric (constant length) conditions. The muscle is firmly fixed at both ends and is unable to shorten. Following stimulation, the muscle contraction is manifested as a force exerted against its restraints (*see* Note 5) *(3)*. Consequently, internal dimension changes in the muscle are minimized, thus avoiding complications in the calculation of contractile activity owing to changes in tissue geometry. A force transducer is commonly used to measure the force of an isometric contraction by detecting the small displacements in response to load as electrical output. The transducers must be able to respond rapidly. Most force transducers are fast enough to follow the normal slow contractile force changes of most smooth muscles. As a general rule, the response of the transducer should be 10 times faster that the most rapid event to be recorded *(3)*.

It is necessary to use dissected muscle preparations that can be directly attached to the recording apparatus. Although the use of isolated muscle preparations reduces complex geometry changes in the muscle observed in vivo, allows greater access of the peptide to the cells, and reduces proteolytic degradation of the peptide, some new problems are introduced. Dissection is always accompanied by some tissue and cellular damage, and the altered physical environment may modify the responses of the preparation.

Selection of a tissue that contains receptors that are specific for the peptide under investigation is of crucial importance. Ideally the tissue chosen should contain receptors specific for the peptide under investigation and be free from responses to homologous peptides. Such monoreceptorial assays, thus, can provide simple systems to analyze in detail the peptide–receptor interactions and serve as bioassays for the development of new ligands with high potency and selectivity (*see* Note 6) *(4)*.

This chapter briefly describes a previously established protocol used to determine the agonistic or antagonistic properties of analogs of Neurokinin A (NKA) *(5)* on hamster tracheal smooth muscle (*see* Note 7). The ability of various analogs of NKA to induce contraction, and/or block the NKA-induced contraction can be easily measured under isometric conditions.

2. Materials

2.1. Apparatus

Several pieces of equipment are required. Gould 8000 chart recorders with built-in bridge amplifiers are used to record the analog signals generated from Hugo Sachs F30 force transducers following muscle stimulation. Also required are a 95% O_2, 5% CO_2 gas supply, a vacuum pump, and a circulating water bath maintained at 37°C. Water-jacketed organ baths (4 mL) should keep the contents of the bath at 37°C and allow continuous aeration (95% O_2, 5% CO_2) of the bath. Plastic tubing, a peristaltic pump, pipets (20, 100, and 200 µL), and dissecting equipment are also necessary.

2.2. Chemicals

1. Nanopure water.
2. Krebs buffer (1 L): 7.00 g (120 m*M*) NaCl, 0.35 g (4.7 m*M*) KCl, 2.10 g (25 m*M*) $NaHCO_3$, 0.31 g (1.5 m*M*) $MgCl_2$, 0.37 g (2.5 m*M*) $CaCl_2$, 0.15 g (1.2 m*M*) NaH_2PO_4, 0.99 g (5.5 m*M*) glucose, 1 mL stock phosphoramidon, and 1.8 mg (5 µ*M*) indomethacin, pH 7.4 (*see* Note 8) *(5,6)*.
3. 80 m*M* isotonic Krebs buffer (1 L): 2.61 g (44.7 m*M*) NaCl, 5.97 g (80 m*M*) KCl, 2.10 g (25 m*M*) $NaHCO_3$, 0.31 g (1.5 m*M*) $MgCl_2$, 0.37 g (2.5 m*M*) $CaCl_2$, 0.15 g (1.2 m*M*) NaH_2PO_4, 0.99 g (5.5 m*M*) glucose, 1 mL stock, and 1.8 mg (5 µ*M*) indomethacin, pH 7.4 (*see* Note 9) *(5,7,12)*.
4. 100 µ*M* stock solution of peptide is made by dissolving 1 mg peptide in 100 µL dimethyl sulfoxide (DMSO) (*see* Note 10) and diluting to the desired concentration with distilled water.
5. 100 µ*M* stock solution of carbachol (18.26 mg/L) (*see* Note 11).

3. Methods

3.1. Preparation of the Hamster Trachea

1. Sacrifice male Syrian hamsters (*see* Note 7) (120–140 g) by CO_2 intoxification in a gassing chamber. Immediately after death, make a vertical incision through the sternum, from the base of the rib cage to the mouth, allowing access to the thoracic cavity.
2. Remove any connective tissue surrounding the trachea and cut through the larynx. Hold the larynx with forceps or the trachea will disappear into the thoracic cavity. Gently tease the trachea from the hamster removing any further connective tissue. Excise the trachea with the larynx and bronchioles attached to avoid damage to the trachea, and place in aerated (95% O_2, 5% CO_2) Krebs buffer at 20°C.
3. Remove the freshly excised trachea from the Krebs buffer. Carefully pin it, through the larynx and bronchioles, to a cork mat. While keeping the tissue immersed in Krebs buffer, remove any remaining connective and/or fat tissues. Remove the bronchioles and cut transverse sections of the trachea to give approximately eight rings about 3 mm in width. Place the unused rings back into

the continuously aerated Krebs buffer at 20°C. Rings must be used during the day of preparation.

4. Position the rings between the tungsten separation wires (as shown in Fig. 1) so that the smooth muscle is directly between the wires. Lower the trachea into the continuously oxygenated organ bath containing Krebs buffer. Place the tissue under 5 m*N* resting tension. After acclimatization and achievement of a steady baseline (1 h), readjust the tension to 5 m*N*. To ensure suitability for testing, contract the rings with 1 μ*M* carbachol (*see* Note 11) at 20-min intervals and wash several times with Krebs buffer until reproducible responses are obtained. Following a 20-min recovery period, determine the agonistic activity of the peptide.

3.2. Assessment of Agonism

1. Serially dilute the stock peptide solution with Krebs buffer so that the final concentration of peptide in the organ bath is in the range of 10^{-10}–10^{-6}*M* (*see* Note 12).
2. Pipet the lowest concentration of peptide solution (10^{-10}*M*) directly into the organ bath and observe the response on the pen recorder.
3. Add each of the successively more concentrated peptide solutions to the organ bath once the previous response reaches a plateau, or, if no response is observed, after 5 min. If the response decreases upon subsequent peptide additions, the tissue may exhibit desensitization *(8)* (*see* Note 13).
4. After the last response to peptide has been recorded, wash the tracheal tissue several times with Krebs buffer and once basal tension is restored, allow the tissue to equilibrate for 30 min.
5. Empty the organ bath and fill it with the 80 m*M* isotonic Krebs buffer. Once the response has reached a plateau, wash the tissue several times with Krebs buffer.
6. Peptide-induced contractions are expressed as the percentage of the reference contraction induced by 80 m*M* isotonic Krebs buffer, and are plotted against the logarithm of the peptide concentration required to elicit the response as shown in Fig. 2.

3.3. Assessment of Antagonism

1. Make 500 mL of a 10^{-8}–10^{-6}*M* solution of the putative antagonist in Krebs buffer. Each tracheal preparation can be used to assess the activity of only one concentration of antagonist.
2. Determine the tissue responses to concentrations of a known full agonist in Krebs buffer as described in Section 3.2.
3. Allow the tissue to recover for 20 min in Krebs buffer.
4. Add the putative antagonist solution (step 1) to the organ bath and let the tissue equilibrate for 40 min. Note all changes in basal tension observed during this equilibration. Differences are owing to the presence of the putative antagonist and must be considered in all subsequent calculations of agonist responses.
5. Prepare solutions of various concentrations of full agonist by serial dilution with solutions of the putative antagonist at selected concentrations. Repeat step 2 in the continued presence of the putative antagonist.

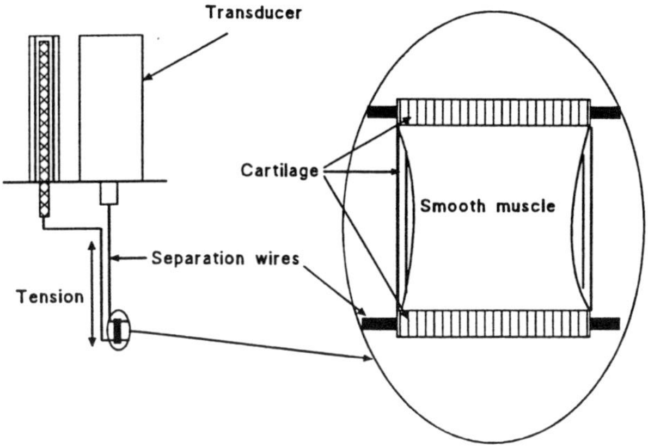

Fig. 1. Apparatus used for the functional experiments. The tracheal ring is loaded onto the wires with the muscle placed directly between the wires. The isometric force produced in response to peptide stimulation of the muscle is detected by the transducer, and the electrical signal generated is recorded by the pen recorder.

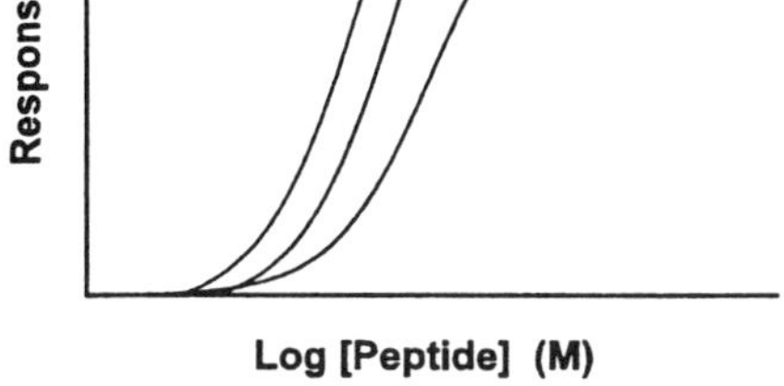

Fig. 2. Concentration–response curve for three representative full and partial agonists, A, B, and C. As the concentrations of A and B are increased, the response increases accordingly until the maximal tissue response is obtained. Thus A and B are full agonists. Although A and B produce the maximal tissue response, a greater concentration of B is required to produce the same response as a particular concentration of A, as shown by the rightward position of B with respect to A. Therefore, A is more potent than B. At high concentrations, however, C does not induce a maximal tissue response and is characteristically a partial agonist.

6. Compare the carbachol-induced contractions in the absence and presence of the putative antagonist. If any difference is observed, the results with this tissue should be excluded from the experiment.
7. Compare the response induced by the 80 m*M* isotonic Krebs buffer in the absence and presence of the putative antagonist; again, differences mandate exclusion.
8. Agonist-induced contractions, in the absence (pooled data) and presence of the putative antagonist, are expressed as a percentage of the reference contraction induced by 80 m*M* isotonic Krebs buffer and are plotted against the logarithm of the agonist concentration required to elicit the response as shown in Fig. 3.

4. Notes

1. Agonists that induce maximal tissue response are regarded as full agonists. They have high affinity for and high efficacy at the receptor as illustrated by the concentration–response curves for peptides A and B in Fig. 2. Partial agonists usually do not induce a maximal response, as shown by the curve for peptide C in Fig. 2. They may have comparable affinity for the receptor to that of a full agonist, but significantly lower efficacy. Thus, a markedly higher concentration of a partial agonist is required to elicit an equivalent response compared with the full agonist.
2. According to the Occupancy Theory *(9,10)*, the magnitude of an effect is directly proportional to the concentration of peptide–receptor complex formed and the maximal effect occurs when all receptors are occupied. Thus, the efficacy is equal to the maximal effect produced. Stephenson, however, modified the theory to explain the results of several laboratories, with the major change that a maximal effect can be produced without full receptor occupancy *(11)*. Therefore, the potency and efficacy of the peptide must be determined by null methods *(1,12)* such as partial irreversible receptor blockade using phenoxybenzamine *(13)*.
3. In the presence of a competitive antagonist, the agonist concentration–response curve is shifted to the right by increasing the concentration of antagonist without significant depression of the maximal response, as shown in Fig. 3.
4. In the presence of a noncompetitive antagonist, the agonist concentration–response curve is slightly shifted to the right, but the maximal response is significantly reduced by increasing antagonist concentration, as shown in Fig. 4.
5. Various materials (silk thread, stainless steel wires, tungsten wires) can be exploited to apply tension to the muscle in the organ bath. The material employed must be resistant to corrosion under the organ bath conditions.
6. Data taken from tissues assumed to be monoreceptorial should be interpreted cautiously *(4)*. If the assay data with a peptide do not fit the criteria for a particular classification of receptor, it is assumed that the peptide crossreacts with another receptor for a related peptide. The possible presence of a novel type of receptor subtype, however, should not be excluded and should be investigated by further functional and radioligand binding studies *(4)*.
7. The hamster trachea was chosen because, in this tissue, the responses to neurokinin A (NKA) are mediated only by the NK_2 receptor. The tachykinin

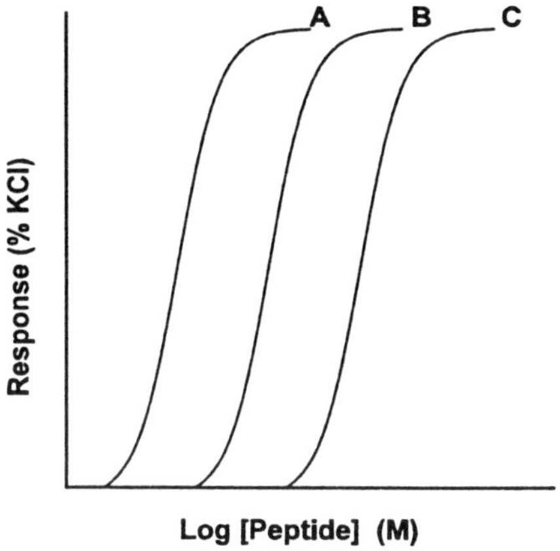

Fig. 3. The effect of a competitive antagonist on the agonist concentration–response curve. As the concentration of agonist is increased, the response increases until the maximal tissue response is obtained (A). In the presence of increasing concentrations of antagonist (B, then C), the agonist concentration–response curve is shifted to the right without depression of the maximal response.

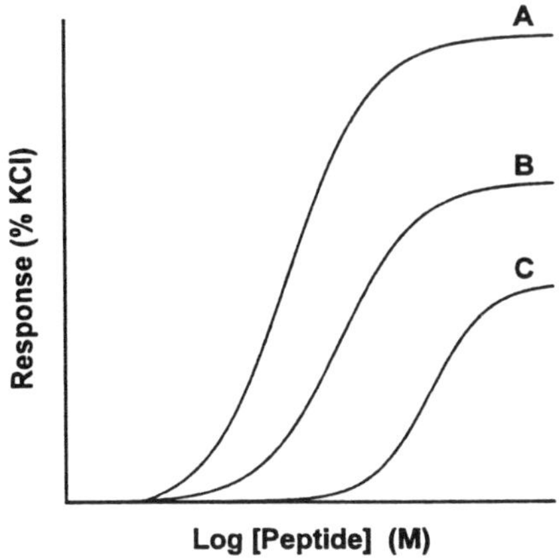

Fig. 4. The effect of a noncompetitive antagonist on the agonist concentration–response curve. As the concentration of agonist is increased, the response increases until the maximal tissue response is obtained (A). In the presence of increasing concentrations of antagonist (B, then C), the agonist concentration–response curve is not significantly shifted to the right. However, the maximal responses are depressed.

family of neuropeptides includes NKA, neurokinin B (NKB), and substance P (SP), all of which share a common C-terminal sequence of Phe-Xaa-Gly-Leu-Met.NH_2. They exert their biological actions via three pharmacologically distinct receptors. SP preferentially binds to the NK_1, NKA to the NK_2, and NKB to the NK_3 receptor *(14)*. Since the C-terminal pentapeptide is the major receptor binding site for each peptide, there is considerable crossreactivity of each of the tachykinins with receptors for the other tachykinins. The response of the hamster trachea to SP, NKB, and specific NK_1 and NK_3 receptor agonists was several orders of magnitude less than to NKA *(5,7)*. Various putative monoreceptorial bioassays have been developed in which the response to the tachykinins is mediated solely by NK_1 (dog carotid artery), NK_2 (hamster trachea), and NK_3 (rat portal vein) receptors *(4)*.

8. Phosphoramidon is an inhibitor of endopeptidase 24.11, which is largely responsible for the degradation of the tachykinins. A stock solution of 100 μ*M* phosphoramidon is made by dissolving 1 mg in 18.4 mL nanopure water. Aliquots (1-mL) of the stock solution are added to 1.5-mL Eppendorf tubes and stored in the freezer until required for inclusion in 1 L Krebs buffer to give a final concentration of 100 n*M* phosphoramidon. Indomethacin inhibits the enzyme cyclooxygenase and thus stops the formation of prostaglandins. Unlike phosphoramidon, indomethacin is insoluble in water. Thus, 1.8 mg is first dissolved in a small volume (200 μL) of ethanol and then made up to the desired concentration with nanopure water. This presence of ethanol has no effect on the muscle preparation. To maintain the ratio of sodium-to-potassium ions in the Krebs buffers, the pH is adjusted to 7.4 with either 1*M* acetic acid or 1*M* ammonium hydroxide.
9. The protocol described in this chapter uses 80 m*M* isotonic Krebs buffer to establish a reference receptor-independent contraction; all peptide-induced contractions are expressed as a percentage of this contraction. Other research groups have also used 80 m*M* isotonic Krebs buffer to produce a reference contraction *(7,12)*, whereas others have used carbachol-stimulated contractions for reference *(6)*.
10. The DMSO dissolves most hydrophobic peptides and the small quantities used have no effect of the eventual results.
11. In this protocol, carbachol is used to screen the tracheal rings for their suitability for peptide testing. 1 μ*M* Carbachol is added repeatedly to each ring before the agonistic/antagonistic properties of each peptide are assessed. If differences are observed between consecutive carbachol-induced contractions, then the tissue is replaced. Prior exposure to carbachol does not reduce subsequent carbachol-induced contractions (i.e., the tissue does not become desensitized). Thus, changes in subsequent carbachol-induced contractions indicate that a problem has developed in the tissue.
12. The stock peptide solution is serially diluted so that the final concentration of peptide in the bath ranges from 10^{-10}, 3×10^{-10}, 10^{-9}, 3×10^{-9}, 10^{-8}, 3×10^{-8}, 10^{-7}, 3×10^{-7}, to 10^{-6}*M*. The volume of concentrated peptide solution added to the organ bath should not exceed 3% of the organ bath volume. The small volumes added

do not substantially change the total volume in the organ bath, and thus the final concentration of peptide. Temperature and pH in the bath are also unchanged by the addition of small volumes of peptide solution.

13. Desensitization of a tissue to stimulation by a peptide is often assumed when the effect of a particular concentration of peptide gradually diminishes when the peptide is added continuously or repeatedly. The cause of desensitization is poorly understood and may involve a change in the receptor structure, receptor internalization, increased metabolic degradation, and exhaustion of intracellular mediators *(2,8)*. If the tissue becomes desensitized, then the peptide cannot be added cumulatively. Instead, the tissue must be thoroughly washed between each individual peptide addition and basal tension must be reached before subsequent additions. Previous studies, however, have shown that the hamster trachea does not display desensitization of its response to NKA *(7)*.

References

1. Kenakin, T. (1993) *Pharmacologic Analysis of Drug-Receptor Interaction.* Raven, New York.
2. Tallarida, R. J. and Jacob, L. S. (1979) *The Dose-Response Relation in Pharmacology.* Springer and Verlag, New York.
3. Meiss, R. A. (1989) Mechanical properties of gastrointestinal smooth muscle, in *Handbook of Physiology: The Gastrointestinal System* (Schultz, S. G., Wood, J. D., and Rauner, B. B., eds.), American Physiological Society, Bethesda, MD, pp. 273–329.
4. Maggi, C. A. (1994) Evidence for receptor subtypes/species variants of receptors, in *The Tachykinin Receptors* (Buck, S. H., ed.), Humana, Totowa, NJ, pp. 395–470.
5. Murnin, M. (1994) Structure-activity studies on some analogues of Neurokinin A. PhD thesis, University of Ulster.
6. Abu Shanab, A. A., Allen, J. M., Guthrie, D. J. S., Irvine, G. B., Murphy, R. F., and Walker, B. (1991) Effects of some neurokinin A analogues on tachykinin-induced contraction of the guinea pig trachea. *Peptides* **12,** 1069–1075.
7. Maggi, C. A., Patacchini, R., Rovero, P., and Meli, A. (1989) The isolated hamster trachea: a new preparation for studying NK-2 receptors. *Eur. J. Pharmacol.* **166,** 435–440.
8. Rang, H. P. and Dale, M. M. (1991) *Pharmacology,* 2nd ed. Churchill Livingstone, London.
9. Clark, A. J. (1933) *The Mode of Action of Drugs on Cells.* Williams and Wilkins, Baltimore, MD.
10. Arëins, E. J. (1954) Affinity and intrinsic activity in the theory of competitive inhibition. *Arch. Int. Pharmacodyn. Ther.* **99,** 32–49.
11. Stephenson, R. P. (1956) A modification of the receptor theory. *Br. J. Pharmacol.* **11,** 379–393.
12. Patacchini, R., Astolfi, M., Quartara, L., Rovero, P., Giachetti, A., and Maggi, C. A. (1991) Further evidence for the existence of NK_2 tachykinin receptor subtypes. *Br. J. Pharmacol.* **104,** 91–96.

13. Furchgott, R. F. (1966) The use of β-haloalkylamines in the differentiation of receptors and in the determination of dissociation constants of receptor-agonist complexes, in *Advances in Drug Research, vol. 3* (Harper, N. P. and Simmonds, A. B., eds.), Academic, New York, pp. 21–55.
14. Van Giersbergen, P. L. M. and Buck, S. H. (1994) Biochemical methods and assays, in *The Tachykinin Receptors* (Buck, S. H., ed.), Humana, Totowa, NJ, pp. 39–68.

31

Measurement of Efflux Rates from Brain to Blood

William A. Banks, Melita B. Fasold, and Abba J. Kastin

1. Introduction

The blood–brain barrier (BBB) regulates the exchange of substances between the fluids of the central nervous system (CNS) and the blood *(1)*. As such, the BBB is actively involved in providing nutrition and maintaining the homeostatic environment for the brain and spinal cord. Recently, the BBB has been postulated to be important in communication between the nervous system and the peripheral tissues through its ability to control the exchange of regulatory substances *(2,3)*. Furthermore, the BBB is increasingly being found to play important and active roles in disease states *(4–9)*.

Most studies have concentrated on blood to CNS movement, or influx, of substances. However, brain-to-blood movement, or efflux, of substances is equally important. Efflux systems have been shown to aid in keeping substances such as potassium within a very narrow range in the cerebrospinal fluid (CSF), in keeping the CNS levels of some amino acids and neurotransmitters low, and in preventing the accumulation within the CNS, and consequently the effects on the CNS, of drugs and toxins *(10)*. Efflux systems are also a route by which materials found within the brain can enter the circulation in amounts sufficient to produce measurable concentrations in the blood and to induce peripheral effects *(11,12)*. This can be of concern in studies in which the hypothesis that a CNS site of action exists for a substance is tested by administration of the substance into the brain. In the presence of robust transport systems, blood levels obtained after intracerebroventricular (icv) injection can mirror those produced after iv injections *(13)*. Efflux systems also may be associated with disease *(4,14)*.

Measurement of efflux rates traditionally has been cumbersome, involving complicated cannulation and instrumentation of large animals. We outline here

From: *Methods in Molecular Biology, Neuropeptide Protocols*
Edited by: G. B. Irvine and C. H. Williams Humana Press Inc., Totowa, NJ

a method that we have developed involving only a free-hand icv injection into an anesthetized rodent. This method has been widely applied in several strains of mice and rats under numerous pathophysiological conditions to study a wide range of substances.

2. Materials

2.1. Equipment

1. One-microliter Hamilton syringe (Model # 7101KH; Hamilton, Reno, NV). This is necessary to ensure the accuracy and reliable repetition of the icv injection (*see* Note 1). With a permanent marker and ruler, make a line 3.0–3.5 mm from the end of the needle.
2. Guarded needle: A standard 26-gage hypodermic needle is used with a length of polyethylene tubing (Intramedic #7406) over it. The tubing is cut to a length of 3.5 mm to ensure that the needle does not penetrate too far into the brain after the skull is punctured.
3. Instruments (Roboz Surgical Instrument, Rockville, MD).
   a. Heavy-duty scissors with at least one blunt edge are necessary for decapitation of the animal.
   b. Dean tonsil scissors (long-handled, with short angled blades) are used to cut along the sides of the skull before removal of the brain.
   c. Four-inch microforceps with slightly curved tips are used to remove the brain from the skull.
   d. Stopwatch or other timing device.

2.2. Chemicals and Injectables

1. Urethane (ethyl carbamate; Sigma, St. Louis, MO): A 40% solution of urethane (add 40 g to 100 mL of 0.5*N* saline) is used as an anesthetic (*see* Note 2). The recommended dose is about 10 µL/g body wt. Therefore, a 20-g mouse would receive 200 µL of urethane or 2 g/kg body wt (*see* Note 3).
2. Lactated Ringer's solution.
3. Radioactive materials: Although the primary radioactive label is ^{125}I, other labels including ^{99m}Tc, ^{3}H, ^{14}C, and ^{35}S also have been used successfully. Dilute the labeled compound to 25,000 cpm/µL in lactated Ringer's solution for injection (*see* Note 4). If the compound is found to adhere to the inside of the Hamilton syringe (as determined by injection checks: Note 5), lactated Ringer's solution containing 1% bovine serum albumin may be used.

3. Methods

3.1. Icv Injection

1. After the mouse is anesthetized, resect the scalp. Remove enough of the fur and skin to expose the suture lines of the skull.
2. Locate the bregma, the intersection of the coronal and sagittal sutures (Fig. 1). This is easier to do if the skull is allowed to dry slightly after the scalp has been resected.
3. Puncture the skull with the guarded needle at a point 1 mm posterior and 1 mm

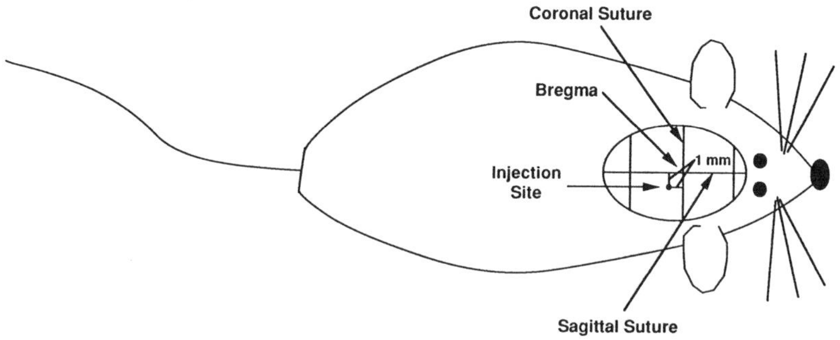

Fig. 1. Landmarks for intracerebroventricular injections.

lateral to the bregma. Whereas it is possible to measure this point exactly, it is more practical to visualize a 1-mm square with the bregma at one corner, then make the hole at the opposite, diagonal corner.

4. Draw 1 μL of labeled material into the Hamilton syringe. Wipe the outside of the needle with a tissue to remove any residual solution. Insert the syringe with a twisting motion, until the mark on the needle is flush with the skull. Inject the material with a single, smooth motion. Leave the syringe in place for 5–10 s after injection (*see* Note 6). Start timing from this point.
5. Decapitate the mouse at a predetermined time after the injection. For most experiments, except disappearance curves, this would be at 10 min. Remove the brain from the skull and place it in a container suitable for the detection of residual radioactivity (*see* Note 7). Count for a suitable time in the appropriate radiation counter; usually 3 min for γ-emitters or 5 min for β-emitters, if 25,000 cpm are used. If fewer cpm are injected, the counting time will need to be increased to allow for the accumulation of sufficient counts for accuracy.

3.2. Experimental Design

3.2.1. Disappearance Curve

To determine the rate at which the labeled material is removed from the brain, decapitate the animals at 0, 2, 5, 10, and 20 min after the icv injection, with 7–10 mice at each time-point (*see* Note 8). If the transport rate is exceptionally slow, the decapitation points may be extended to include 30, 45, 60, 90, and 120 min after the icv injection.

3.2.2. Inhibition with Unlabeled Material

To determine the saturability of the transport system, inject labeled material with and without unlabeled material at a dose of 10 n*M*/mouse, with 7–10 mice for each group (*see* Note 8). If this dose inhibits the transport system, additional higher doses can be tested such as 30 and 100 n*M*/mouse and lower doses such as 3, 1, 0.3, and 0.1 n*M*/mouse (or even lower concentrations; *see* Notes 9 and 10).

3.2.3. Manipulations of the Transport System

If a saturable transport system is found to exist for the compound being tested, other experiments may be done to further define the parameters of the system. These could include pretreatment with aluminum *(15,16),* determination of cross-inhibition with related compounds *(17–21),* and cross-inhibition with compounds of previously defined transport systems *(2).*

3.3. Calculations

3.3.1. Half-Time Disappearance ($t_{1/2}$)

The half-time disappearance of the labeled compound is calculated by a plot of the log (mean cpm per whole brain) against the time after icv injection (Fig. 2). The half-time disappearance (in min) is the inverse of the slope multiplied by 0.301. This can range from about 15 min or less for a compound with an active transport system (such as Tyr-MIF-1) to more than 60 min for one removed by CSF reabsorption (such as albumin). Substances sequestered by brain tissues have a longer $t_{1/2}$. The antilog of the intercept of the line is taken as the amount of material available for transport (*see* Note 8) and used in the calculation of the transport rate.

3.3.2. Transport Rate

The transport rate (T) expressed in mol/g of brain per min is calculated from the equation:

$$T = (A - M)C/itw$$

where M is the cpm remaining in the individual brain, C the amount of material expressed in moles, i the amount of injected material expressed in cpm, t the time in min from injection to decapitation, and w the weight of the brain in grams. A is the amount of material available for transport, as determined from the half-time disappearance curve, expressed in cpm (*see* Note 11).

3.3.3. Percent Inhibition

Any inhibition of the transport system by either unlabeled material or other candidate inhibitors can be expressed as a percent of the transport rate with the equation:

$$\%T = 100(A - E)/(A - C)$$

where E is the cpm remaining in the brains of mice receiving the inhibitor.

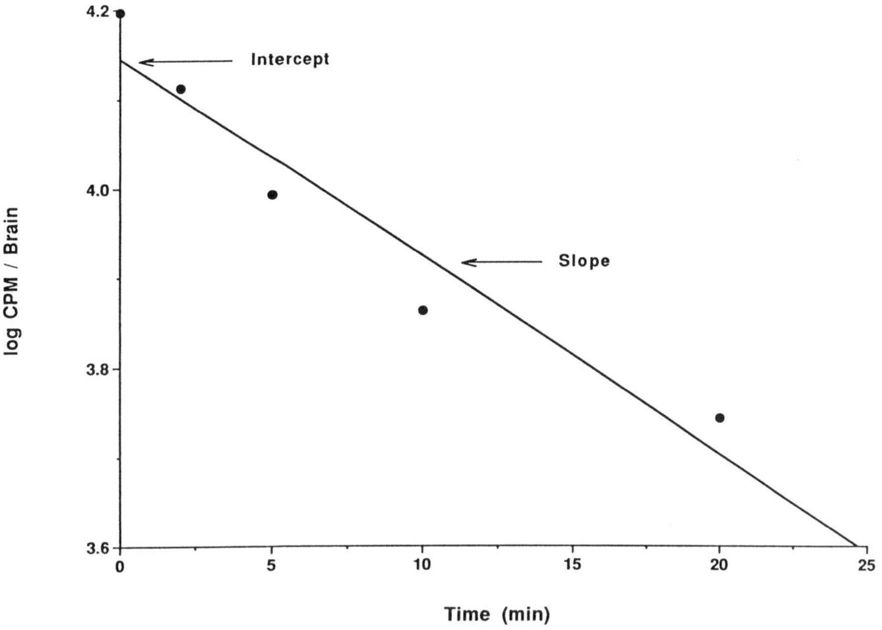

Fig. 2. Idealized disappearance curve for ^{125}I-Tyr-MIF-1.

3.3.4. Statistics

Regression curves are calculated by the least squares method. Computer software programs or hand-held calculators can be used. A significant correlation between log(M) and t is taken as evidence for disappearance from brain. Two or more regression lines can be compared for statistically significant differences by a variety of statistical methods or packages such as the BMDP1R program (BMDP Statistical Software, Los Angeles, CA). Means are reported with their standard errors. Groups are initially compared by analysis of variance (ANOVA) and, if a statistical difference exists ($p < 0.05$), further compared by a multiple comparisons test.

4. Notes

1. This particular model of Hamilton syringe is outfitted with a needle with a #3 point. It has a 90° bevel and an electrotaper. The point of this needle does little damage to brain tissue upon entering, and the taper simplifies entry through the skull. The injection volume of the syringe is contained entirely within the needle; however, the wire of the plunger does travel through the barrel and is extremely fragile. Avoid bending or breaking the plunger wire, as this renders the syringe

inoperable. A syringe guard is available, if desired, to stiffen the plunger and avoid this potential problem.

2. Urethane produces deep, long-lasting anesthesia more reliably than other agents, with only minor alterations in physiological parameters. It is best used for nonrecovery procedures. Although labeled as a carcinogen, this potential may be limited to genetically predisposed strains. There are currently no case reports or epidemiological studies on urethane as a human carcinogen *(22,23)*. Some precautions are recommended:
   a. Wear gloves and mask when working with the crystalline form.
   b. Work under a fume hood; cover the solution while it is dissolving.
   c. Store in sealed bottles; do not reuse the empty bottles.
   d. Wear gloves while working with the solution.
3. Because of differences in each animal's metabolism, a mouse may require varying amounts of urethane to be completely anesthetized. It is better to begin with a lower dose, then administer more as necessary. Complete anesthesia should result in a lack of reaction to a toe-pinch, as well as rapid, shallow respiration and loss of righting reflex.
4. 25,000 cpm/μL of labeled material is our standard concentration. The concentration used may range from 10,000–50,000 cpm but should not go outside this range, owing to variability in counting.
5. Checks consist of 1-μL injections of labeled material made into empty test tubes before, during, and after injection into each group of animals. The mean of these three injections is used to normalize the counts remaining in the brain to the standard 25,000 cpm that had been intended to be injected. This allows direct comparison of groups receiving unequal amounts of labeled material, as well as day-to-day comparisons of experiments.
6. This helps prevent the labeled solution from refluxing through the skull. If reflux occurs despite this precaution, record as a partial injection. This will aid in the possible elimination of widely outlying values for the final calculations.
7. It is not important to remove the brain from the skull in intact form. The number of pieces of brain tissue does not affect the counting of the radiation, as long as each piece of the entire brain is placed in the container. Use a 12 × 75-mm test tube for γ-emitters or a 20-mL glass scintillation vial for β-emitters. Centrifuge the brain for 5 min at 3000*g* to pull the brain tissue to the bottom of the 12 × 75-mm test tube. This ensures that the gamma counter can detect all of the radioactivity.
8. To decapitate a mouse at a 0-min time-point, overdose the animal with urethane before the icv injection. The animal must be dead for 10–15 min before the injection. This ensures that any active transport processes are eliminated and gives an accurate measure of the material available for transport (less than the amount injected).
9. To inject labeled and unlabeled material simultaneously, make up the materials to twice the desired concentration (50,000 cpm and 20 n*M*/mouse) and combine equal portions of each; alternatively, the 25,000 cpm/μL solution may be made first, and then used to dilute the unlabeled material to the desired concentration.

10. If the material of interest has limited solubility, use the highest soluble concentration possible. If it is not water soluble, alternative solvents may be used if the labeled material is also diluted in the solvent and the solvent itself does not affect the transport system (*see* Section 3.2.3.).
11. Transport rate can also be calculated from the rate of appearance of radioactivity in blood as previously described *(24)*. However, the statistical variability is greater with this method and peripheral pharmacokinetic parameters must be determined in a separate group of animals.

References

1. Davson, H. (1967) The blood-brain barrier, in *Physiology of the Cerebrospinal Fluid,* Churchill, London, pp. 82–103.
2. Banks, W. A. and Kastin, A. J. (1990) Editorial review: Peptide transport systems for opiates across the blood-brain barrier. *Am. J. Physiol.* **259,** E1–E10.
3. Banks, W. A. and Kastin, A. J. (1993) Physiological consequences of the passage of peptides across the blood-brain barrier. *Rev. Neurosci.* **4,** 365–372.
4. Banks, W. A. and Kastin, A. J. (1989) Inhibition of the brain to blood transport system for enkephalins and Tyr-MIF-1 in mice addicted or genetically predisposed to drinking ethanol. *Alcohol* **6,** 53–57.
5. Banks, W. A. and Kastin, A. J. (1988) Review: Interactions between the blood-brain barrier and endogenous peptides: emerging clinical implications. *Am. J. Med. Sci.* **295,** 459–465.
6. Rapoport, S. I. (1976) Pathological alterations of the blood-brain barrier, in *Blood-Brain Barrier in Physiology and Medicine,* Raven, New York, pp. 129–152.
7. Knobler, R. L., Marini, J. C., Goldowitz, D., and Lublin, F. D. (1992) Distribution of the blood–brain barrier in heterotopic brain transplants and its relationship to the lesions of EAE. *J. Neuropathol. Exp. Neurol.* **51,** 36–39.
8. Stitt, J. T. (1990) Passage of immunomodulators across the blood-brain barrier. *Yale J. Biol. Med.* **63,** 121–131.
9. Cserr, H. F. and Knopf, P. M. (1992) Cervical lymphatics, the blood-brain barrier and the immunoreactivity of the brain: a new view. *Immunol. Today* **13,** 507–512.
10. Davson, H., Welch, K., and Segal, M. B. (1987) Blood-brain-CSF relations, in *The Physiology and Pathophysiology of the Cerebrospinal Fluid,* Churchill Livingstone, Edinburgh, pp. 375–451.
11. Jones, P. M. and Robinson, I. C. A. F. (1982) Differential clearance of neurophysin and neurohypophysial peptides from the cerebrospinal fluid in conscious guinea pigs. *Neuroendocrinology* **34,** 297–302.
12. Mens, W. B. J. and Van Wimersma Greidanus, T. B. (1983) Penetration of neurohypophysial hormones from plasma into cerebrospinal fluid (CSF): half-times of disappearance of these neuropeptides from CSF. *Brain Res.* **262,** 143–149.
13. Davson, H., Welch, K., and Segal, M. B. (1987) The return of the cerebrospinal fluid to the blood. The drainage mechanism, in *The Physiology and Pathophysiology of the Cerebrospinal Fluid,* Churchill Livingstone, Edinburgh, pp. 485–521.

14. Banks, W. A. and Kastin, A. J. (1993) The potential for alcohol to affect the passage of peptide and protein hormones across the blood–brain barrier: a hypothesis for a disturbance in brain-body communication, in *NIAAA Research Monograph 23, Alcohol and the Endocrine System* (Zakhari, S., ed.), National Institutes of Health; National Institute on Alcohol Abuse and Alcoholism, Bethesda, MD, pp. 401–411.
15. Banks, W. A. and Kastin, A. J. (1983) Aluminium increases permeability of the blood-brain barrier to labelled DSIP and β-endorphin: possible implications for senile and dialysis dementia. *Lancet* **ii,** 1227–1229.
16. Banks, W. A., Kastin, A. J., and Fasold, M. B. (1988) Differential effect of aluminum on the blood-brain barrier transport of peptides, technetium and albumin. *J. Pharmacol. Exp. Ther.* **244,** 579–585.
17. Barrera, C. M., Kastin, A. J., Fasold, M. B., and Banks, W. A. (1991) Bidirectional saturable transport of LHRH across the blood-brain barrier. *Am. J. Physiol.* **261,** E312–E318.
18. Banks, W. A., Ortiz, L., Plotkin, S. R., and Kastin, A. J. (1991) Human interleukin (IL) 1α, murine IL-1α and murine IL-1β are transported from blood to brain in the mouse by a shared saturable mechanism. *J. Pharmacol. Exp. Ther.* **259,** 988–996.
19. Banks, W. A., Kastin, A. J., and Ehrensing, C. A. (1993) Endogenous peptide Tyr-Pro-Trp-Gly-NH_2 (Tyr-W-MIF-1) is transported from the brain to the blood by peptide transport system-1. *J. Neurosci. Res.* **35,** 690–695.
20. Ferguson, A. V. and Marcus, P. (1988) Area postrema stimulation induced cardiovascular changes in the rat. *Am. J. Physiol.* **255,** R855–R860.
21. Banks, W. A. and Kastin, A. J. (1985) Peptides and the blood-brain barrier: lipophilicity as a predictor of permeability. *Brain Res. Bull.* **15,** 287–292.
22. Field, K. J. and Lang, C. M. (1988) Hazards of urethane (ethyl carbamate): a review of the literature. *Lab. Animals* **22,** 255–262.
23. Wood, E. M. (1956) Urethane as a carcinogen. *The Progressive Fish-Culturist,* 135,136.
24. Banks, W. A. and Kastin, A. J. (1989) Quantifying carrier-mediated transport of peptides from the brain to the blood, in *Methods in Enzymology,* vol. 168 (Conn, P. M., ed.), Academic, San Diego, pp. 652–660.

32

Assay of Neuropeptidases Using Fluorogenic Substrates

Carvell H. Williams

1. Introduction

Among the more sensitive methods available for the assay of peptidases are those that use fluorogenic substrates, which are of three general types depending on the substrate specificity of the peptidase of interest:

1. Those in which a fluorescent aromatic amine, typically 7-amino-4-methyl coumarin (AMC) (I, Fig. 1) or 4-methoxy-2-naphthylamine (II, Fig. 1) (Note 1), is linked by an amide bond to the C-terminal amino acid of a peptide substrate. The fluorescence is largely suppressed when the aromatic amino group of the fluorophore is so linked. Specificity is confined to P subsites since no P' amino acids are present (Note 2). Enzymic cleavage of the amide bond linking the peptide to the fluorogenic group liberates the fluorescent molecule. This type of substrate was introduced for proteases such as trypsin and chymotrypsin *(2),* but is also extensively used for the assay of aminopeptidases, in which case a single amino acid is linked to the fluorogenic group.
2. A second method utilizes an internally quenched substrate that contains a fluorescent group and some other molecular species (most usually, one containing a nitro group, NO_2) that is capable of absorbing the fluorescence energy (intramolecular quenching). In these circumstances, the substrate itself either does not fluoresce, or has only weak fluorescence. The design of an internally quenched substrate is such that when it is cleaved by the peptidase under study, the quencher and the fluorescent moiety become separated, resulting in relief of the quenching effect, and hence to a substantial increase in fluorescence. However, the liberated fluorescent species may be subject to some quenching by the substrate (i.e., intermolecular quenching). This can be minimized by using low concentrations of substrate in the assay, but it is still usually necessary to carry out appropriate calibration experiments to take account of the effect. This type of assay allows

From: *Methods in Molecular Biology, Neuropeptide Protocols*
Edited by: G. B. Irvine and C. H. Williams Humana Press Inc., Totowa, NJ

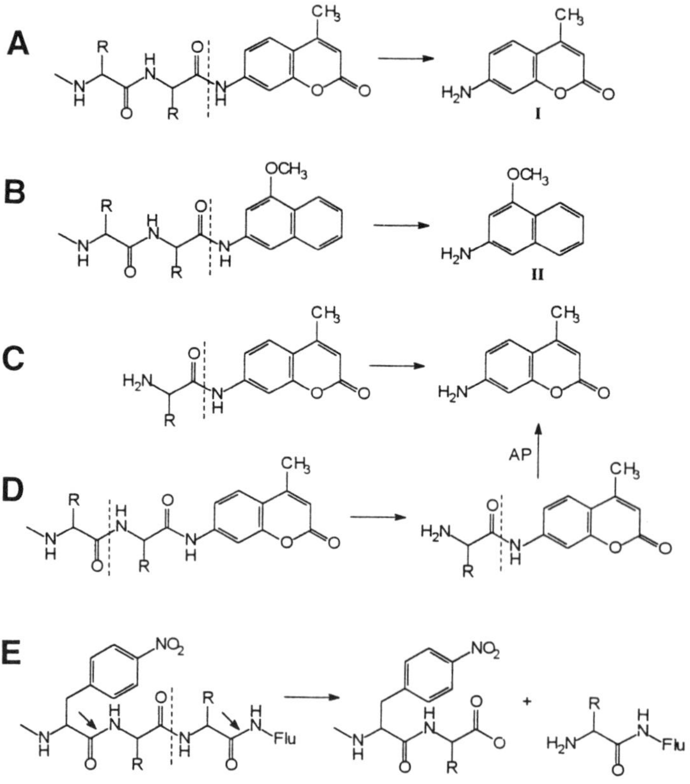

Fig. 1. Some typical schemes for generation of fluorescence by action of peptidases on fluorogenic substrates. Vertical dashed lines indicate sites of peptide bond cleavage. **(A,B)** Action of an endopeptidase on fluorogenic substrates, liberating the fluorescent 4-methyl-7-amino coumarin (I) and 4-methoxy-2-naphthylamine (II), respectively. **(C)** As (A), but in this case, the enzyme is an aminopeptidase. **(D)** Coupled assay for an endopeptidase that generates a substrate for aminopeptidase (AP, present in excess), which then liberates the fluorescent molecule (cf c). **(E)** Structure III contains a fluorescent group (designated Flu) and *p*-nitro-Phe as a putative internal quencher. Their separation by peptidase action (dashed line) generates the fluorescent fragment IV. Alternative cleavages (arrows) would achieve a similar effect. Note that in (E), the fluorescent portion (Flu) can be any convenient fluorophore provided that it be compatible with the substrate specificity requirements of the enzyme and that it be effectively quenched.

more versatility in the design of substrates because, provided that the quenching and fluorescent groups become separated from each other by the action of the enzyme, such a substrate can have either or both P and P' specificity determinants in its amino acid sequence.

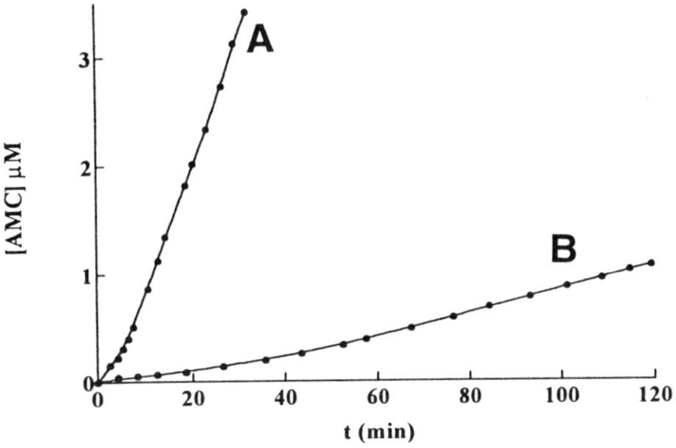

Fig. 2. Progress curves for (A) coupled assay of endopeptidase using leucine aminopeptidase 3.4.11.1 as coupling enzyme; (B) the aminopeptidase alone indicating the presence of contaminating endopeptidase activity. Note the much more prolonged lag phase in (B) as compared to that in (A). Substrate: Succinyl-Ala-Ala-Phe-AMC. Excitation, 380 nm; emission, 460 nm.

3. A variation of method 1., the coupled assay, can sometimes be used when a direct, continuous assay is not available for a particular enzyme. For example, it can be applied to enzymes that cleave the penultimate peptide bond of substrates such as those referred to in method 1., generating an aminoacyl fluorophore. The latter is cleaved in turn by an aminopeptidase that is included in the assay system. Such assays are characterized by an initial lag phase (rate of reaction increases with time) as the concentration of the aminoacyl fluorophore increases from zero to a steady state, when the rate should become linear. This condition exists when the rate of formation of the product of the first enzymic reaction (v_1) equals the rate of consumption of this product by the second (coupling) enzyme (v_2). It is necessary to ensure that this condition (or something close to it) is reached within a time when v_1 is still essentially linear. In practice, this is achieved by having a sufficient amount of coupling enzyme activity present.

The three types of assay referred to above are exemplified in Figs. 1 and 2.

Discussions of the theory of coupled assays can be found in refs. *3–5*. Although fluorogenic substrates for some peptidases are available from commercial sources, it may be necessary in other cases to design and synthesize a substrate. Some recent examples (by no means an exhaustive list) can be found in refs. *6–11*.

This chapter exemplifies the direct assay for an aminopeptidase by describing the measurement of the Michaelis constant for the fluorogenic substrate L-leucyl-7-amino-4-methyl coumaryl amide (Note 3) and a coupled assay for endopeptidase activity using succinyl-Ala-Ala-Phe-AMC as substrate (Scheme 1). The use of this substrate in a coupled assay was first described by Mumford et al. *(12,13)*.

$$\text{HOOCCH}_2\text{CH}_2\text{CO-Ala-Ala-Phe-AMC} \xrightarrow{1} \text{Phe-AMC} + \text{HOOCCH}_2\text{CH}_2\text{CO-Ala-Ala}$$

$$\text{Phe-AMC} \xrightarrow{2} \text{Phe} + \text{AMC}$$

Scheme 1. 1, endopeptidase; 2, aminopeptidase.

Both assays are carried out using a recording spectrofluorometer. Other assay methods have been reported, some requiring specially synthesized substrates. The methods described here make use of materials that are available commercially, except for the endopeptidase itself.

2. Materials

2.1. Equipment

1. A spectrofluorometer capable of measuring fluorescence at a wavelength of 460 nm, together with a suitable pen recorder or similar. We currently use a Perkin Elmer Model MPF 44B (Norwalk, CT), but there are several suitable instruments available from other manufacturers.
2. Quartz cuvet(s) designed for fluorimetric use (*see* Note 3).
3. A thermostatically controlled water bath and associated pump for maintaining constant temperature in the cuvet compartment. (Other methods of temperature control fitted to some instruments are equally suitable.)

2.2. Aminopeptidase Assay

1. 50 m*M* Tris-HCl, pH 7.5. Dissolve 3.03 g of Tris in about 450 mL of water. Bring the pH to 7.5 by adding 2*M* HCl. Make up to 500 mL. Re-check pH.
2. 7-Amino-4-methyl coumarin (AMC) solution. Dissolve 1.76 mg (10 μmol) of AMC in 2 mL of 0.02*M* HCl (*see* Note 5). Dilute this to 50 mL with 50 m*M* Tris-HCl buffer, pH 7.5. This gives a stock solution of $4 \times 10^{-4}M$ (*see* Notes 6 and 7).
3. N-(L-Leucyl)-7-amino-4-methylcoumarin hydrochloride (Leu-AMC). Dissolve 2.6 mg in 1 mL of water to give an 8 m*M* stock solution (*see* Note 7).
4. Aminopeptidase N (*see* Note 8).

2.3. Coupled Assay

1. Triton X-100 detergent.
2. 50 m*M* Tris-HCl, pH 7.5, prepared as in Section 2.2., but containing 0.1% v/v Triton X-100.
3. 7-Amino-4-methyl coumarin solution, prepared as in Section 2.2.
4. Succinyl-Ala-Ala-Phe-4-methyl-coumaryl-7-amide (succinyl-Ala-Ala-Phe-AMC) solution. Dissolve 5.64 mg in 1 mL of dimethyl sulfoxide, giving a stock solution of 10 m*M*.

5. Porcine cytosolic leucine aminopeptidase Type III (Sigma, St. Louis, MO). Dilute a sample of the enzyme preparation with ice-cold 50 m*M* Tris-HCl buffer to give a protein concentration of 6 mg of protein/mL. Keep on ice.
6. Neutral endopeptidase EC 3.4.24.11. Prepare the enzyme sample at a concentration of 25 µg of protein/mL in Tris buffer containing 0.1% (v/v) of Triton X-100 (*see* Note 9). Keep on ice.

3. Method

3.1. Assay of Aminopeptidase (Note 8)

All references to buffer mean 50 m*M* Tris-HCl, pH 7.5 (Section 2.2.).

1. Prepare a series of dilutions in water of the stock solution of the substrate to give the following range of concentrations: 0.4, 0.2, 0.1, 0.08, 0.06, 0.04, 0.03, and 0.02 m*M* (*see* Note 10).
2. Switch on the thermostatically controlled water bath and associated pump.
3. Place the Tris buffer solution in the water bath (*see* Note 11).
4. Switch on the fluorimeter and allow at least a 15-min warm-up period.
5. Set monochromators as follows: excitation, 380 nm; emission, 460 nm (*see* Note 12).
6. Place a cuvet in the sample compartment and add 1 mL of Tris buffer.
7. Open the shutters for incident and emitted light and set chart reading to zero. Close shutters.
8. Dilute the AMC.HCl standard (Section 2.2.) ×100 with buffer to give a solution of concentration 1 µ*M* (*see* Note 13). Place 1 mL in a cuvet into the sample compartment, open the shutters, and set full scale deflection using appropriate gain controls. Close shutters.
9. Pipet 0.94 mL of buffer into a cuvet in the fluorimeter. Then add 50 µL of one of the substrate solutions. Leave for 5 min to ensure thermal equilibration.
10. Open the incident light shutter. Start the chart recorder. Start the reaction by adding 10 µL of enzyme solution to the cuvet and immediately open the shutter for the emitted fluorescent light (*see* Note 14). Allow the reaction to proceed for about 5 min.
11. Close the shutters, remove, and clean the cuvet. Repeat the assay twice more at the same concentration of substrate.
12. Repeat steps 8–10 for the remaining concentrations of substrate (*see* Note 15).
13. Calculate the mean rates of reaction for each substrate concentration.

3.2. Method for Coupled Assay of Endopeptidase 3.4.24.11

All references to buffer mean 50 m*M* Tris-HCl buffer, pH 7.5, containing 0.1% Triton X-100 (Section 2.3.).

1. Switch on the thermostatically controlled water bath and associated pump.
2. Place the Tris buffer solution in the water bath (*see* Note 11).
3. Switch on the fluorimeter and allow at least a 15-min warm-up period.

4. Set monochromators as follows: excitation, 380 nm; emission, 460 nm (*see* Note 12).
5. Place a cuvet in the sample compartment and add 1 mL of Tris buffer.
6. Open the shutters for incident and emitted light and set chart reading to zero. Close shutters.
7. Dilute the AMC.HCl standard (Section 2.2.) ×100 with buffer to give a solution of concentration 1 μ*M* (*see* Note 13). Place 1 mL in a cuvet into the sample compartment, open the shutters, and set full scale deflection using appropriate gain controls. Close shutters.
8. Pipet 0.975 mL of buffer into a cuvet in the fluorimeter. Then add endopeptidase (10 μL, 0.25 μg of protein, *see* Note 9) and leucine amino peptidase (10 μL, 60 μg of protein, *see* Note 16).
9. Open the incident light shutter. Start the chart recorder. Start the reaction by adding 5 μL of substrate solution to the cuvet and immediately open the shutter for the emitted fluorescent light (*see* Note 14). Allow the reaction to proceed for 15–20 min.
10. Repeat the assay twice or three times more.
11. Repeat the coupled assay as described in steps 1–10 but at step 8 omit the endopeptidase and add 10 μL of buffer instead. Allow the reaction to proceed until the observed increase in fluorescence (if any) becomes linear (*see* Note 17 and Fig. 2).
12. Calculate the mean rate of reaction from the measured slopes of the linear portions of the fluorescence vs time traces on the chart recorder.
13. Subtract the mean rate of endopeptidase activity (if any) obtained in step 11.

4. Notes

1. 4-Methoxy-2-naphthylamine has largely replaced the carcinogenic 2-naphthylamine as a fluorophore. 7-Amino-4-methyl coumarin is not known to be carcinogenic. However, it is wise to treat all aromatic amines with caution. Wear gloves when handling them, and take steps to avoid inhalation of the powder.
2. The nomenclature of Schecter and Berger *(1)* is used here. Amino acid residues on the carbonyl side of the scissile peptide bond (-CO-NH-) are designated P_1, P_2, and so on, numbering from the residue carrying the carbonyl group toward the N-terminus. On the NH side of this bond, residues are numbered outward toward the C-terminus, as P_1', P_2', and so on.
3. There are numerous ways of calculating K_m values from kinetic data. Reference *3* is recommended for details of several of these. In the author's laboratory, a K_m value of 0.08 m*M* was obtained for the L-leucyl-AMC under the conditions described here.
4. We use 10-mm square cuvets, with a total volume of approx 1.4 mL, in which an assay volume of 1 mL is used. These cells have two parallel walls that are considerably thicker than the other pair. Because the emitted fluorescence is detected at right angles to the incident beam, it is important to ensure that such a cuvet is always placed in its holder in the same orientation. It will be found that when the cuvet is placed so that the incident beam (excitation) passes through the thinner wall, the measured fluorescence is greater than if the cell is turned through 90°.

5. Because AMC (free base) is not very soluble in water, it is sometimes recommended to dissolve it in, e.g., dimethyl sulfoxide before diluting into buffer. The method used here avoids organic solvent and is just as effective in ensuring initial dissolution of the amine.
6. This solution can be stored for at least 4 wk at 4°C without noticeable loss of fluorescence.
7. In practice, it is simpler to accurately weigh out approximately the required amount and adjust the volume of solvent accordingly to give the required concentration.
8. This method can be adapted as it stands for the assay of other aminopeptidases, provided that an appropriate substrate be used, e.g., Asp-AMC for aminopeptidase A; Arg-AMC for aminopeptidase B, but *see* Note 10.
9. The concentration of enzyme used will depend on its specific activity. That quoted here is based on the specific activity obtained by purification of the enzyme as described by Relton et al. *(14)*.
10. These concentrations of substrate are suitable for aminopeptidase N. The range for other aminopeptidases will depend on the anticipated K_m value and must be found by trial and error. Preliminary assays, at three or four well-spaced concentrations of substrate, should provide a rough guide as to where on a plot of rate vs [s] these values lie. A range of concentrations can then be chosen that span the K_m value (from a maximum concentration of about 3–5 times the anticipated K_m to about one-third to one-quarter of its value).
11. This ensures that the buffer is at approximate working temperature, minimizing thermal equilibration times when samples are added to the cuvet.
12. These wavelengths are not the emission and excitation maxima for AMC, but are used to minimize interference from the native fluorescence of the substrate.
13. With a 1 μ*M* standard, formation of 0.1 nmol of product will produce a deflection of 10% of full-scale. For higher sensitivity, a more dilute standard can be used but this will cause a decrease in the signal-to-noise ratio.
14. The resulting slight delay in the onset of measurement of the enzymic reaction should not present a problem unless the rate differs noticeably from linearity in the early stages. If this is the case, it is likely to be because the enzyme concentration is much too high. A small upward kick on the recorder trace at the start of the reaction is owing to the inherent fluorescence of the substrate.
15. It is advisable to randomize the order of substrate concentrations used in case of systematic error such as instrument drift, minor changes in enzyme activity, and so on, during the experiment. In a prolonged experiment, it is advisable to check the response to the AMC standard occasionally.
16. This is based on a specific activity of 200 U/mg as defined by the suppliers. The lag phase is no more than 5 min when coupled with the suggested amount of endopeptidase activity, during which time about 1% of the substrate is consumed.
17. The aminopeptidase coupling enzyme may contain traces of contaminant endopeptidase activity. If so, there will be an increase in fluorescence, even in the absence of added endopeptidase activity. This will be small and

accompanied by a substantial lag period (*see* Fig. 2). By allowing the reaction to continue until it becomes linear, the amount of the contaminant endopeptidase activity can be ascertained and subtracted from value obtained in step 8 (Section 3.2.).

References

1. Schecter, I. and Berger, A. (1968) On the active site of proteases. III. Mapping the active site of papain; specific peptide inhibitors of papain. *Biochem. Biophys. Res. Commun.* **32,** 898–902.
2. Zimmerman, M., Ashe, B., Yurewica, E. C., and Patel, G. (1977) Sensitive assays for trypsin, elastase and chymotrypsin using new fluorogenic substrates. *Anal. Biochem.* **78,** 47–51.
3. Cornish-Bowden, A. (1995) *Fundamentals of Enzyme Kinetics.* Portland, London, pp. 58–62.
4. Storer, A. C. and Cornish-Bowden, A. (1974) The kinetics of coupled enzyme reactions. *Biochem. J.* **141,** 205–209.
5. Brooks, S. P. J. and Suetler, C. H. (1989) Practical aspects of coupling enzyme theory. *Anal. Biochem.* **176,** 1–14.
6. Tisljar, U., Knight, C. G., and Barrett, A. J. (1990) An alternative quenched fluorescence substrate for Pz-peptidase. *Anal. Biochem.* **186,** 112–115.
7. Molineaux, C. J., Yu, B., and Ayala, J. M. (1991) Distribution of endopeptidase 24.15 in rat brain nuclei using a novel fluorogenic substrate: comparison with endopeptidase 24.11. *Neuropeptides* **18,** 49–54.
8. Knight, C. G. (1991) A quenched fluorescent substrate for thimet peptidase containing a new fluorescent amino acid, DL-2-amino-3-(7-methoxy-4-coumaryl) propionic acid. *Biochem. J.* **274,** 45–48.
9. Butenas, S., Orfeo, T., Lawson, J. H., and Mann, K. G. (1992) Amino-naphthalenesulfonamides, a new class of modifiable fluorescent detecting groups and their use in substrates for serine protease enzymes. *Biochemistry* **31,** 5399–5411.
10. Wang, G. T. and Krafft, G. A. (1992) Automated synthesis of fluorogenic protease substrates: design of probes for Alzheimer's disease-associated proteases. *Bioorganic Med. Chem. Letts.* **2,** 1665–1668.
11. Goudreau, N., Guis, C., Soleilhac, J.-M., and Roques, B. P. (1994) Dns-Gly-(p-NO_2)Phe-β-Ala, a specific fluorogenic substrate for neutral endopeptidase 24.11. *Anal. Biochem.* **219,** 87–95.
12. Mumford, R. A., Strauss, A. W., Powers, J. C., Pierzchala, P. A., Nishino, N., and Zimmerman, M. (1980) A zinc metalloprotease associated with dog pancreatic membranes. *J. Biol. Chem.* **255,** 2227–2230.
13. Mumford, R. A., Pierzchala, P. A., Strauss, A. W., and Zimmerman, M. (1981) Purification of a membrane-bound metalloendopeptidase from porcine kidney that degrades peptide hormones. *Proc. Natl. Acad. Sci. USA* **78,** 6623–6627.
14. Relton, J. M., Gee, N. S., Matsas, R., Turner, A. J., and Kenny, A. J. (1983) Purification of endopeptidase 24.11 ("enkephalinase") from pig brain by immunoadsorbent chromatography. *Biochem. J.* **215,** 519–523.

33

Characterization of Neuropeptidases Using Inhibitors

Nigel M. Hooper

1. Introduction

It is now well established that the major mechanism for the termination of a neuropeptide signal is not internalization of the peptide, but rather its metabolism by one or more neuropeptidases (reviewed in refs. *1* and *2*). By the very nature of their action in hydrolyzing peptides that have been released into the synaptic cleft, these neuropeptidases are ectoenzymes; that is, they are integral proteins of the plasma membrane, asymmetrically oriented with the bulk of the protein, including the active site, exposed at the extracytoplasmic surface *(3)*. In addition to the termination of peptide signals, some neuropeptidases are involved in the activation or modulation of certain peptides, in which case the enzyme may be localized either extracellularly or intracellularly. Although 100 or more potential neuropeptides have now been discovered, there is a substantially smaller number of neuropeptidases (Table 1). The reason for this is that most of the neuropeptidases can act on more than one peptide substrate. For example, neprilysin (neutral endopeptidase-24.11; EC 3.4.24.11) can hydrolyze the enkephalins, tachykinins, cholecystokinin, natriuretic factors, bradykinin, and so on, whereas peptidyl dipeptidase A (angiotensin converting enzyme; EC 3.4.15.1) can hydrolyze bradykinin, the enkephalins, substance P, and so on, in addition to angiotensin I *(1)*. Whether a peptide is a substrate for a particular neuropeptidase in vivo will depend on whether the two are colocalized in the same neuronal pathway and whether the kinetics of hydrolysis are favorable. Thus, although neprilysin can hydrolyze luteinizing hormone-releasing hormone (LH-RH) in vitro, the kinetics for this reaction are so poor that even if the enzyme and peptide colocalized in the brain, it is unlikely that neprilysin would contribute in any appreciable way to the in vivo metabolism of LH-RH. As well as these broad-acting neuropeptidases, there are a few sub-

From: *Methods in Molecular Biology, Neuropeptide Protocols*
Edited by: G. B. Irvine and C. H. Williams Humana Press Inc., Totowa, NJ

Table 1
Neuropeptidases and Their Inhibitors

Neuropeptidase (EC no.)	Inhibitor	K_i, n*M*	I_{50}, μ*M*
Membrane alanyl aminopeptidase (3.4.11.2)	Actinonin		1.0
	Amastatin	19	0.05
	Bestatin	4100	3.0
	Kelatorphan	7000	0.7
	Probestin		0.05
Arginyl aminopeptidase (3.4.11.6)	Arphamenine A	5.0	
	Arphamenine B	5.0	
Glutamyl aminopeptidase (3.4.11.7)	Amastatin	250	2.0
	Probestin		19.9
X-Pro aminopeptidase (3.4.11.9)	Apstatin		4.0
	Enalaprilat[a]		2.2
X-Trp aminopeptidase (3.4.11.16)	Amastatin		2.0
	Bestatin		6.0
	Probestin		5.0
Membrane dipeptidase (3.4.13.19)	Cilastatin	700	0.11
Dipeptidyl peptidase IV (3.4.14.5)	Diprotin A	1000	3.2
	Diprotin B	5000	
Peptidyl dipeptidase A (3.4.15.1)	Captopril	1.7	0.021
	Enalaprilat	0.2	0.003
	Lisinopril	0.1	0.011
Carboxypeptidase M (3.4.17.12)	DL-mercaptomethyl-3-guanidinoethylthio-propanoic acid (MGTA)		0.3
Pyroglutamyl peptidase I (3.4.19.3)	Oxoprolinal	20	
Prolyl oligopeptidase (3.4.21.26)	Cbz-Pro-prolinal	14	
Neprilysin (3.4.24.11)	Phosphoramidon	2	0.013
	Thiorphan	4.7	0.013
	Kelatorphan	1.4	0.002
Thimet oligopeptidase (3.4.24.15)	N-[1-(R,S)-carboxy-2-phenylpropyl]-Ala-Ala-Tyr-pAB	16	
Neurolysin (3.4.24.16)	Pro-Ile		90
Endothelin converting enzyme-1	Phosphoramidon		0.5

Further information can be found in refs. *1* and *2*.
[a]Only effective toward certain substrates.

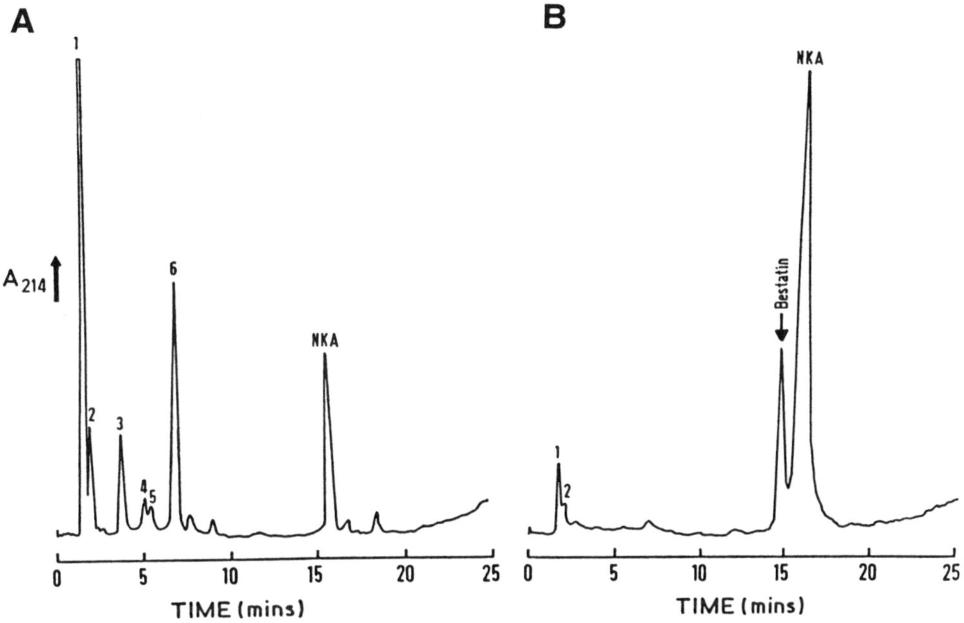

Fig. 1. Hydrolysis of neurokinin A by porcine striatal synaptic membranes. *See* legend to Table 2 for experimental details. **(A)** No inhibitors present; **(B)** incubation in the presence of phosphoramidon (1 μ*M*) and bestatin (0.1 m*M*). The products were resolved by RP-HPLC and monitored at 214 nm. The peptides are numbered as in Table 2. NKA, neurokinin A (His-Lys-Thr-Asp-Ser-Phe-Val-Gly-Leu-MetNH$_2$). Note that the inhibitor bestatin absorbs at 214 nm.

strate-specific enzymes such as thyrotropin releasing hormone-degrading enzyme *(4)* and endothelin converting enzyme *(5)*. This latter group of enzymes are potentially more difficult to identify because of the necessity to have the specific substrate available first.

The most widely used method for identifying which particular neuropeptidase(s) is involved in the metabolism of a peptide is to observe the effect of selective inhibitors on the metabolism of the peptide by brain preparations (*see* refs. *6–8*). Figure 1 and Table 2 show the results of such a study to investigate the effect of inhibitors on the metabolism of neurokinin A by porcine striatal synaptic membranes *(9)*. From this study, it was clear that captopril, an inhibitor of peptidyl dipeptidase A (Table 1), had no significant effect on the metabolism of the peptide by the striatal synaptic membranes. In contrast, phosphoramidon, an inhibitor of neprilysin (Table 1), essentially completely blocked the metabolism. Bestatin, a relatively

Table 2
Effect of Inhibitors on the Hydrolysis of Neurokinin A by Porcine Striatal Synaptic Membranes

Peak	Retention time (min)	Peptide	Effect of inhibitors on peak area (control [no inhibitors] = 100 for each peak)				
			P	B	C	P + B	K
1	1.60	His-Lys-Thr-Asp-Ser (NKA 1-5)	14	50	89	0	9
2	2.00	Val-Gly (NKA 7-8)	22	19	100	0	0
3	3.74	Leu-MetNH_2 (NKA 9-10)	0	60	96	0	0
4	5.08	Not determined	0	63	110	0	0
5	5.44	Phe-Val-Gly (NKA 6-8)	0	342	69	0	0
6	6.76	His-Lys-Thr-Asp-Ser-Phe-Val-Gly (NKA 1-8)	9	100	127	0	7

Synaptic membranes from porcine striatum were prepared as described in ref. *12*. Samples of membranes (60 μg of protein) were incubated with neurokinin A (0.5 m*M* final concentration) in the absence and presence of the indicated inhibitors at 37°C. The reaction was stopped by heating at 100°C for 4 min, and the substrate and products separated and quantified by reversed-phase HPLC (*see* Section 3.3. for details). Products were identified either by using synthetic marker peptides or by collection of the individual product peaks and amino acid analysis of the corresponding peptides. The concentrations of the inhibitors were: P, phosphoramidon, 1 μ*M*; B, bestatin, 0.1 m*M*; C, captopril, 1 μ*M*; K, kelatorphan, 10 μ*M*.

broad acting aminopeptidase inhibitor (Table 1), had some effect on its own but, when combined with phosphoramidon, led to a total block in the metabolism of neurokinin A by the striatal synaptic membranes. Kelatorphan, designed to block enkephalin metabolism in the brain by inhibiting neprilysin and membrane alanyl aminopeptidase (aminopeptidase N; EC 3.4.11.2) *(10)* (Table 1), also completely blocked the metabolism of neurokinin A. Thus, from these studies it would appear that neprilysin is the major neurokinin A hydrolyzing activity in the striatal synaptic membranes, with an aminopeptidase, possibly membrane alanyl aminopeptidase, contributing to secondary hydrolysis. The pattern of product peaks produced upon metabolism of neurokinin A by purified porcine kidney neprilysin was essentially the same as that produced by the striatal synaptic membranes, and the hydrolysis of neurokinin A by neprilysin displayed one of the highest k_{cat}/K_m values of any peptide tested *(1)*, adding further support to this conclusion.

Although the above approach has been used successfully to determine the neuropeptidase(s) involved in the metabolism of numerous peptides, considerable caution is required in the interpretation of the results from such an experiment. This mainly arises because many of the available inhibitors can affect the action of more than one neuropeptidase (*see* Table 1). An inhibitor may only appear to be selective for a particular enzyme because it has not been screened against other neuropeptidases. For example, bestatin, widely viewed as a selective inhibitor of membrane alanyl aminopeptidase, is actually more potent against X-Trp aminopeptidase (aminopeptidase W; EC 3.4.11.16) (Table 1), and amastatin, originally described as an inhibitor of glutamyl aminopeptidase (aminopeptidase A; EC 3.4.11.7) is essentially equipotent against this enzyme, membrane alanyl aminopeptidase and X-Trp aminopeptidase (Table 1) *(11)*. More recently, it has been shown that phosphoramidon, long regarded as a relatively selective inhibitor of neprilysin, also inhibits endothelin converting enzyme *(5)*. As some of the neuropeptidases also have overlapping substrate specificities, further caution must be exercised when interpreting the results from such inhibition experiments. One of the only truly reliable ways around this problem is to check that a purified preparation of the neuropeptidase in question can actually hydrolyze the peptide (*see* above). Alternative approaches such as immunodepletion of a neuropeptidase from the membrane sample with a specific antibody can provide important information but again need to be performed with care as closely related enzymes may also be recognized by the antibody. If, when all the necessary control experiments have been included and the results interpreted correctly, the pattern of inhibition of the activity being studied does not fit with any of the known neuropeptidase activities, it may be that one is dealing with either a previously uncharacterized or a poorly characterized enzyme, as undoubtedly there are still numerous neuropeptidases to be discovered.

This chapter describes approaches to the characterization of neuropeptidases using inhibitors that have been used in our laboratory for a number of years. First, Sections 2.1. and 3.1. describe a method for the simple preparation of striatal synaptic membranes from brain tissue that is enriched in a number of the membrane-bound neuropeptidases. A more lengthy procedure that results in a higher purity of membranes has been described *(12),* although the results obtained with both preparations have been found to be almost identical (Hooper, N. M., unpublished results), and in the first instance much time and effort can be saved by following the former procedure. Obviously, other membrane or soluble preparations from other regions of the brain may be more applicable to the particular peptide under investigation. Second, in Sections 2.2. and 3.2., a protocol is detailed to assay for the enzyme activity with the enzyme preparation in the absence and presence of inhibitors. If one is starting from the perspective of not knowing anything about the neuropeptidase activity, the first step is to evaluate the effect of class specific inhibitors (Table 3) in order to determine the mechanistic class (metallo, serine, cysteine, or acidic) of enzyme that the activity belongs to. More selective inhibitors of individual neuropeptidases (or groups of neuropeptidases) (Tables 1 and 4) can then be used to characterize the particular activity in more detail. Probably the most common method for assessing the metabolism of a peptide is to separate and quantitate the substrate and products by reversed-phase high performance liquid chromatography (RP-HPLC) (*see* Figs. 1 and 2). In Sections 2.3. and 3.3., details are provided of the HPLC system that has been used extensively in our laboratories for this purpose.

2. Materials

2.1. Isolation of Synaptosomal Membranes

1. Pig striatum (approx 10 g) (*see* Note 1).
2. Buffer A:10 m*M* Tris-HCl, pH 7.0, 0.32*M* sucrose.
3. Buffer B: 10 m*M* Tris-HCl, pH 7.0, 0.2*M* NaCl.
4. Glass/glass hand-held homogenizer.
5. High-speed ultracentrifuge with fixed angle rotors.

2.2. Enzyme Assays

1. Microcentrifuge tubes (1.5-mL, hinge-capped) and an appropriate rack.
2. Buffer C: 0.1*M* Tris-HCl, pH 7.4 (*see* Notes 2 and 3).
3. Appropriate enzyme, inhibitor, and substrate samples (*see* Section 3.2.).
4. A water bath at 37°C with a rack to hold the microcentrifuge tubes.
5. A boiling water bath with a rack to hold the microcentrifuge tubes.

Table 3
Class Specific Protease Inhibitors

Class of protease inhibited	Inhibitor	Mode	Stock concentration	Effective concentration	Notes
Metallo	EDTA	R	50 m*M*	1–5 m*M*	
	EGTA	R	50 m*M*	1–5 m*M*	
	1,10-phenanthroline	R	100 m*M*	1 m*M*	Dissolve in methanol.
Serine	Aprotinin	R	1–10 mg/mL	2–10 μg/mL	
	Di-isopropylfluorophosphate (DipF)	I	1*M*	0.1 m*M*	**Extremely toxic.** Dissolve in propan-2-ol. Half-life in aqueous solution 1 h at pH 7.5.
	Phenylmethanesulfonylfluoride (PMSF)	I	100 m*M*	1 m*M*	Dissolve in propan-2-ol and store at 4°C. Half-life in aqueous solution 1 h at pH 7.5.
Cysteine	E-64 (trans-epoxysuccinyl-L-leucylamido-[4-guanidino] butane)	I	1 m*M*	10 μ*M*	
	Iodoacetamide	I	100 m*M*	1 m*M*	Prepare fresh as required.
	p-Hydroxymercuri-phenylsulfonic acid	I	100 m*M*	1 m*M*	
Acidic	Pepstatin	R	1 m*M*	1 μ*M*	Dissolve in methanol.

The mode of inhibition is indicated as either reversible (R) or irreversible (I). Unless otherwise indicated: the inhibitor should be dissolved in water to the indicated stock concentration and then stored at –20°C; the stock solution is stable when stored at –20°C for up to 6 mo; and the inhibitor is available from Sigma (St. Louis, MO).

Table 4
Properties of Neuropeptidase Inhibitors

Inhibitor	Stock concentration, m*M*	Effective concentration, μ*M*	Notes
Actinonin	1	10–100	
Amastatin	1	10–100	Dissolve in methanol, 30–60 min preincubation required for maximal inhibition.
Arphamenine A	1	10	
Arphamenine B	1	10	
Bestatin	1	100	Dissolve in methanol.
Captopril	100	10	
Cilastatin	10	10–100	Available from Merck, Sharp and Dohme (West Point, PA).
Diprotin A	1	10–50	
Diprotin B	1	50–100	
Enalaprilat	1	10	Available from Merck, Sharp and Dohme.
Lisinopril	1	10	
DL-mercaptomethyl-3-guanidinoethylthio-propanoic acid (MGTA)	1	10	Available from Calbiochem.
Phosphoramidon	1	10	
Pro-Ile	10	100–500	
Thiorphan	10	10	

Unless otherwise indicated: the inhibitor should be dissolved in water to the indicated stock concentration and then stored at –20°C; the stock solution is stable when stored at –20°C for up to 6 mo; the inhibitor is available from Sigma; and the mode of inhibition is reversible.

2.3. RP-HPLC

1. An HPLC system configured with two pumps, a UV 214 nm detector, and an automatic injector. We use a Waters/Millipore (Bedford, MA) Maxima 2 Workstation with two pumps, a variable wavelength detector, and an automatic sample injector (WISP) (*see* Note 4).
2. A μBondapak C_{18} column either 8 × 100 mm Radial Pak or 3.9 × 300 mm steel column.
3. High grade far UV acetonitrile (*see* Note 5).
4. Analar H_3PO_4.
5. Solvent A: 0.08% (v/v) H_3PO_4 at pH 2.5 (*see* Note 2).

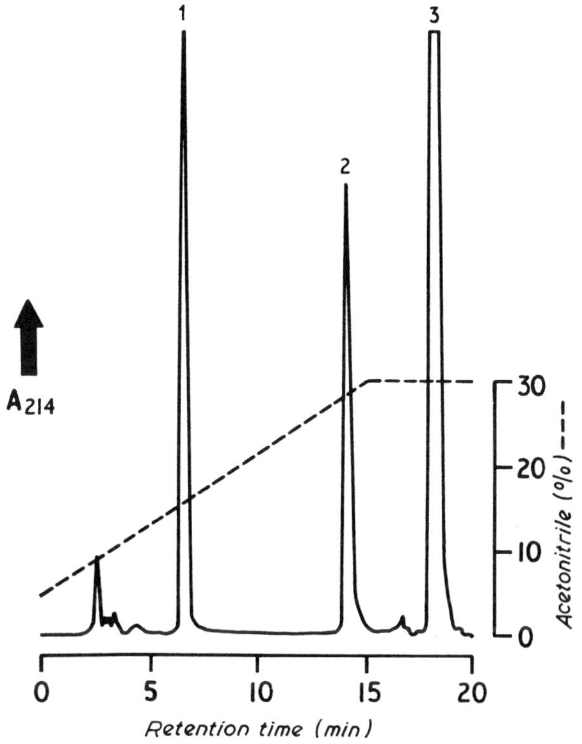

Fig. 2. RP-HPLC analysis of [D-Ala^2, Leu^5]enkephalin hydrolysis. [D-Ala^2, Leu^5]enkephalin (0.5 m*M* final concentration) was incubated with purified porcine kidney neprilysin (EC 3.4.24.11) (30 ng) for 45 min at 37°C. The reaction was terminated by boiling for 4 min and the substrate and hydrolysis products separated and quantitated by RP-HPLC (*see* Section 3.3. for details). The absorbance peaks represent: 1, Tyr-D-Ala-Gly; 2, Phe-Leu; 3, [D-Ala^2, Leu^5]enkephalin.

6. Solvent B: 30% (v/v) acetonitrile in 0.08% (v/v) H_3PO_4 at pH 2.5 (*see* Notes 2 and 6).
7. Degas all solvents by filtering through 0.45-µm solvent-resistant filters using a 1-L glass suction filtration unit attached to the water mains.

3. Methods

3.1. Isolation of Synaptosomal Membranes

Perform all steps at 4°C. All g-forces are g_{av}.

1. Cut 10 g of pig striatum into smaller pieces with a pair of scissors and then homogenize in 100 mL Buffer A using a glass/glass homogenizer with 12 up and 12 down strokes.

2. Centrifuge the homogenate at 1000*g* for 3 min in a fixed angle rotor.
3. Collect the supernatant, resuspend the pellet in 100 mL Buffer A, and repeat step 2.
4. Collect the supernatant and combine with the first supernatant. Then centrifuge both at 17,000*g* for 20 min in a fixed-angle rotor.
5. Discard the supernatant and resuspend the pellet in 100 mL ice-cold water using the glass/glass homogenizer. Centrifuge at 100,000*g* for 30 min.
6. Repeat step 5.
7. Discard the supernatant and resuspend the pellet in 100 mL Buffer B. Centrifuge at 100,000*g* for 30 min.
8. Discard the supernatant and resuspend the pellet in a minimum volume (2–5 mL) of Buffer B. Store in aliquots at –70°C.

3.2. Enzyme Assays (see Note 7)

Use ice-cold buffers or set up the tubes on ice prior to incubation at 37°C.

1. In a 1.5-mL microcentrifuge tube, pipet 70 μL of Buffer C (*see* Note 8).
2. Add 10 μL of a tenfold concentrated solution of an appropriate inhibitor to give the final effective concentration as indicated in Tables 3 and 4 (*see* Notes 9 and 10).
3. Add 10 μL of the synaptosomal membrane sample or other enzyme sample to be assayed. If necessary (*see* notes in Table 4), preincubate the enzyme/inhibitor/buffer mix (*see* Note 11).
4. Add 10 μL of a tenfold concentrated solution of the peptide substrate to start the reaction, mix, cap, and incubate at 37°C for the appropriate length of time (*see* Note 12).
5. Terminate the reaction by heating the tubes in a boiling water bath for 4 min (pierce the cap of each tube with a syringe needle to prevent the cap blowing open upon heating). Then microfuge the tubes at 10,000*g* for 5 min to pellet particulate material and transfer 80 μL of the supernatant into the appropriate HPLC vials.

3.3. RP-HPLC

1. Equilibrate the HPLC column in 4.5% (v/v) acetonitrile in 0.08% (v/v) H_3PO_4 at pH 2.5 at a flow rate of 1.5 mL/min until the baseline is steady (this will take approx 20–30 min).
2. Inject 60 μL of the supernatant from the enzyme assay onto the column.
3. Elute bound peptides with a 15-min linear gradient of 4.5–30% (v/v) acetonitrile in 0.08% (v/v) H_3PO_4 at pH 2.5, followed by 5 min elution at the final conditions at a flow rate of 1.5 mL/min (*see* Fig. 2) (*see* Note 13).
4. Detect the peptides at 214 nm (*see* Note 14).
5. Re-equilibrate the column for a minimum of 8 min in the starting conditions before injecting the next sample.

4. Notes

1. Approximately 5–10 pig brains will be required in order to obtain 10 g of striatum.
2. Use high-purity water, e.g., from a Milli-Q purification system (Waters/Millipore), in order to avoid impurities binding to and eluting from the HPLC column.
3. A pH 7.4 buffer is probably a sensible pH to start with if nothing is known about the neuropeptidase being assayed, but, obviously, alternative pH and buffers may be more appropriate for certain enzymes.
4. Several other HPLC systems are on the market, any of which would be suitable. A manual injection unit can be used instead of an automated sample injector but with obvious limitations if multiple assays are to be chromatographed.
5. It is imperative to use high-grade acetonitrile, which is designated "far UV" when employing detection at 214 nm, otherwise a high baseline is observed that will increase on increasing the acetonitrile concentration gradient and obscure lower absorbing peaks.
6. If the absorbance at 214 nm of solvents A and B is substantially different, the baseline will alter during the chromatogram. If this is a problem, the two solvents can be brought to the same absorbance at 214 nm by adding 10–20-µL aliquots of high-grade acetic acid to the solvent with the lower absorbance.
7. Remember to include appropriate controls, in particular a tube with enzyme sample, peptide substrate, and buffer (no inhibitor). When using a new assay system for the first time, it is also advisable to include a substrate blank (i.e., substrate alone), a buffer blank, an enzyme sample blank, and an inhibitor blank as any one of these components may absorb at 214 nm, producing absorbance peaks on the HPLC chromatogram that may be mistaken for products. In particular, several of the neuropeptidase inhibitors are peptide analogs and absorb at 214 nm (for an example of this, *see* Fig. 1B, in which bestatin clearly shows up on the HPLC chromatogram).
8. The final reaction volume should be 0.1 mL, so the amount of buffer used will need to be adjusted depending on the volumes of inhibitor(s), enzyme sample, and substrate used.
9. Tables 3 and 4 are not exhaustive lists of neuropeptidase inhibitors. Other lists of protease inhibitors with details of their properties can be found in ref. *13* and a useful leaflet entitled "The easy way to customize your protease inhibitor cocktail," produced by Calbiochem (La Jolla, CA).
10. When diluting an inhibitor from the stock solution to the working concentration, use buffer C or an appropriate alternative (*see* Note 3) in order to minimize the effects of pH or solvents on the enzyme assay.
11. Some of the inhibitors are classified mechanistically as "slow, tight-binding" inhibitors. Experimentally, this means that a preincubation of the inhibitor with the enzyme in the absence of substrate is required in order to achieve maximal inhibition (*see* Table 4 for those inhibitors for which this applies). This preincubation is best performed at 4°C to minimize unwanted proteolysis of the enzyme sample.

12. The length of time of incubation at 37°C will vary for each combination of peptide and neuropeptidase and will have to be determined experimentally.
13. Very hydrophobic peptides, or peptides that are hydrolyzed to hydrophobic products, may require a higher concentration of acetonitrile to be eluted from the C_{18} column.
14. At 214 nm, peptide bonds absorb and so any substrate or product with two or more amino acid residues will be visible on the HPLC chromatogram at this wavelength. In addition, the aromatic amino acids Trp, Tyr, and Phe, as well as His, absorb at 214 nm to different extents. By this same argument, all other amino acids are invisible at 214 nm; thus, if these are released individually upon hydrolysis of the peptide, they will not appear on the chromatogram. A quantitation of the absorbance at 214 nm of individual amino acids and peptide bonds (amide and imido) can be found in ref. *14*.

References

1. Kenny, A. J. and Hooper, N. M. (1991) Peptidases involved in the metabolism of bioactive peptides, in *Degradation of Bioactive Substances: Physiology and Pathophysiology* (Henriksen, J. H., ed.), CRC, Boca Raton, FL, pp. 47–79.
2. Checler, F. (1993) Neuropeptide-degrading peptidases, in *Methods in Neuropeptide and Neurotransmitter Research* (Parvez, S. H., Naoi, M., Nagatsu, T., and Parvez, S., eds.), Elsevier, Amsterdam, pp. 375–418.
3. Hooper, N. M. (1993) Ectopeptidases, in *Biological Barriers to Protein Delivery* (Audus, K. L. and Raub, T. J., eds.), Plenum, New York, pp. 23–50.
4. Bauer, K. (1995) Inactivation of thyrotropin-releasing hormone (TRH) by the hormonally regulated TRH-degrading ectoenzyme. A potential regulator of TRH signals? *Trends Endocrinol. Metab.* **6,** 101–105.
5. Turner, A. J. and Murphy, L. J. (1996) Molecular pharmacology of endothelin converting enzymes. *Biochem. Pharmacol.* **51,** 91–102.
6. Matsas, R., Fulcher, I. S., Kenny, A. J., and Turner, A. J. (1983) Substance P and (Leu) enkephalin are hydrolysed by an enzyme in pig caudate membranes that is identical with the endopeptidase of kidney microvilli. *Proc. Natl. Acad. Sci. USA* **80,** 3111–3115.
7. Hooper, N. M. and Turner, A. J. (1985) Neurokinin B is hydrolysed by synaptic membranes and by endopeptidase-24.11 ("enkephalinase") but not by angiotensin converting enzyme. *FEBS Lett.* **190,** 133–136.
8. Bourne, A. and Kenny, A. J. (1990) The hydrolysis of brain and atrial natriuretic peptides by porcine choroid plexus is attributable to endopeptidase-24.11. *Biochem. J.* **271,** 381–385.
9. Hooper, N. M., Kenny, A. J., and Turner, A. J. (1985) The metabolism of neuropeptides. Neurokinin A (substance K) is a substrate for endopeptidase-24.11 but not for peptidyl dipeptidase A (angiotensin converting enzyme). *Biochem. J.* **231,** 357–361.
10. Fournie-Zaluski, M.-C., Chaillet, P., Bouboutou, R., Coulaud, A., Cherot, P., Waksman, G., Costentin, J., and Roques, B. P. (1984) Analgesic effects of

kelatorphan, a new highly potent inhibitor of multiple enkephalin degrading enzymes. *Eur. J. Pharmacol.* **102,** 525–528.

11. Tieku, S. and Hooper, N. M. (1992) Inhibition of aminopeptidases N, A and W. A re-evaluation of the actions of bestatin and inhibitors of angiotensin converting enzyme. *Biochem. Pharmacol.* **44,** 1725–1730.
12. Fulcher, I. S., Matsas, R., Turner, A. J., and Kenny, A. J. (1982) Kidney neutral endopeptidase and the hydrolysis of enkephalin by synaptic membranes show similar sensitivity to inhibitors. *Biochem. J.* **203,** 519–522.
13. Beynon, R. J. and Bond, J. S., eds. (1989) *Proteolytic Enzymes: A Practical Approach,* IRL, Oxford, p. 259.
14. Stephenson, S. L. and Kenny, A. J. (1987) Metabolism of neuropeptides. Hydrolysis of the angiotensins, bradykinin, substance P and oxytocin by pig kidney microvillar membranes. *Biochem. J.* **241,** 237–247.

Index